EMPIRE AND CATASTROPHE

FRANCE OVERSEAS: STUDIES IN EMPIRE AND DECOLONIZATION

Series editors: A. J. B. Johnston, James D. Le Sueur, and Tyler Stovall

Empire and Catastrophe

*Decolonization and Environmental Disaster in
North Africa and Mediterranean France since 1954*

Spencer D. Segalla

University of Nebraska Press | Lincoln

Library of Congress Cataloging-in-Publication Data
Names: Segalla, Spencer D., author.
Title: Empire and catastrophe: decolonization and
environmental disaster in North Africa and Mediterranean
France since 1954 / Spencer D. Segalla.
Description: Lincoln: University of Nebraska Press, 2021. |
Series: France overseas: studies in empire and decolonization |
Includes bibliographical references and index.
Identifiers: LCCN 2020040356
ISBN 9781496219633 (hardback)
ISBN 9781496222138 (epub)
ISBN 9781496222145 (mobi)
ISBN 9781496222152 (pdf)
Subjects: LCSH: Environmental disasters—Political aspects—
Africa, French-speaking. | Environmental disasters—Political
aspects—Africa, North. | Decolonization—Africa,
French-speaking. | Decolonization—Africa, North.
Classification: LCC GE146 .S44 2021 |
DDC 363.700961—dc23

LC record available at https://lccn.loc.gov/2020040356

For Maya and Ryan

CONTENTS

MAPS
_____

ACKNOWLEDGMENTS

This book would not have been possible without the tireless support of my partner and spouse, Prof. Amanda Bruce, and the foundation provided by my parents, Wendy and David Segalla. I also thank Charles McGraw Groh for tackling the enormous challenge of serving as the founding chair of our new department so that I could focus on this book. This book has also been made possible by support from an American Institute for Maghrib Studies research grant, University of Tampa David Delo research grants and Dana Foundation grants, and by Fred and Jeanette Pollock Research Professor grants. Digital access has been supported by a Mellon Foundation grant to the Sustainable History Monograph Pilot.

A historical book such as this inevitably builds upon the scholarship of others, and the chapters that follow are particularly indebted to the work of El Djamhouria Slimani Aït Saada, Valentin Pelosse, Paul Rabinow, Janet Abu-Lughod, Marie-France Dartois, Thierry Nadau, and Yaël Fletcher. I am also grateful for the models and mentorship provided by Herman Lebovics and for the input provided by Brock Culter, Mitch Aso, Joomi Lee, Ahmed Sabir, Daniel Williford, Julia Clancy-Smith, Stacy Holden, Mohammed Daadaoui, James Mokhiber, Moshe Gershovich, and Shana Minkin, and by the members of the Florida Maghreb workshop group: Ann Wainscott, Adam Guerin, Amelia Lyons, and Darcie Fontaine. In Morocco, I received invaluable assistance and insight from Lahsen Roussafi, Yazza Jafri, Mohamed Bajalat, Rachid Bouksim, Jamila Bargach, Suad Kadi, Abdallah Aourik, Hassan Bouziane, Jacques Lary, and Mohamed Mounib. I am also immensely grateful to the archivists, librarians, and staff at archives and repositories of sources on three continents, without whom I would have been lost.

I offer special thanks to the anonymous peer reviewers for the University of Nebraska Press, whose invaluable input on various manuscript drafts helped to improve this work, to UNP editors-in-chief Alisa Plant and Bridget Barry for their support of this project, to Erin Greb of Erin Greb Cartography, to Ihsan Taylor of Longleaf Services, and to copyeditor Bridget Manzella. Chapter 3 of this volume expands on "The 1959 Moroccan Oil Poisoning and U.S. Cold War Disaster Diplomacy," originally published in the *Journal of North African Studies* 17 (2012), 315–36, available online at https://tandfonline.com/. I am grateful to the journal editors for their support of my early work on this project. I thank Todd Shepard and Patricia Lorcin for their close readings and detailed comments on my contribution to their anthology, *French Mediterraneans: Transnational and Imperial Histories* (University of Nebraska Press, 2016), a piece which eventually grew into chapters 5 and 6 of the present volume.

Finally, I thank the design and production teams at the University of Nebraska Press, and the indexer, for their work on the print edition of this book.

CHAPTER I

Introduction

We have always lived in the shadows like obscure cockroaches, powerless before the reading of a newspaper, babbling before the morose enemy who stole our crops, stripped our land, converted them from wheat into vineyards useless for our hunger. We receive the natural verdict with relief. It would finally break the monotony of servitude. And it is the native soil that will shake the rocks and that now opens, to bury us with our denouncers, these arrogant masters who know the pain that they inflict.
—"Old Man," in Henri Kréa, *Le séisme: Tragédie*

ALMOST FOUR YEARS AFTER the French Algerian city of Orléansville was devastated by an earthquake in September 1954, the Franco-Algerian playwright Henri Kréa published a play that presented the seismic disaster as a harbinger of a painful but necessary decolonization. Kréa, a.k.a. Henri Cochin, son of a French father and Algerian mother, was an advocate of Algerian independence, and the struggle for decolonization was still underway when Kréa wrote *Le séisme: Tragédie*.[1] The Algerian Revolution, which had begun just weeks after the earthquake, would not achieve the independence of Algeria from French rule until 1962. In the intervening years, North Africa and France would be wracked by a series of disasters: seismic aftershocks and years of brutal warfare in Algeria, a dam collapse in France, a mass poisoning, and another catastrophic earthquake in Morocco.

Kréa's play, begun in 1956 and published in 1958 in both Paris and Tunis, was explicitly anti-imperialist. The play builds on the synchronicity of the Orléansville earthquake and the nationalist revolution that began a few weeks later. Yet Kréa's play purports to render chronology irrelevant: *Le séisme* opens with a recitation of major earthquakes of the 1900s, from Calabria in 1905 to Orléansville in 1954, and then juxtaposes these disasters, particularly the Orléansville earthquake, with an ancient narrative of anti-colonial resistance in North Africa: the second-century BC rebellion of the Numidian king Jugartha against Roman occupation.[2] The geoenvironmental disaster of 1954 was linked to colonial violence through the category of *malheur* (misfortune, woe); the revolution of 1954 was linked to antiquity through the theme of oppression and anti-colonial rebellion.

At the outset, the play's portrayal of natural disasters is intertwined with its portrayal of Roman/French colonialism in North Africa. The prologue begins with a voiceover explaining geological theories of earthquake production and then turns to antiquity, with the Romans invading North Africa like "locusts that periodically swoop down to bring famine to fertile Numidia."[3] After the prologue, however, the ancient setting abruptly dissolves, and the focus of the first act of the play ("Episode 1") shifts to the 1954 earthquake as the play turns to the relationship between the seismic and political events in modern Algeria. Revising a trope that geographer El Djamhouria Slimani Aït Saada has traced back to the colonial discourses of the nineteenth century, Kréa portrayed a country battered by misfortune. Earthquakes, war, floods, locusts, and the tragedy of death in childbirth all converge in the suffering of the Algerian people. The choir chants in the final passage of the play:

> This country, crucible of men of all origins of all poetic destinations
> Collides
> With clatterings of fire
> With the deaf rhythm of blood
> Flowing in streams
> Like a flooding river
> Breaking the dikes of the narrow valleys. . . .
> The eternal wave of generations . . .
> Crushed by the cosmic pestle
> Of misfortune.[4]

However, this grim finale, as well as the play's subtitle—*Tragédie*—contradicts the dominant narrative of the play, which portrays the earthquake as a vehicle of salvation.[5]

At first, the earthquake seems to sweep away the injustices of colonialism, offering death as the only liberation. However, the pessimism of the Old Man's interpretation of the disaster is superseded by a vision of hope through catharsis, communicated at the end of "Episode 1" in a voiceover that reinforces the relationship between the seismic and political upheavals in Algeria: "Understand that the earth shook at the same time that the people arose from their torpor."[6] Thus the moral righteousness of the struggle for liberation gives meaning to all other forms of suffering. In Kréa's play, the earthquake is not just a portent of the coming revolution, it is also retribution:

> All the dead howl and stir in concert, these fields and these villages that were stolen from them. The trees open to battle the convoy of sacrilege. Nothing different, the insects, stones, man. Their planes like sharks are drowned by the welcome invasion of locusts. Their arms are of no use.[7]

One disaster avenges another, and if the colonized suffer, the colonizer is also weakened. The earthquake is a disaster not only for the long-suffering Algerians but also for the French; as such, it is the first strike against empire. The varieties of catastrophe that afflict humanity blur into one another: sharks and aerial bombardments; childbirth and epidemics; earthquakes and locusts. Yet there is order and meaning in this litany of suffering:

> The great day has arrived
> With its procession of misfortunes
> But misfortunes are good for something
> For example the general suppression of misery
> The resurrection of grand sentiments
>
> Certainly
> One must die of hunger . . . to be human[8]

The earthquake, and the war, serve a purpose: the liberation and redemption of the Algerian people.

Kréa's association of French colonialism with natural scourges like sharks or locusts portrays individual human actors as overwhelmed by a situation they neither understand nor control. The French occupiers are confused, struggling to follow an imperial ideology but bewildered by the realities of colonial North Africa. Believing in the good works brought about by colonialism but pelted by children throwing potatoes "larded with razor blades," the soldiers ask why their beneficence is welcomed with such hostility, even from the earth itself: "the stones themselves ruminate with menace."[9] Ultimately, the colonial situation leads the occupiers to extremes of evil. A soldier who was a surgeon in France, saving lives, becomes a torturer in Algeria, disemboweling prisoners. Another soldier speaks with regret of two children he slaughtered "like partridges";

another gleefully recounts the story of a humanist professor who spoke of ethics until he came under fire, at which point he became "as savage as a cannibal," getting drunk and "thinking only of raping little girls." This brutality, the play suggests, stems from colonial occupation; French intentions and the idea of a civilizing mission are irrelevant: they cannot mitigate the brutal nature of colonialism any more than they can alter the movement of tectonic plates.[10]

The dramatic turning point in the play is the radicalization of the older generation of Muslim Algerians. This occurs in the last section of the play, when the Old Man and the Old Woman break free from their tragic flaw, their resigned accommodation to colonialism, which Kréa portrays as characteristic of the older generation. Awakened from their "hypnosis"[11] by the earthquake, the old couple becomes politicized, and the Old Man leaves to take up arms for the nationalist cause. The sound of the rumbling earth fills the theater, and an image of a mask of Jugartha, symbol of anti-colonial rebellion, is projected onto a screen. Rather than representing the earthquake as an inexplicable, meaningless bringer of suffering, Kréa's characters interpret it as a clarion call to revolution and as a sacrifice for the cause. In Kréa's play, the earthquake-inflicted suffering of the people, like the violence of rebellion, is the price of salvation. In this vision of human politics and the physical environment, the two become one: the earthquake and the Algerian Revolution form an inseparable cataclysm.[12]

Decolonization and Disaster

Historians have increasingly recognized the role of environmental disasters in the expansion of colonial empires and the development of colonial states.[13] In the historiography of decolonization in North Africa, however, environmental events, no matter how catastrophic in scale or transformative for those involved, are often relegated to the background.[14] For the average historian, "Algeria 1954" is shorthand for one thing: the beginning of the Algerian Revolution, a war which finally led to Algerian independence over seven years later. For the survivors of Orléansville, however, "Algeria 1954" invokes not only the revolution but also the earthquake. For those who experienced environmental disasters in Morocco, Algeria, and France between 1954 and 1960, the consequent horrors were major events, not mere footnotes to the military and political upheavals of those years. However, the experience of empire and decolonization did not cease when environmental disasters erupted into the social and political lives of humans. The inseparability of the human and the environmental, of disaster and decolonization, was inscribed across a range of historical texts that can be examined by the historian: in archival documents, in architecture, in fiction, and in memoir. The interpenetrations of disaster and decolonization in these texts have often been obscured by the tendency of popular and academic historiography to separate narratives of political and military events and narratives of human culture and society from the history of the inanimate environment. Yet the evidence examined here reveals ways in which the environmental is lived and understood by humans through the experience of the political and the social.

The earthquake portrayed by Kréa was one of several environmental catastrophes that were bound up in the history, literature, and memory of decolonization in the French empire. This book focuses on four major environmental disasters that afflicted France, Algeria, and Morocco. These disasters occurred in the period of French Africa's transition to independence, which came, formally, to Morocco (and Tunisia) in 1956; to much of sub-Saharan Africa in 1960; and, finally, to Algeria in 1962. Two of the disasters examined here are earthquakes: the September 9, 1954, earthquake and its seismic aftershocks in Algeria's Chélif Valley, and the February 28, 1960, earthquake in Agadir, Morocco. The other two are of overtly anthropogenic origins: the flooding of Fréjus, France, due to the collapse of the Malpasset Dam in 1959, and a mass outbreak of paralysis in 1959 Morocco, caused by the contamination of the food supply with jet engine lubricant from an American airbase.

These four disasters were interrelated in multiple ways. Refugees from the Orléansville earthquake found themselves in Fréjus when the Malpasset Dam collapsed, and Orléansville became a model for state responses to disaster in both Fréjus and Agadir. The experience of the 1959 poisoning altered the political calculus of both the US State Department and the Moroccan political opposition following the 1960 earthquake. The Fréjus flood was invoked by French, American, and Moroccan actors in the diplomatic wrangling, domestic politics, and great-power machinations that followed both the 1959 poisoning and 1960 earthquake. Due to these interconnections, these four events constitute a single object of study, impacting Algerian struggles for independence, French reckoning with the loss of empire, and Moroccans' endeavors to extricate themselves, even after formal independence, from the continuing military and cultural legacy of French occupation.

These four disasters were not the only four horsemen of the French empire's apocalypse in North Africa. Indeed, drought, famine, disease, torture, and massacres are all mentioned in the chapters that follow. Elinor Accampo and Jeffrey

Jackson have defined "disasters" as "extraordinary circumstances that force individuals, governments, and organizations to operate in unusual and stressful ways."[15] There were many such circumstances amidst the tumult of imperialism and decolonization. Nevertheless, these four catastrophes were distinct from the other calamities of the era in their interconnectedness and in that they were large-scale transformative events produced by "rapid-onset hazards," the onsets of which were not intentionally initiated, thus fitting a narrower scholarly definition separating disasters from other infelicitous events.[16]

No great importance is placed here on the distinction between "nature-induced" and anthropogenic disasters.[17] Rousseau said of the 1755 earthquake that devastated Lisbon and Meknes that, if people did not build homes made of heavy materials, earthquakes would cause little damage.[18] Rousseau's argument becomes particularly relevant to the present study when one learns of the injuries to earthquake-stricken rural Algerians caused by collapsing tile roofs, an innovation that had replaced the mud-and-thatch used in decades past. More than two centuries after Rousseau, scholars now widely accept that the effects of so-called "natural disasters" are determined by historical processes and human actions. As Jonathan Bergman has put it, natural disasters constitute "a meeting ground or 'human ecology' between human and non-human worlds."[19] This is also true of unintentionally produced disasters such as the Malpasset Dam collapse in France and the mass chemical poisoning in Morocco, which were triggered by anthropogenically manufactured hazards.

Ted Steinberg has argued that the modern idea of a "natural" disaster is a technology of power, allowing elites to obscure the processes that produce disproportionate suffering among the disempowered.[20] In all four of the disasters addressed here, the colonized suffered more than did the colonizers, and the poor more than the rich, due to the unequal distribution of resources both before and after the disaster. In the case of the two earthquakes, the idea of "natural disasters" helped elites' efforts to obscure the social determinants of this suffering. But all four of these disasters revealed the inequities and injustices of colonialism.

However, while recognizing the revelatory aspect of disasters, we must also recognize that disasters can also be transformative, creating abrupt changes in political, social, and cultural landscapes as well as physical ones. The present volume takes up the challenge posed by environmental historians to consider seriously the impact of the environment on human history.[21] Arguing against the notion of "natural" (or "accidental") disasters in favor of the idea that the results of disasters are determined by human decisions and social arrangements, one runs the risk of denying the importance of the environment in history, repeating the notion of modern history as a story of humans shaping their own destinies using ingredients provided by nature.[22]

In the mythology of modernity, the environment begins as a problem but becomes a tool. This myth is shadowed by its inverse: the Frankenstein narrative of human hubris that places (inadequate) human rationality in opposition to (uncontrollable) nature. However, sources close to these events reveal a multivalent relation between the human history of decolonization and the environmental history of floods, chemical compounds, and earthquakes. As Timothy Mitchell has argued, the opposition between inanimate nature and active human rationality is problematic, a discursive tradition which itself ought to be an object of study, imposed upon a messy web of causality involving both human and non-human actors. In contrast to this hegemonic discourse, survivors' memoirs and literary representations such as Kréa's represent counter-hegemonic visions of what Mitchell calls "the ambivalent relation between the nonhuman and the human."[23] Archival documents reveal how disasters disrupted colonial projects and undermined the propaganda of imperialist "civilization" that promised material and moral benefits for the colonized; disasters likewise undercut triumphalist nationalist narratives promising that political independence would bring salvation to the poor and oppressed. In other instances, however, the new circumstances created by disasters were mastered by the powerful, who forged new tools of colonial oppression, international diplomacy, anti-colonial propaganda, or authoritarian nation-building.

In Kréa's creative vision, humans are not the only actors, and the distinction between the human and inanimate fades. Kréa's narrator informs us that "earthquakes are subject to the same causes whether they be human or telluric."[24] Henri Kréa's portrayal of the 1954 earthquake's relation to the struggle for independence might be dismissed as "mere" metaphor, a literary device, if it were not for other types of documentary evidence revealing the effects of these disasters, as "both ecological facts and representational spaces,"[25] on the experience of decolonization, while also revealing the impact of decolonization on the experience of these disasters. This evidence includes French, Algerian, Moroccan, and American sources ranging from diplomatic cables and political polemics to memoirs and architectural reviews. These sources demonstrate that the intersections of disaster and decolonization affected not only memoirs and imaginative writing but also Cold War diplomacy, humanitarian medical missions, and the

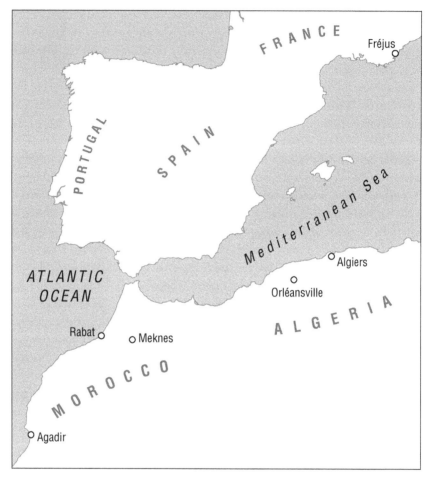

MAP 1. The western Maghreb and southwestern Europe. (Erin Greb Cartography.)

architecture and urbanism of post-disaster reconstruction. This archival record is replete with evidence that the processes of decolonization and disaster were not just related by a coincidental similarity in their effects on human bodies and built environments but that disaster and decolonization impacted each other in multiple ways, as the legacy of colonialism and the politics of decolonization shaped the distribution of harm and the distribution of aid, while the destruction of cities created new landscapes for the imagination and implementation of post-imperial visions.

Calamity, Empire, and Locality

The events described here all took place within a particular trans-Mediterranean region of the waning French empire, encompassing the western Maghreb and the Côte d'Azur. This book situates environmental disasters in the context of the processes of decolonization through which France, Algeria, and Morocco negotiated new relationships as separate entities rather than as components of an empire. However, the focus here is on localities rather than nation-states. To be more precise, the book centers on the study of disaster-afflicted provincial cities and the areas around them: Orléansville and the Chélif Valley, Fréjus and the Reyran Valley, Agadir and the Souss Valley, and, to a lesser degree, Meknes, the initial epicenter of the 1959 poisoning that afflicted much of Morocco. Scholars Gregory Mann and Emmanuelle Saada, among others, have urged their colleagues to integrate the study of particular localities into our understanding of imperialism in both metropole and colony and to explore the relation of localities to each other and to global and imperial institutions and networks.[26] In an article focusing on Fréjus, Mann seeks to push colonial studies to see beyond the colonial cultures and discourses of the imperial administrative centers (Paris, Dakar) to the varied and specific places that constituted the empire. This book follows Mann by working to "disaggregate"[27] the waning French empire by focusing on specific provincial localities—but it also explores what might be illuminated by re-aggregating these localities through the trans-local category of "disaster" (*sinistre, cataclysme, catastrophe, karitha*), a category neglected in academic historiography but dynamic in the popular press of the time, the archival record, and the remembered experience of those involved. Mann writes: "The town [Fréjus] and places like it offer immense potential for understanding the cultivation and evolution of historically grounded social and political formations, as well as the emergence of new ones."[28] If we read "places like it" to mean other locales both transformed by decolonization and afflicted by disaster—such as Orléansville, Meknes, Agadir—we can then examine how disasters catalyzed the "emergence" of new "social and political formations" and how imperialism and decolonization shaped how people in these locales interpreted and responded to disaster.

The historiographic turn toward locality parallels efforts by Algerian scholars and memoirists to explore and remember local pasts, a trend which has produced complex treatments of the colonial era. Mostefa Lacheraf, in his 1999 preface to Dr. Belgacem Aït Ouyahia's memoir *Pierres et lumières*, describes the attempt by Algerian writers to liberate Algerian history from the dominance of "abstract ideologies, ahistorical and without relation to reality past or present" through the writing of local histories of village life and of cities. Writing during the civil war of the 1990s, Lacheraf argued that this localist turn could serve to counter not only "stereotyped ethnography" of Algerians (a legacy of colonialism) but also visions of Algeria promoted by Islamist radicals.[29] This historiography of the local also offers an alternative to nationalist narratives of heroic resistance to a uniformly destructive colonizer.[30] As chapter 7 of the present volume will discuss, Aït Ouyahia's memoir explicitly seeks to provide such an alternative through his self-deprecatory representations of his own tentative relation to the struggle for independence, and, less explicitly, through his ambiguous portrayals of French colonial education and medicine.

The turn toward local history in Algeria has also brought to light the relationships between local disasters and experiences of colonialism and decolonization. El Djamhouria Slimani Aït Saada has made the region of the Chélif Valley, including Orléansville (later known as El Asnam and today as Chlef), the focus of a study which "seeks to unravel the ties between physical and human geography and space, and economic, social, and cultural life" through an examination of memoirs, geographical works, and literature written from the time of French conquest to the late twentieth century.[31] Aït Saada's examination of the Chélif Valley details an enduring discourse about the violence of the environment in Algeria that has persisted from the time of French conquest through half a century of Algerian independence. French sources from the nineteenth century reveal that the Orléansville region was defined, in the minds of many French writers, by the environmental hardships inflicted on its inhabitants, as "the heat, the floods, and the earthquakes contributed to the construction of a strongly deprecatory imagination of the region."[32] Since independence, Algerian writers have continued to reflect on the brutality of this environment. This discourse of the violent environment resembles the persistent narrative of human violence as the defining feature of Algerian history recently deconstructed by James Mc-Dougall in both its racist imperialist form ("imagined native 'savagery'" requiring repressive violence from the colonizer) and its nationalist reincarnation ("inflexible, unreformable, total oppression" requiring revolutionary violence from the colonized).[33] But these discourses about political and environmental violence are not separate, parallel strands. Aït Saada's work reveals, with local specificity, a double helix of connections between discourses about calamity, colonial rule, and decolonization in the Chélif region.

Aït Saada explains that the association of natural disaster with the uprising of the Algerian people appeared not only in the work of Henri Kréa but also in that of other Algerian writers such as Habib Tengour and Belgacem Aït Ouyahia (both discussed at length in this book), as well as Jean Millecam and Mohamed Magani.[34] Aït Saada could also have included the leftist dissidents Boualem Khalfa, Henri Alleg, and Abdelhamid Benzine, who echoed Kréa in their memoir of the dissident newspaper *Alger républicain*: "The autumn of 1954 opened with a cataclysm. As if nature wanted to be the herald of the hurricane that, for more than seven years, would tear apart and convulse the country."[35] For these writers, as Aït Saada argues, the "political and telluric tremors" of the period could not be separated.[36]

The intersections of colonialism and decolonization with environmental calamity are equally evident in writings about the disasters in Fréjus, Agadir, and Meknes. In Morocco, there has been much interest in distinct histories of localities long neglected in nationalist historiographies—particularly in the Tashelhit-speaking south, where Agadir is located. This interest is due both to the widespread perception, discussed in chapter 6, that the 1960 earthquake had a deleterious and disorienting effect on Agadir's relation to Moroccan heritage and identity and to the countervailing narrative of the Berber cultural movement that posits Agadir as the "capital" of Morocco's Berber culture.[37] In Fréjus, meanwhile, local notions of a particular identity have undergirded resistance to more inclusive notions of Frenchness from the time of the 1959 Malpasset Dam collapse to the present. The chapters that follow make the argument that, in these localities, the experience of catastrophe was inseparable from the upheavals of decolonization, and the shape of decolonization was crafted by catastrophe.

The Long and the Short of It

Decolonization and disasters were often tied together in survivors' and observers' interpretations and memories of events—that is to say, in the lived experience of events, for we live in realms of memory and interpretation. Was this merely an ahistorical psychological or linguistic phenomenon—a tendency to associate suffering with suffering, sometimes including other, more personal traumas? Does Kréa's metaphor of disaster and decolonization have dramatic force only by means of a trick of the mind or linguistic sleight of hand through which the reader accepts the equation of one sort of woe with another? In Kréa's play, the casualties of war and earthquake are intermingled with childbirth, death, and the suffering of famine.[38] In other accounts of disaster discussed in this book, an automobile accident, a geriatric medical crisis, and a childhood sexual assault mingle with the violence of decolonization and of disasters. Meanwhile, the equation of environmental disaster with war had become commonplace in the twentieth century, as new technologies and methods of warfare—bomber planes and basket bombs—now leveled buildings and created carnage in a way that resembled the effects of earthquakes and floods.

This book argues that the survivors, novelists, and memoirists who associated the disasters of the period with decolonization were not merely suffering from a painful mental illusion or capitalizing on a coincidental resemblance or useful metaphor. They were invoking a literary trope, to be sure, but the power of that literary trope was based, as good literature is, on astute observation. These individuals recognized that distinctions between the natural and the social, the seismic and the political, are illusory. No environmental event is experienced outside of sociopolitical contexts, and no human story takes place in an environmental vacuum. The archival record reveals that the upheavals in Algeria, Morocco, and France that were triggered by the disasters of 1954–1960 were inseparable from the upheavals produced by the violence of colonialism and decolonization. The human relation to the environment is not separate from political history, and one cannot write the history of these disasters without writing the history of decolonization.

Unlike Mitchell's *Rule of Experts* and Gregory Clancey's *Earthquake Nation*, works which have deftly treated intersections of environment and Western imperialism, this book does not focus primarily on "experts," although experts are certainly included here. Much of this book examines the disruption caused by disasters and the efforts to conceptualize and instrumentalize disasters among those whose claims of expertise lay elsewhere: mayors, legislators, bureaucrats, diplomats, journalists, novelists, dissidents, a fisherman, an obstetrician, a filmmaker. Some of the sources used here conform to the traditional definition of "hard" primary sources: written close to the time of the event, for purposes seemingly other than the representation of events to posterity, but collected by archivists for later use in the construction of historiography. However, the present volume also examines, as source evidence, texts written months, years, or decades later. These texts include not only imaginative writing and personal memoirs but also local histories. However, while written long after the onset of the disasters discussed here, such sources are no less "primary" to an examination of the long-term effects of these disasters and the meanings that humans ascribed to them.

This book examines disasters and decolonization as temporally extensive phenomena that do not occur in a "catastrophic instant" but that unfold over the course of time—not only weeks and years but also decades.[39] Cities, once destroyed, remain transformed evermore, even if rebuilt. Bodies that become paralyzed remain impaired for decades, even if some recovery occurs. And the dead remain dead, even if they do not all remain in their original graves. Therefore, sources that would usually be considered "secondary"—produced long after the initiating "event"—become "primary" in this analysis, revealing the disaster and decolonization in their temporally extended forms. Conversely, the "primary" sources produced soon after the onset of catastrophic events are no less representational than texts produced decades later, but they can represent only a temporally truncated version of the event; other sources must be consulted to understand catastrophic events as they unfold over a longer duration and are inscribed into built environments, histories, and memorializations.[40]

Decolonizations, like disasters, are also extended affairs. The term decolonization includes the formal recognition of national political independence through a treaty or accord (e.g., Morocco and France in 1956; Algeria and France in 1962); it can also refer to the mass exodus of colonists and descendants of colonists from a particular locality (e.g., Agadir in 1960; Orléansville in 1962). Decolonization also includes the long chronology of revolutionary and counter-revolutionary violence that helped to bring about formal independence and continued thereafter. Sometimes, decolonization involves the departure not only of the living but also of the dead. Sometimes, the end of military occupation comes long after formal independence and after the outmigration of colonists and their corpses. In certain contexts, decolonization can mean the displacement of European imperialism by Cold War geopolitics and American (or in other cases, Soviet) neo-imperialism. Decolonization can also mean the end of the hegemony of Western imperialist cultural values, aesthetics, and epistemologies—a process that, if not unattainable, is far slower, more fraught, and much less complete than the others. Consequently, the synchronicity and interpenetration of the history of decolonization with the four disasters examined here does not consist solely in the fact that the *onset* of these disasters began during the years of the Algerian Revolution or within a few years of Morocco's emergence as an independent nation. The disasters that began in 1954–1960 were long, and so were the decolonizations, and their interconnectedness has spanned decades.

The enduring temporality of both disasters and decolonization has been explored within the field of trauma studies. Approached from a psychoanalytic perspective, traumatic events are identified as such because the individual cannot comprehend them as they occur; consequently, they are experienced only later, through flashbacks, emotions, and sensory experiences that disrupt the normal experiencing of time.[41] Scholars of colonialism and decolonization have applied the concept of trauma to describe the social and psychological effects of colonial oppression on both the colonized and the colonizer.[42] Paralleling the ideas of Henri Kréa, scholars Kai Erikson and Andy Horowitz have problematized the distinction between sudden traumas and the slow grind and quotidian violence of social injustice or political repression, while applying this expanded concept of trauma to the experience of putatively natural disasters. Erikson has argued that not only sudden events but also "chronic conditions" can produce the traumatic psychological and social reactions that, in his view, define disasters.[43] Horowitz has argued that the trauma of disaster develops within the context of longstanding social injustice and political violence. In such contexts, the sudden "traumatizing agency" of the short-term event cannot be separated from the suffering caused by the long history and enduring legacies of oppression and conflict.[44] Human experiences of the earthquakes of Orléansville and Agadir, the 1959 poisoning in Morocco, and the flooding of Fréjus were indeed inseparable from the "chronic conditions" of injustice and violence brought about by colonialism and decolonization.

However, the enduring experience of environmental disasters, like that of decolonization, cannot be wholly subsumed within the category of trauma.[45] Disasters and decolonization involve not only traumatized sufferers but also resilient survivors—and triumphant opportunists. In the weeks, months, and years following the onset of the disasters studied here, the effects of colonialism, decolonization, and catastrophe manifested, for some, in the struggle to survive extraordinarily difficult times; for others, in the struggle for power. For many— diplomats and political leaders, rulers and dissidents, antagonists and allies in the era of the Cold War and decolonization—disaster was not trauma but rather was war by other means: the attempt to make Algeria French; the attempt to make it clear that Algerians were not French; the attempt to maintain Morocco's dependence on France; the attempt to extend American influence in Morocco; the attempt to demonstrate Morocco's independence from France. The effects of disasters were tied to questions ranging from whether French military bases might remain after independence to what a modern, independent North African city ought to look like.

In their short and their long manifestations, the disasters discussed here shaped international relations, urban landscapes, and the attitudes of French, Moroccan, and Algerian individuals toward the colonial, pre-disaster past and toward the post-disaster, post-independence future. The breakup of the French empire shaped how these disasters were conceptualized, how historical actors responded to them, and how the suffering was distributed. The ongoing lives of disasters and decolonizations extended through the decades in the physical environment of cities and ruins, buildings and cemeteries, and in the culture and politics surrounding memory and the built environment. Rebuilding the cities of Orléansville, Fréjus, and Agadir took years, and these urban environments were forever transformed by their destruction and reconstruction, and by the political and demographic upheavals of decolonization. Survivors would forever have to cope with the absence of what had been destroyed, and with the meaning of living in, or in exile from, a city that had been transformed. However, post-disaster urban landscapes also provided an opportunity for new assertions about culture, identity, and power. For some, recovery from disaster offered opportunities to break from the past and impose visions of a glorious future; for others, disaster offered an opportunity to lament how much had already been lost. The documentary record produced after these catastrophes reveals competing elites, dissidents, and disaster survivors jockeying to advance competing and protean visions for Algeria, France, and Morocco in spaces fundamentally altered by political and environmental events.

This book argues that the interconnections between these disasters and the decolonization of Morocco and Algeria are evident in both the long and the short term, as evinced by sources produced throughout this temporal scale, ranging from diplomatic dispatches and radio broadcasts to memoirs, novels, architectural plans and urban landscapes. Chapters 2 and 3 examine the Chélif Valley earthquake and the Malpasset Dam collapse in both their short and long manifestations, through the lenses of archival documents and through works of long-term memory and representation. Chapter 4 concentrates on the Morocco oil poisoning in the short term, examining the expanding impact of the United States and the Cold War on the political and diplomatic environment in the waning French Empire. Chapters 5 and 6 both examine the Agadir earthquake, providing an extended case study of the short and long temporality of a particular disaster. Chapter 5 treats the short term, investigating the manifestations of the political and diplomatic struggles of decolonization and the Cold War in controversies concerning the treatment of survivors, the burial of the dead, and the reconstruction of the city. Because the built environment of Agadir's post-earthquake urban landscape and the legacy of disaster remained central to debates about the meaning of Moroccan identity in ways not paralleled in Fréjus or in Orléansville-Chlef, chapter 6 extends the treatment of the urban reconstruction of Agadir into the longer term, examining the politics of decolonization in relation to architecture and urban planning. Chapter 7 returns to the topic of memory and literary representation, examining the long process of making meaning of catastrophe among survivors and observers of disasters and decolonization in both Morocco and Algeria.

In these chapters, I have sought to demonstrate that the 1954 Algeria earthquake, the 1959 Malpasset Dam collapse, the 1959 Morocco oil poisoning, and the 1960 Morocco earthquake were not just momentary ruptures in human history, but were enduring phenomena intimately connected with a decolonization process that continued long after the 1962 Évian Accords brought formal independence to the last of France's North African holdings. It may not be precisely true that "earthquakes are subject to the same causes whether they be human or telluric."[46] But Henri Kréa was right to see human and environmental history, decolonization and disaster, as interwoven both in the immediate aftermath of catastrophic events and through the long "eternal wave of generations."

CHAPTER 2

Algeria, 1954

A T 1:06 A.M. ON SEPTEMBER 9, 1954, Algeria's Chélif Valley was struck with an earthquake measuring 6.7 on the Richter scale. Two hundred kilometers northeast of the epicenter, the news reached a young Muslim doctor, Belgacem Aït Ouyahia, sitting in his new Renault 203 in the town of Haussonvillers (today Naciria). Aït Ouyahia had just left the Chélif the week before after finishing his surgical internship in Orléansville; he was on his way to take up a position in the colonial health service as a "médecin de colonization de la région" in his native Kabylia, while he finished his thesis. It was eight in the morning, and he had just stopped for gas, coffee, and a *beignet*. Aït Ouyahia recounted that moment in his 1999 memoir:

> I got back behind the wheel and turned on the radio:
> "... [*sic*] has shaken the region of Orléansville. Numerous buildings have collapsed. Already there are known to be many victims, and the hospital is inundated with the injured. This is the largest earthquake ever known in Algeria..."
> —My God! My God!
> And I surprised myself; I, who was not too observant—even not observant at all—I surprised myself by reciting the *shahada* in a loud voice:
> "There is no God but God, and Mohammed is his prophet."[1]

Aït Ouyahia immediately turned the car around and returned to Orléansville to rejoin the medical team at the hospital there. Aït Ouyahia was fortunate to have been far from the center of the earthquake when it struck, but upon his return to Orléansville, a city of thirty thousand people just thirty kilometers from the epicenter, he would soon be confronted with the horrors that this sudden movement of the inanimate had inflicted on human bodies.

In his 1955 memoir, a French official, René Debia, described the experience of the city's inhabitants:

Those who were not crushed immediately were thrown to the ground "like the fruits of a fig tree bent by a storm," like sailors in the midst of a storm, of which the same din arose around them; it was a deafening noise, formed of an extraordinary rumbling that rose from the subsoil of the earth, of the crashing of walls, and their cracking, like that of a ship rocked by waves, of the dull thud of buildings that crashed on their neighbors like water on a bridge. And the fathers, the mothers who gathered together their children that night, in the darkness, did so instinctively with the idea of awaiting together a death which seemed to them inevitable; but the shaking of the ground abated, little by little; without realizing it, they tried to grope their way to an exit or a stairwell, they climbed over the piles of rubble and twisted iron; surprised to be still alive, they reached a courtyard, a garden, or a street; and there, breathing, breathed finally an odor of sulfur that came, they didn't know from where, and this opaque dust cloud that enveloped the city, adding to the thickness of the dark.[2]

When the sun rose, wrote Debia, Orléansville resembled a bombed out city, a "landscape of death."[3] Debia stated that four thousand homes had been destroyed in the city; in the entire affected region, he counted eighteen thousand ruined "houses," plus the destruction of thirty-five thousand *gourbis*—small, windowless structures made of earthen bricks, packed earth (*pisé*), or sticks or stones cemented with mud, that were home to the vast majority of the rural Muslim population. In Debia's words, the gourbis were obliterated "as if by an explosion. Often the roof was intact but had collapsed in one piece on top of the rest of the structure; from this debris one pulled out the cadavers and the injured."[4]

Scientists would later attribute the disaster's onset to the slow collision of tectonic plates in the Dahra mountains near the Mediterranean coast.[5] An aftershock measuring 6.2 on the Richter scale struck farther north near the coastal city of Ténès the next day, bringing down more buildings. Countless more aftershocks followed, including significant tremors on September 16 and October 19 and 21. The earthquake and its aftershocks killed at least twelve hundred people throughout the region, injured about fourteen thousand, and left as many as two hundred thousand homeless. The vast majority of the dead—over 90 percent—were Muslim Algerians, reflecting the overall population of the affected area as well as the quality of housing.[6] As in most earthquakes, it was human construction that did most of the actual killing. The deadly pancaking of gourbis described by René Debia was most likely due to the adoption of tile roofing material weighed down and held in place by large rocks. In urban areas, the widespread use of masonry in building construction proved similarly lethal.[7] Neither metropolitan nor Algerian France had building codes specifically developed for areas of seismic risk, and the general French building code was only applied to larger buildings constructed after 1946 from steel-reinforced concrete.[8] On the other hand, reinforced concrete was insufficient to save an almost-completed nine-story, low-income apartment building (*habitation à loyer modéré* or HLM) in Orléansville, which "collapsed like a house of cards, crushing the workers who were living on one of its floors."[9]

Those who died there left no memoirs, of course. If disaster victims are a particular subcategory of subaltern, then only the survivors speak.[10] However, the voices of the most disempowered survivors are also muted: the destitute, the illiterate, those too traumatized by disaster or terrorized by war, or too occupied by the struggle to survive to provide testimony for posterity. René Debia and Belgacem Aït Ouyahia were in many ways typical of those who were able to provide such testimony. As the French subprefect for the region of Orléansville and an educated surgeon and future professor of obstetrics at the Algiers School of Medicine, respectively, these men were sufficiently privileged to get their own representations of the disaster published, even if this event, so enormous in scale and transformative in their own lives, would be pushed to the margins of the dominant historical narratives of the era.[11] Both Debia and Aït Ouyahia saw the disaster as intimately related to the question of decolonization in Algeria, but the two men held sharply contrasting views of this relationship. For Debia, the earthquake both revealed and augmented the commonality of interests between the French and the Muslim Algerians of the Chélif Valley, demonstrating the necessity of continued French rule; for Aït Ouyahia, the disaster was a turning point that decisively demonstrated the oppressive nature of French colonialism. Their writings demonstrate the fractured experience of these events as well as the inseparability of the environmental catastrophe from the experience of political turmoil.

Like the memoirs of Debia and Aït Ouyahia, sources written in the days and weeks following the 1954 earthquake also reveal divergent understandings of these events as well as the linkages between the earthquake and the possibility of decolonization. Archival sources demonstrate that the catastrophe became a tool for the French colonial state to use disaster response to counter nationalist narratives and to defend a vision of benevolent colonialism. Yet, because the earthquake wreaked enormous destruction on the built environment, especially on the humble edifices housing most of the region's Muslim population, it dramatically increased the financial cost, technical difficulty, and bureaucratic complexity of the colonial state's efforts to make plausible its narratives of civilization, development, and solidarity. Consequently, the disaster provided an opportunity for dissidents to offer counter-narratives decrying the hypocrisy and futility of the colonial project. The earthquake thus became an important part of the struggle for and against Algerian independence, and the political struggle in Algeria shaped human responses to the seismic shocks that rocked the Chélif Valley in 1954.

A Department of France, a Valley in North Africa

French histories of Algeria's Chélif Valley typically begin with the founding of Orléansville in 1843 as a military camp by General Thomas-Robert Bugeaud. Bugeaud's occupation of Orléansville, and of the Mediterranean town of Ténès, forty kilometers to the north, was intended to create a bulwark against the return of the forces of Amir Abd al-Qadir, the Algerian resistance leader whom French forces had driven out of the Chélif Valley.[12] The new French outpost was near the site of an ancient Roman colony of Castellum Tingitanum, a fact that proved useful to the narratives of French imperialists seeking to link the French conquest of Algeria to the heritage of Rome. The local name for the site (which may have been the location of a weekly souk) was Lasnab or in classical Arabic, al-Asnam, meaning "idols," a name possibly derived from the memory of the Roman presence or of their ruins. El Asnam became the official name of the city after independence until it was renamed "Chlef" after another earthquake devastated the city in 1980.[13] In 1848, the northern part of what is today Algeria, including the Chélif Valley, was declared part of France by the Second Republic.

Orléansville's location, isolated by the Dahra mountains to the north and the Ouarsenis to the south, was perceived as inhospitable by the French due to its climate and due to the ongoing resistance of the inhabitants of the two mountain ranges.[14] Nevertheless, the European settlement at Orléansville grew into a thriving town and became the seat of a subprefecture administering a district, or *arrondissement*, extending north to Ténès, within the larger French department of Algiers, *département* 91 among the departments of France.[15] (The *départements* are roughly akin to a North American state or province.)

To begin the history of this valley in North Africa with the arrival of Europeans reinforces an obviously Eurocentric historical metanarrative. If the scope of the narrative is narrowly confined to the city of Orléansville, this approach might seem, at first, to be satisfactory. Geographer Valentin Pelosse points out that, unlike Algerian cities such as Algiers, Constantine, and Tlemcen, the

French post at Orléansville was not grafted onto an Algerian Muslim town, and there was only a limited presence of Muslim Algerians in Orléansville during the early years of the settlement's growth.[16] For René Debia, the subprefect and administrative head of the arrondissement in 1954, Orléansville, though inhabited by migrants "from all shores of the Mediterranean," grew into a typical "small bourgeois city" of the Third Republic.[17]

A transformation was underway, however: the city started to become what Debia called "a great indigenous city, one of the most significant Arab cities in all of Algeria."[18] By 1900, the city of Orléansville was already home to approximately three thousand Muslim Algerians, half the total population.[19] These Muslims did not materialize out of thin air. Even if Orléansville was a creation of the French, the Chélif Valley was not—there was an earlier history of the region. The Chélif Valley was inhabited, prior to the French arrival, by the *Awlad Qasir*, among others. By 1863 the French had driven the *Awlad Qasir* out of twelve thousand hectares of the best agricultural land.[20] For the next several decades, Muslims often sought to escape the rural poverty caused by French conquest of the Chélif by emigrating to Tunisia or other Muslim countries rather than to European-dominated Algerian cities. Around the turn of the century, however, migratory flows shifted toward the city, and the history of the rural Chélif merged with the history of French-built Orléansville.[21]

This growing city, like all of Algeria, was dominated by its European inhabitants, who enjoyed the full rights of French citizenship, while the Muslims were French subjects ruled according to the *indigénat*, or native-status, laws.[22] Nevertheless, certain leading Muslim families were able to maintain their prominence by switching their allegiance to the colonial state, which needed collaborators who could facilitate control of the Muslim population as French-appointed rural notables, or *caïds*. The most significant of these families in the Chélif was the Saïah family.[23] The Saïahs became a focus of controversy and criticism after the Orléansville earthquake. As propertied elites with close ties to the French regime, they represented a symbolic fulcrum in the struggle for post-disaster justice in the region. Though members of the Saïah family positioned themselves as advocates for the Muslim population, they were portrayed by critics as oppressors, criminally implicated in an unjust system.

The Saïahs' relationship with the French was shaped, over the years, by French reform measures intended to justify, mitigate, or occlude the arbitrary and oppressive character of colonial rule. While the most dramatic of these reforms were enacted after the Second World War, historians have argued that this process stretched back into the nineteenth century and included the 1898 establishment of the Délégations Financières, an assembly that included a limited number of seats for privileged Muslim Algerians. Reform accelerated after the First World War, when almost half of adult Algerian men were exempted from the *indigénat* and given voting rights in local elections. After World War II, the French regime affirmed the citizenship of Muslim Algerians and, in 1947, replaced the Délégations Financières with a 120-seat Algerian Assembly, elected by voters who were divided into two "colleges," one composed of Europeans and a small number of select Muslim Algerians, and one composed solely of Muslim Algerians.[24]

These postwar reforms maintained a system of governance firmly under the control of the French of European origin and their chosen Muslim allies. Although the *indigénat* was eliminated, and adult male Muslim voters could now elect some representatives to the French National Assembly and the Algerian Assembly, in Algeria the influence of the sixty Algerian Assembly delegates chosen by the Muslim "second college" voters was checked by the power of the sixty chosen by the overwhelmingly European voters of the "first college." Consequently, as has often been pointed out, delegates representing a European population of less than a million and less than sixty thousand of the most privileged Muslims (those who had been given "first college" voting status) could block the will of the delegates who represented a population of almost eight million. The ability of the settler population to prevent the Muslim majority from expanding their power was reinforced by the requirement of a two-thirds supermajority for certain proposed reforms. When an upswell of Muslim political activity led to nationalist successes in municipal elections in 1947, the authorities further subverted the limited democratic potential of this assembly through intimidation and election-tampering in favor of approved Muslim candidates.[25]

This "reformed" political system prevented real democratization in Algeria, but it provided opportunities for a well-positioned few. Closely tied to the French colonial system, the Saïah family rose to top of the reformed political structure. At the time of the earthquake, members of the family held important positions in French Algeria: Saïah Abdelkader was a member and former president of the Algerian Assembly, Bouali Saïah was a member of the Assembly, and Saïah Menouar was a representative to the French National Assembly in Paris. Locally, several members of the family still held posts of *caïd* in the Chélif region.[26] The vast majority of Muslim Algerians were less fortunate, and it was the poor Muslim population that would suffer the most after their homes were destroyed by the earthquake, in the growing shantytowns around Orléansville housing migrants from the countryside and in the vast rural areas housing an impoverished and scattered population. The relation of the Saïahs to this population—as oppressors or advocates—would be a matter of controversy when the earthquake struck.

City and Country on the Eve of Destruction

In sharp contrast to Agadir, destroyed by an earthquake less than six years later, controversy and political struggle following the 1954 earthquake in the Chélif would center as much on rural as on urban areas, while the destruction of Orléansville's architectural legacy, such as it was, would provoke little comment. Orléansville's twentieth-century landscape still bore the imprint of its military origin. A grid of streets was surrounded by a wall and a belt of military land, with military buildings dominating the western part of the city. Outside of the walls, the metropolitan area (*agglomération*) of Orléansville included two suburbs (*faubourgs*), La Ferme in the north and La Bocca Sahnoune in the south. These suburbs became increasingly Muslim as they absorbed migrants from the countryside. Europeans engaged in "white flight" from the faubourgs into city proper, while many Muslim Algerians lived in improvised mudbrick and *bidonville* (shanty) housing without piped water or sanitation infrastructure.[27]

As Benjamin Stora explains, many Muslims across Algeria had been driven to cities by French expropriation of land and water, the disruption of pre-colonial social and economic networks, and population growth. In the nineteenth century, communal tribal lands, properties of religious brotherhoods, and lands of the defeated Ottoman governor had been partitioned into private plots, leaving the average Muslim Algerian farmer with only seven acres, barely enough for subsistence. Meanwhile, traditional systems of water management were disrupted, along with communal landholding and charitable religious foundations. By 1919, *colons*, farmers of European descent with French citizenship, possessed a million acres of land in the département of Algiers alone, although, it should be noted, there was enormous inequality of wealth within the European population, and agricultural consolidation in the twentieth century also led to the urbanization of the European population.[28] In the Chélif Valley, rural poverty and further economic disruption in the period of the world wars prompted an acceleration of rural-to-urban migration, and by 1948 the official census in Orléansville counted 13,693 Muslim Algerians, out of a population of 17,223.[29] The rate of migration was such that, by 1954, the majority of the city's population had been born in the countryside. By 1960, when the European population of Orléansville reached its apex, the Muslim population had grown to almost thirty-eight thousand, and people classified as "European" constituted only 16 percent of the population.[30]

These transformations of the city created some unease among the French elite. A report by the département's Service of Urbanism described the population growth of the suburbs as "disordered."[31] For Sub-Prefect Debia, an advocate of colonialism through economic development, the growth of the city's Muslim population was, in part, a positive development, insofar as it included Muslim merchants and functionaries and later some doctors and lawyers, of Kabyle and Arab backgrounds, who "adopted our way of life, if not in terms of dress then at least in the realm of habitat," moving into villas and apartments formerly occupied by Europeans. "Less encouraging," lamented Debia, "because it was a sign of poverty and because it often resulted in social uprootedness," was the much more numerous settling of Muslims in the faubourgs "where they often lived as if they were in the *douars* [rural villages]."[32] Debia described the exponential growth of unregulated housing in these suburbs as a demographic battle that threatened to overwhelm the legacy of the planned French city, protected by the "solid corset" of its ramparts.[33] Debia's modernist, imperialist fear of unregulated Muslims was echoed by another French official who described the faubourgs as "two popular quartiers constructed in an anarchic fashion, in violation of the most elementary rules of hygiene and urbanism."[34]

Reinforcing this vision of a city divided between realms of European progress and Muslim disorder, the European sections of Orléansville had by 1954 become home to dazzling monuments to Europeans' belief in their own modernity. Debia described the architectural innovations: "Here, an ultra-modern building; there, a gigantic school, the largest in France [*sic*]; an administrative *hôtel*; a ten-story building under construction; a magnificent subprefecture in the hispano-mauresque style." It was a city "glittering with light at night" and a city of motorcars.[35] The city walls were undergoing demolition, and the glittering modernity of the city extended westward beyond the old city limits, exemplified by the construction of a "sumptuous" new building to house the administrative offices of the subprefecture.[36]

All this was far removed from the lives of most inhabitants of the region, however. Beyond the immediate environs of the city, the arrondissement administered by the subprefecture of Orléansville, which included most of the zone affected by the earthquake, was overwhelmingly rural. This area included eight sizable communities, with a total population of about eighty-five thousand, that were classified as *communes de plein-exercice,* endowed with elected municipal governments due to their significant European populations. In addition to the city of Orléansville, these included Ténès, on the coast (population ten thousand); and Oued Fodda (population twelve thousand), east of Orléansville, and five smaller towns. Nevertheless, most of the arrondissement's three hundred thousand inhabitants lived in rural districts classified as *communes mixtes*, where the almost entirely Muslim residents were administered by appointed officials.[37] In these rural areas, "extreme dispersion characterized the distribution of the rural population," who survived through a combination of pastoralism and agriculture.[38] This "extreme dispersion," over an area of 450,000 hectares (more than 1,700 square miles), would make disaster response slow and difficult.

Disaster Response

The effectiveness, earnestness, and equity—or lack thereof—of the French disaster response effort became a central focus of public contestation in Algeria in the autumn of 1954, as competing voices struggled to frame the shortcomings of the disaster response as either the fruit of an intrinsically unjust system, or as the inevitable result of the sheer magnitude of the "natural" disaster amid the putatively primordial backwardness of the Algerian people, or simply as the result of organizational failures that could be corrected for future disasters through technocratic adjustments.

Those involved in the disaster response in its early stages testified to the magnitude of the challenges they faced. On the night of the first earthquake, "total confusion, in darkness" reigned in Orléansville until sunrise; one early report described an "atmosphere of war (presence of numerous soldiers) and of post-bombardment."[39] The seismic shock had destroyed the city's means of telecommunication with the outside world and had interrupted its electrical supply. A gendarme was dispatched to the town of Oued Fodda, twenty kilometers away, where he was able to reach Algiers by telephone, forty minutes after the disaster.[40] The radio transmitter of the French military subdivision in Orléansville had been damaged but was repaired within two hours, enabling Subprefect Debia to send three messages requesting tents and food as well as civil engineers, trucks, and bulldozers. As the hours passed, news of destruction in Ténès and in several towns in the Chélif Valley trickled into Orléansville, and Débia requested helicopters.[41]

The city was home to a volunteer crew of thirty or so firefighters (*sapeurs-pompiers*). They were reportedly unable to organize as a team in the early hours of the disaster, but were instead drawn individually or in small groups into rescue efforts in their immediate vicinities; the fire chief and several pompiers became engaged in efforts to extricate victims buried in the ruins of the Hotel Baudoin.[42] The two hundred legionnaires housed at the military garrison and the thirty or so *police d'État* were able to respond in more organized fashion within hours. By the afternoon, their numbers were augmented by the arrival of an additional five hundred troops, with trucks and bulldozers, and several sapeur-pompier units from Relizane, to the southwest, and Algiers, to the northeast.[43]

The medical staff at the hospital constituted another indispensable group of first responders, reinforced by health professionals arriving from other areas. In a chapter titled "Orléansville 54," Dr. Aït Ouyahia described the scene in the hospital just after his arrival the next day, when additional tremors struck. Patients sustained additional injuries as they were thrown from their beds by the aftershocks. Beds were moved to the garden as new patients were brought in with fresh injuries. Amid the fear and chaos that ensued, the medical staff, led by Aït Ouyahia's mentor Dr. Kamoun, kept working.[44]

As crucial to the disaster response as the functioning of the hospital was the fact that the city possessed not only a railway depot but an airport; both were damaged but still usable. The morning after the first earthquake, the first relief shipment, containing medical supplies, arrived by air.[45] The Cold War presence of the US Air Force in Europe and in neighboring Morocco proved valuable, as six American C-119s joined ten French army planes in the airlift of goods from France to Algiers and Algiers to Orléansville.[46] By afternoon, flights were arriving in Orléansville every twenty minutes, bringing supplies and evacuating the seriously injured; 117 were evacuated the first day. Hunger and thirst were rampant, but in the afternoon, the arriving army trucks brought bread and two cisterns of water.[47] Shipments of tents also began to arrive immediately, but the supply was grossly inadequate. People had no choice but sleep in the open, although some found refuge in train cars still on the tracks of the train station, slowing the arrival of shipments by rail.[48]

On September 11, a national disaster relief committee, the Comité National de Secours aux Victimes du Séisme de la région d'Orléansville, was created by the Ministry of the Interior in Paris, and the Government-General of Algeria established a parallel committee.[49] A large role in the post-disaster relief effort was played by the metropolitan Service National de Protection Civile (SNPC), which dispatched a team to Orléansville, including seventeen members of the Paris *Sapeurs-Pompiers* to supplement the firefighters from Orléansville and from Algiers. The SNPC team leader, Lieutenant Colonel Curie, arrived in

Algiers on September 11 at what is today the Houari Boumedienne airport, along with a shipment of 16 large tents and 245 beds. That night, Curie flew to Orléansville, where he met with Subprefect Debia and Mr. Freychet, director of the departmental relief service (Service de Secours). The rest of the metropolitan SNPC team arrived within the next forty-eight hours, accompanied by an engineer named Marius Hautberg, who had been appointed to serve as an assistant (*adjoint*) to Colonel Curie, with a mission to conduct a study of structural damage, methods of clearing debris, and the organization of the disaster response.[50]

Those engaged in disaster response were not immune to the stress created by the carnage and destruction that surrounded them. Once assembled in Orléansville, the sapeurs-pompiers slept in tents at the military garrison, where food supplies were inadequate, while Curie and Hautberg joined Subprefect Debia, his staff, and his family in tents near the slightly damaged subprefecture building.[51] Hautberg recounted that, within the SNPC team, "overwork, fatigue, and a kind of necro-psychosis caused an ambiance of nervousness," and tempers flared.[52] Aït Ouyahia, too, referred to his own shock and emotional distress upon viewing the carnage.[53] As the days passed, response workers undertook the grisly task of excavating the ruins to recover bodies. Soon, "in the stifling heat of September, the atmosphere was permeated with the odor of decay."[54] In the blazing heat, workers began to use the stench to help them locate the bodies, which had to be painstakingly extricated from the rubble and then coated with quicklime in an attempt to prevent outbreaks of disease.[55] DDT was sprayed liberally over the city by helicopter and from trucks.[56] A school was converted into a makeshift morgue, where the bodies were placed in coffins, which were being shipped in from throughout the region.[57] After another major aftershock on September 16 brought still more damage to the city, eighty-four more firefighters arrived from Paris, bringing the total size of the SNPC team to about one hundred.[58] However, the dispatch of these reinforcements from the metropole was rushed after the new tremor struck. Consequently, they arrived without adequate advance planning or materials—they lacked sufficient food supplies for themselves and were not accompanied by the fifty tons of tents they had been expected to bring. As a result, the SNPC team was, an according to Curie, ill-equipped to respond to the new wave of disaster.[59]

Beni Rached

In the weeks and months following the earthquake, much controversy surrounded the dire conditions and scarcity of relief aid in rural areas, where the population was overwhelmingly Muslim. For the first few days, however, officials had initially assumed that the epicenter of the earthquake was in Orléansville[60] where initial counts of the dead ranged from 153 (including 23 "Europeans" and 130 "français musulmans") to 168.[61] There were also reports from other towns describing death and damage throughout an area extending from the Chélif Valley towns of Oued Fodda (163 dead) and Pontéba ("total destruction—numerous dead"), to Ténès on the coast.[62] However, officials were slow to recognize the extent of the disaster in rural areas, and no effort was made in the first forty-eight hours to extend disaster aid into the smaller villages, or *douars*, in hard-to-reach areas not served by roads, where most of the thousands of casualties had in fact occurred.[63]

It was not until September 11 that aid workers in Orléansville became aware of the enormity of the devastation of the village of Beni Rached, forty kilometers to the east at the true epicenter of the earthquake, where 300 residents had been killed.[64] Sources provide conflicting accounts of the discovery of the tragedy there. Official reports neglected the role of Algerian Muslim agency in uncovering and treating the suffering in Beni Rached, emphasizing instead the vigorous state response that followed. Colonel Curie's report from September 27, 1954, stated that the discovery of Beni Rached on September 11 was made "by chance" by a gendarme.[65] According to Colonel Curie's concise report, the morning after the discovery of Beni Rached, a US Air Force helicopter then flew reconnaissance missions in the area, returning with one of the injured. More helicopter evacuations followed, and a systematic effort was undertaken to identify affected rural communities, with ten medical teams sent out to canvass the region.[66] Colonel Curie's report on the SNPC mission to Orléansville was followed, in December 1954, by a report on the organization of the disaster response written by Philippe Kessler, a recent graduate of the elite École nationale d'administration who had been conducting an administrative traineeship near the Chélif Valley in September, and by Marius Hautberg's report addressing both disaster response and the structural effects of the earthquake on buildings.

Kessler's report credited the medical service's staff as being the first to address the full extent of the rural disaster. Like Curie and Hautberg, Kessler emphasized the importance of helicopters.[67] Kessler was impressed by the heroic drama of the aerovac: "It is thus that in certain places where, in the memory of man, no 'European' had ever passed, families affected by the disaster could see one of these providential machines descend from heaven, land at their door, from which would disembark '*toubib*' [doctor] or nurse. This medical penetration, provoked by the event, brought a royal and marvelous path to the unhappy people who benefitted from it."[68] This story of miraculous, technological "penetration" (a term of colonial conquest) by European saviors is redolent of the mythology of colonialism—the "providential" machines a modern version of Columbus's ships, appearing as gods, as Europeans liked to claim, to the inhabitants of a New World. Frantz Fanon informs us that Algerian Muslims often saw the colonial doctor as threatening and humiliating rather than "marvelous."[69] While there is no doubt that the helicopter evacuations saved lives, Kessler's version of the narrative emphasizes the importance of the colonizer's military technology, and erases the agency of Muslims—both outsiders and residents of Beni Rached—who responded to the disaster.

Dr. Aït Ouyahia tells a very different story. According to Aït Ouyahia, the medical staff at the Orléansville hospital, finally taking a dinner break on the evening of September 10, were joined by local notable Saïah Menouar, a deputy (representative) from Orléansville to the National Assembly in Paris. According to Aït Ouyahia, the young doctor turned to his supervisor, Dr. Kamoun, and said, "I have noticed, sir, that all the injured who have come to us come from the farms and villages that are along the roads. I wonder, in what condition are the isolated *douars* and *mechtas* [villages and hamlets]?"[70] Saïah Menouar offered the use of his jeep, and, after a few hours of sleep, Menouar and Aït Ouyahia left, still in the dark of night.

Aït Ouyahia may have downplayed Menouar's role in initiating the expedition. French records indicate that Saïah Menouar was born in Beni Rached and that six members of the Saïah family died there during the earthquake. This suggests that Saïah Menouar played a more active role in initiating the expedition and determining its destination than Aït Ouyahia indicated: Saïah Menouar was very likely the driving force of the expedition, if not the actual driver, as in the doctor's memoir.[71] Aït Ouyahia's account, published in 1999, reflects some ambivalence about Saïah Menouar, mentioning that Menouar had been "elected" by the Muslim population only after being handpicked "by the administration and *colons* of the Chélif," in consultation with the head of the Saïah family, Saïah Abdelkader.[72] Aït Ouyahia's depiction of Saïah Menouar's role in this story may have been influenced by nationalist condemnations of those who, like the Saïah, collaborated with French rule. Nevertheless, the doctor's memoirs granted Saïah Menouar a role, unlike the reports of Curie and Kessler.

According to Aït Ouyahia's memoir, he and Menouar drove about thirty kilometers on the road, through the town of Oued Fodda. (There, the ten-year-old Ali Bouzar, who would later write his own memoir of the earthquakes of 1954 and 1980, had just survived the disaster and was fearing for the life of his father, a medical worker in Orléansville—likely one of Aït Ouyahia's colleagues.)[73] Past Oued Fodda, they left the road and turned north, following a trail along a dry riverbed. At dawn, they reached a pair of collapsed dwellings. Under a fig tree lay, still alive, a woman, seven months pregnant, the skin on her bloody abdomen torn back as if "scalped," along with a man and a small child. Around them lay corpses: their three sons, and the man's parents. Aït Ouyahia applied sulfa and bandages to the woman's wounds and promised the man they would soon return to take the woman and child with them to the hospital. They then pressed on for another dozen kilometers to the village of Beni Rached. There, they found that "not a single house had resisted the earthquake; Beni Rached was nothing more than a gigantic cluster of earth and stone, planted here and there with torn up walls."[74] The survivors recognized "Si Menouar" and kissed his hand. They reported that there were several dead in every household; the mosque had been converted into a morgue; survivors were still trying to dig out the dead from the ruins. Dr. Aït Ouyahia worked for several hours treating the injured, until he ran out of supplies. Aït Ouyahia and Saïah Menouar were forced to return to Orléansville to summon more assistance. On the way, they came to the first family they had encountered by the fig tree. The woman and child were still there. The man was on his way to bury the dead. His donkey and mule were laden with corpses; his parents and two of his sons were stuffed into the saddle bags, his third son lay across the back of the mule.[75]

The contrast between Aït Ouyahia's account and those of the French reports raises certain questions about sources. The 1999 publication date of Aït Ouyahia's book makes it different from Debia's 1955 memoir and from other sources used in this chapter such as contemporaneous press reports and archival documents such as cablegrams and official government reports: it is inflected by a greater passage of time and by the knowledge that the turmoil of the Algerian rebellion would lead to independence in 1962 (and then to an imperfect polity in independent Algeria). Can Aït Ouyahia's memoir of his life and family history published many decades later, in 1999, be useful in understanding events following the disaster in 1954? Or can it only be used as evidence of the long-term, retrospective intermingling of understandings of decolonization and the 1954 disaster in imagination and memory (a purpose to which it will be put in chapter 7)? Certainly, Aït Ouyahia's memoir cannot be considered entirely reliable. Yet the early genesis of the reports available in French archives does not necessarily make them more reliable than the memories of Aït Ouyahia or those of writers such as Mohammed Khaïr-Eddine, Habib Tengour, and Ali Bouzar, whose work will be discussed in chapter 7. Historians of colonialism are accustomed to reading primary archival sources "against the grain" and with an awareness that authors' depictions of events may be shaped by the cultures of colonialism; we are equally aware, in dealing with post-independence memoirs, that depictions of events may be shaped by cultures of anti-colonial nationalism, by the preoccupations of later decades, and by the desire to tailor memorialization to the needs of a specific audience.

However, exclusively privileging early archival documents when examining the events of 1954 would privilege the French who were in a position to write official accounts, skewing our historical understanding in ways that would reflect the distribution of power in 1950s Algeria. Though Aït Ouyahia's account of "Orléansville, 1954" is separated from events by the passage of more time than are Curie's, Hautberg's, and Kessler's, it must be recognized that the French accounts, even those written just days after the events, are also works of memory and representation for a specific audience. The historian must also approach those accounts skeptically, in recognition of their neglect (both ideologically conditioned and individually self-serving) of Muslim agency, and in recognition of their echoes of imperialist narratives. Hitherto unexploited archives in Algeria may eventually reveal additional perspectives on these events, but in any case, our understanding of history will remain an ongoing work of construction out of the "disparate and multiple" memories (both long- and short-term) and representations by those involved.[76]

The Second Phase

French archival documents provide much detail about the disaster response as the French state's efforts shifted from the initial phase of rescuing victims, treating the injured, and retrieving the dead, to "interventions of secondary urgency": housing the displaced and beginning the process of reconstruction. However, contemporaneous descriptions of events by leftist and nationalist journalists called into question official representations of this second phase of disaster response. Central to the public debate and political struggle in the Chélif Valley in late 1954 were divisions over not only whether the French state was acting with equity to assist both the Muslim and the European survivors but also whether the remoteness and inaccessibility of rural villages such as Beni Rached was part of a status quo ante that French colonialism had to confront and overcome or if the vulnerability of rural Algerians was a product of colonial neglect or exploitative harm—in other words, whether the French colonial state, as it then existed, was the solution to the disaster that afflicted Muslim Algerians or its deepest cause.

The second phase of relief efforts included both direct state intervention through the work of the SNPC and services such as Ponts et chaussées (Bridges and Roads), and donations from private individuals and from organizations such as the Red Cross and Secours Catholique (Catholic Relief). The Interior Ministry's Comité National de Secours organized a "solidarity campaign" to solicit donations, beginning with a "National Day of Solidarity" on September 26. These funds were to be applied toward the purchase and transport of tents, blankets, and other goods to meet the immediate needs of survivors. Throughout metropolitan and overseas France, as well as Morocco and Tunisia, tens of thousands of fundraising posters and hundreds of thousands of solidarity badges were distributed. The total amount collected throughout France and its empire eventually rose to more than 1.5 billion francs (over four million dollars).[77] However, raising the money was one thing; getting aid to the people in need was another.

Monetary donations from throughout France and the overseas French departments were turned over to the treasury office (Trésorier Payeur Géneral) of each department and then consolidated in Paris by the Trésorier Payeur Géneral de la Seine.[78] Donations in kind, however, were sent directly to the Governor-General of Algeria by each department, resulting in a diverse plethora of goods that had to be counted and sorted prior to distribution in the affected areas.[79] Cultural and religious differences produced some glitches in the trans-Mediterranean solidarity effort, most notably an excess of food donations containing pork, and a shortage of clothing for Muslim women,[80] although fifty million francs from the September fundraising were earmarked for the purchase of cloth for such clothing.[81]

It was housing, however, that presented the greatest problem, a fact agreed upon by all sources. If a major rationale for French rule in Algeria in the twentieth century was the ability of the French to improve the material well-being of Muslim Algerians, the earthquake had just made this vastly more difficult.[82] Sources within the National Service of Civil Protection (SNCP) presented the difficulties as largely logistical. By September 13, the Ministry of the Interior in Paris had arranged shipments, with the help of the SNPC, the French Army of the Air, and the US Air Force, of 316 large tents capable of housing over six thousand people.[83] Radio broadcasts in France urged citizens to donate their old camping tents, declaring that "the tent that you have been keeping in the attic and that the grandchildren never use would constitute an undreamed-of solution for an affected family."[84] As more tents arrived, the SNPC team led by Curie and Hautberg took charge of sorting and distribution. However, the tents were of a variety of sizes and types, and most were ill-suited to housing families. It became a nightmare to sort and count the component parts to assure that each recipient obtained a complete kit. Hautberg's log also indicated that there were some enormous American military tents, "worthy of a circus," that neither the SNPC nor the legionnaires could figure out how to assemble.[85] Meanwhile, displaced residents resisted the efforts of aid workers to move them into large tent cities, preferring to camp in front of their damaged homes and keep watch over their goods.[86]

By September 22, Hautberg reported that 2,371 tents had been received and 2,066 of them distributed, leaving 305 still in reserve, with the arrival of another 930 anxiously anticipated.[87] The leaders of the disaster response were well aware that this was insufficient and requested more. A daily report of the Algerian emergency committee estimated that only approximately half of the need had been met.[88] Given the enormity of the disaster, with over forty thousand homes destroyed,[89] this was a gross underestimation of need, reflecting the assumption that the rural *gourbi* dwellers would not receive tents.

In the long term, more than tents would be needed to rebuild Orléansville and to save French Algeria from the political and environmental disasters that threatened. In late September, Hautberg, seeing tents as an unsatisfactory remedy to the housing crisis, traveled to Paris, and communicated "to diverse individuals interested in the events in Orléansville" the urgent need to construct temporary housing, referred to as "barracks." Hautberg pointed out that "the more quickly these are installed, the less need there is for tents." Hautberg also warned, ominously and accurately, that "the rains in the Chélif Valley are said to be torrential."[90] The first step beyond tents was the requisitioning of fourteen thousand square meters of asbestos-reinforced, fiber-cement sheets—roofing material for temporary housing.[91]

With the winter rains coming, temporary housing was an urgent need. However, barracks, like tents, were mainly destined for the cities and towns. Those who lived in gourbis were expected to quickly rebuild their homes themselves, supported by grants of materials and cash payments of ten thousand francs (about thirty US dollars in 1954) to each household, to be followed by an additional ten thousand later.[92] The proponents of this response argued that an illiterate population in desperate need, many living in areas not served by roads, required a process that would be simple and, it was hoped, quick—quick enough to obviate the need for tents or barracks. In practice, however, rebuilding gourbis was not as simple as officials had initially hoped. On September 22, Saïah Abdelkader and the mayor of Orléansville, Ange Bisgambiglia, met with the Governor-General of Algeria, Roger Léonard, during Léonard's visit to Orléansville, and both local officials complained of inadequate efforts to help the inhabitants of gourbis. They denounced the slow pace of distribution of building supplies (specifically, poles to provide a lattice for the roofs) and complained of delays in the distribution of the promised first installment of ten thousand francs for these families without shelter.[93] When the Algerian Assembly convened several days later to address the crisis, one representative (M. Francis) pointed out that the gourbi policy ignored the many rural poor who lived in houses made of stone that could not be rebuilt with some wooden poles and twenty thousand francs. Similarly, representative Bentaieb objected to the use of the term "traditional" in the budget line for "improvements for traditional rural habitats," essentially agreeing with Francis that aid should not be based on an arbitrary distinction between what was modern and what was traditional in the dwellings of the rural poor. The term "traditional" was duly deleted from the legislation, but the Assembly maintained the policy that rural populations would be expected to rebuild their own homes with the assistance of some materials and the fixed payment of twenty thousand francs. Future long-term improvements were promised, but the advocates of the plan claimed that the inhabitants of gourbis preferred to rebuild their own homes.[94]

Solidarity and Division

For the French state, Algeria was France, and the message was one of national solidarity; flags were flown at half-mast on public buildings throughout France.[95] Hautberg believed that this sentiment was sincere and widely shared, referring to "this Algeria, so dear to all the French."[96] Governor-General Léonard proclaimed, "I say above all that the French government considers the Orléansville catastrophe a national catastrophe, which France takes charge of because it affects a French department, French citizens." This statement was meant to be inclusive of Muslim Algerians, who were technically French citizens, albeit unequal ones. Léonard declared, "There has never been, and never will be, discrimination of any kind regarding the victims."[97] Given the obvious divisions in colonial Algerian society amid intermittent nationalist insurgency since 1945, such assurances were aspirational, and had to be made explicit, if only in hopes of minimizing political discord. Léonard's declaration prefigured a more deliberate policy promoting the idea of Muslim equality after 1958.[98]

   As Valentin Pelosse has pointed out, official public declarations from the state—in Paris, Algiers, and Orléansville—presented an image of unity across ethnic divisions, but this "phraséologie officielle"[99] was undermined both by the preexisting inequities of colonialism and by the official disaster response, which treated poor rural Muslims very differently from the rest of the population. Nevertheless, it was frequently claimed that the earthquake had the effect of unifying the population across class and ethnic lines. Raymond LaQuière, president of the Algerian Assembly, declared, "All distinctions, all hierarchies, were leveled in single blow: there remained only brothers, animated by a single and identical desire to help their neighbor with sublime devotion."[100] Orléansville mayor Bisgambiglia echoed this sentiment, declaring to the Assembly, "The Chélif contains two ethnic elements: the Muslims, who are the more unfortunate, and the Europeans. One should not oppose one to other, because they have shown, after the earthquake, that they consider each other as brothers."[101] Several months later, René Debia explained the process by which he believed the earthquake had furthered this inter-ethnic solidarity. From the limited vantage point of the subprefecture, Débia described the first night after the disaster: "An empty lot, across from the subprefecture, was transformed into a city of canvas where, indistinctly, and taking into account only the situation of the family, were settled Europeans and Muslims . . . it never ceased to bring together the ethnic strata of the population so that everyone, rich and poor, and whatever their origins, knew the same hardship and started again together from zero."[102] In Debia's account, it was as if the earthquake had resolved the fundamental contradiction of France's "Impossible Republic," reconciling in one cataclysmic moment the aspirations of French universalism with imperial rule in Algeria.[103] But Debia's optimistic vision, like the broader hope of reconciling imperialism with democracy, was a fantasy. The winds were shifting. By late September, diverse voices, both within the state disaster response effort and in the press, were pointing out the imminent arrival of the rainy season that portended fresh misery for the many thousands sleeping outdoors or in tents.[104]

Organized Protest

Sources contemporaneous with the earthquake response reveal that depictions of the disaster quickly became a field of struggle over the future of Algeria. Even as official French sources promoted a narrative of solidarity and promises of improvements, alternative narratives were being offered within the framework of Algerian nationalism, on the one hand, and leftist calls for class struggle, on the other. Within weeks of the first earthquake, organized opposition groups began to openly denounce the French colonial authorities in Orléansville, and the provision of humanitarian aid became a field of political and ideological struggle in the Chélif Valley.[105] Active post-disaster public relations campaigns and relief aid operations were carried out by various groups in Algeria: the Algerian Communist Party, the communist-linked Confédération Général du Travail (CGT) labor union—and also by Ferhat Abbas's moderate nationalist Union Démocratique du Manifeste Algérien (UDMA) and the less moderate Mouvement Pour le Triomphe des Libertés Démocratiques (MTLD), the nationalist party originally founded by Messali Hadj. The earthquake created an opportunity for these groups to challenge the state's narrative of the disaster response and to present their own alternatives.[106]

The UDMA's newspaper, *La République algérienne*, denounced the authorities' efforts at disaster response and the propaganda of "solidarity" that accompanied it. The authorities were accused of "criminal negligence and scandalous discrimination" based on race. The paper also took note of the manner in which official sources and the mainstream press emphasized the destruction of the urban centers where most Europeans lived, and how they invariably reported the number of European dead separately from casualties among *français musulmans*. *La République algérienne* portrayed the paucity and tardiness of disaster aid in rural areas as a product of racial discrimination. The paper rejected official claims that the lack of roads was to blame for these shortcomings, and argued that transportation infrastructure never seemed to be a problem when the army wanted to send "trucks full of troops" to crush rural disturbances, as they had in the village of Sidi Ali Bounab three years before. Moreover, the paper argued, the flimsy construction of gourbis and the absence of roads and medical facilities only demonstrated the emptiness of the imperialist promises associated with the "civilizing mission."[107]

The authorities were denounced in slightly different terms by dissident political groups of the far Left that included both Muslims and Europeans, most notably the CGT trade union (associated with the French Communist Party) and the Algerian Communist Party (PCA). Like the nationalist UDMA, these groups offered material and political support to the victims of the disaster while portraying the French state as callously indifferent to the needs of the people. Some of their criticisms seemed to echo the UDMA almost verbatim.[108] However, as historian Yaël Fletcher has demonstrated, these non-nationalist groups promoted a class-based vision of colonial oppression that deemphasized ethnic divisions.[109] A major vehicle for this vision was the daily newspaper *Alger républicain*, whose Muslim Algerian and European editors and writers, though predominantly affiliated with the PCA, sought to provide a platform for diverse opposition groups and, as they put it, to "unite, as broadly as possible, all those who—regardless of their political orientations and their origins—want to end colonial oppression."[110] In their criticisms of the state response and their appeals for donations from their members, these groups offered their own vision of "solidarity" between the European French and Muslim and non-Muslim Algerians, based on class identity.[111] This alternative vision criticized the inequities of French colonialism and castigated the French authorities but also hoped to mitigate the "feudal" elements of Algerian nationalism by persuading nationalists to see the French working class as their comrades in the struggle against colonial tyranny.[112]

In early October, articles in *Alger républicain* denounced the empty promises and slow pace of the state's response to the disaster, contrasting the generosity and goodwill of the people who had donated to the solidarity fund with the anemic official efforts to deliver help to the people. Particularly contemptible, in this view, was the state's expectation that the rural population should rebuild their own dwellings, with no help from the state except the paltry payments of twenty thousand francs (less than sixty US dollars)—not even tents for temporary shelter. The earthquake had exposed the falseness of officials' claims about the material benefits the French state had brought to Algeria. The suffering of rural people—rarely identified as Muslim or Arab—was the direct consequence of the failures of the state; villages like Beni Rached had been "abandoned" and left without access roads or medical facilities.[113]

Meanwhile, rural and urban people began to register their discontent, sometimes organized by the dissident political groups. On October 2, *Alger républicain* reported that a hundred "paysans" (peasants or country folk) from the douar Bouarouys had marched in protest of a local official or *caïd* who had demanded bribes from families wishing their names to appear on a list of those to receive the aid allowance for rebuilding—a recurring complaint that the leftist press used to demonstrate the complicity of Muslim elites with the oppressive French state.[114] A day or two later, women from the douar of Oued Larbi, who had organized a "Committee of Disaster Victims," presented themselves at the subprefecture in Orléansville, accompanied by Baya Allaouchiche, secretary of the "Union of the Women of Algeria," and demanded the distribution of tents.[115] The CGT's disaster relief committee organized a delegation of three hundred rural "fellahs" who marched to the town hall in Orléansville, where some of them were able to gain an audience with Saïah Abdelkader's personal secretary, to whom they complained of the lack of tents and the practice of providing only one reconstruction allowance in cases of multiple families living in a single dwelling.[116]

On October 9, the rain began to fall, and the need for shelter became urgent. The CGT responded with its own relief efforts and organized a march of five hundred "fellahs and rural workers" to the town of Oued Fodda, led by syndicate leaders Gessoum Dahmane, Mohammed Marouf, and Zaidi. On October 14, Dahmane led another march—of seven hundred people, according to *Alger républicain*—to the subprefecture in Orléansville, where Debia's reassurances that all would soon be housed were found unconvincing.[117] Yet another march of over seven hundred women took place in Orléansville on October 28.[118] In *Alger républicain*, André Ruiz appealed to international class solidarity: "Brothers and sisters, workers and peasants, of the regions of Orléansville, Ténès, Duperré, you can count on the support of the working class of Algeria and of France, and the support of the international working class. . . .It is incontestable that this catastrophe highlights the misery of our lands, due in the first place to the regime of colonial exploitation."[119]

Critiques of the state response also emerged in the metropole. On October 8, the Catholic Resistance newspaper *Témoignage chrétien* (*Christian Witness*), which would later voice important critiques of French tactics in the Algerian War, published an article titled "Orléansville: Racism is not dead! Does the Mayor only want to feed the Europeans?" The paper quoted a September 15 message, allegedly sent by Bisgambiglia, mayor of Orléansville, to the Red Cross: "Please do not feed Pontéba, the villages Menassis, Maizia, El-Douabed, Gulaftia, Kafafsa, Cheklil and Chouiat, where the men and children did not come to work this morning."[120] This piece of damning evidence was later reprinted in Algeria in the CGT's *La Vie ouvrière* (*The Worker's Life*) and in its local monthly newsletter, *La voix des sinistrés du Chéliff* (*The Voice of the Disaster Survivors of the Chélif*).[121]

Whereas *Témoignage chrétien* had focused on Bisgambiglia, *Alger républicain* and the CGT paired Bisgambiglia's villainy with that of privileged Muslims. *Alger républicain* pointed out that the first cement building to be constructed, in October 1954, was a shed to house Bisgambiglia's horses, but it also addressed continuing demands for bribes from rural Muslim caïds.[122] The CGT's *La voix des sinistrés du Chéliff* paired Bisgambiglia with Saïah Abdelkader, describing the two as "The Profiteers of Misery." Both Saïah and Bigambiglia, it was implied, were guilty of skimming from donations intended for disaster victims; Saïah would later be accused of profiting from the disaster through his family's stake in a cement company which was contracted as a supplier in the construction of HLM housing.[123]

The archival record suggests that this demonization of Bisgambiglia and Saïah Abdelkader was not fully justified; in September 1954 the pair had pressed the

Algerian government to speed the distribution of materials for the reconstruc-tion of gourbis, and in 1955 Saïah would lobby the government in Paris to expand construction of permanent HLM apartment housing in Orléansville for homeless Muslims who had migrated to the city after the earthquake.[124] However, in the Algerian Assembly, it would be the PCA representative René Justrabo who would speak out for the needs of the rural poor. Bisgambiglia and Saïah Abdelkader, in contrast, would focus on maximizing indemnifications for property owners.[125] For the CGT, this dastardly duo constituted a perfect foil to demonstrate that ethnic-ity and religion were irrelevant to the class struggle against capitalist oppression. The oppressors, it was made clear, had no ethnic identity.

This message was reinforced by a complementary message of worker solidarity across ethnic lines. Parisian syndicalists visited Beni Rached in October 1954,[126] and *Alger républicain* contrasted the empty words of Saïah Abdelkader and the inaction of the Algerian Assembly with the successful effort of councilman Ra-chid Dali Bey, a communist, to persuade the Algiers *Conseil Général* to allocate one hundred million francs for disaster relief.[127] Meanwhile, Ruiz was organiz-ing local Muslim Algerian elected officials, who formed a "Comité National algérien d'aide aux sinistrés," which addressed complaints to the Minister of the Interior about the lack of tents and barracks and about the extortion of bribes from disaster survivors by rural caïds.[128]

The situation seemed to be explosive. Faced with signs of popular agitation, the authorities assigned gendarmes to Beni Rached and other villages.[129] As the rains intensified, so did the protests. In late October, *Alger républicain* reported crowds as large as two thousand.[130]

Shortcomings and Deep Causes

The force of seismic waves had produced a dramatic intervention in human history, transforming another environmental factor, the seasonal rains, from a routine and predictable event into a catastrophe: a humanitarian catastrophe for the people of the Chélif Valley and a political disaster for the French state. These catastrophes were also products of late colonialism: a century of impover-ishment and neglect left the rural Algerian population exposed to the elements in the autumn of 1954, while the growing vitality of the anti-colonial opposition made the suffering of poor Muslim Algerians an urgent political concern for the colonial regime.

The French state's response was seen as inadequate not only by the regime's opponents but also by those responsible for the disaster response. By early 1955,

the solidarity campaign had collected over 1.5 billion francs, with donations arriving from across Europe, the Middle East, the United States, and the Soviet
Union. This did not, however, translate into robust action in the Chélif Valley.[131]
Within the SNPC, the shortcomings of the immediate disaster response were
acknowledged, and the event became a case study in unpreparedness and sub-
optimal organization. Although the immediate response of the military units,
sapeurs-pompiers, and especially the medical staff seem to have been universally
applauded, the response from the local government and from Algiers and Paris
was inefficient.[132] As Colonel Curie's report on the disaster response effort would
explain, the local staff of the Service de Santé had performed admirably, but the
local authorities, including Debia, Bisgambiglia, and the mayors of the other
affected towns lacked the "means of communication" to organize an effective
local governmental disaster response operation. Only the military troops in the
area had been able to respond immediately. The subprefecture building had itself been damaged, as had the gendarmerie, and the local officials had themselves
been traumatized by the disaster. The Service de Protection Civile d'Algérie had
been slow to respond to the disaster; the "designated director of disaster relief"
arrived in Orléansville on Friday September 11, only to return to Algiers that day,
and when he returned on the 12th, he possessed no more means of communication or response than did the subprefect or the mayor.[133]

As Kessler noted in his report on the disaster response, the SNPC team that
set up operations on September 20 in the subprefecture fell short of the organization and infrastructure called for in the Plan ORSEC (Organisation de la
réponse de sécurité civile),[134] the guidelines for disaster response promulgated by
the French state in 1952. The Plan ORSEC specified that a team of "specialized
functionaries and technicians" needed to be sent to the disaster site with the authority to respond to the variety of urgent problems that might arise. This team
would have both the skills and the "psychological distance" necessary to confront the disaster, but it was to work closely with the local authorities in order
to benefit from local knowledge. The Orléansville response, however, suffered
from poor coordination. Although the Civil Protection workers from outside
of Orléansville shared space in available buildings with the local authorities,
this resulted in the "dispersion of services" of the SNPC staff while producing
confusion, rather than coordination, between the hierarchy of the SNPC and
that of the subprefecture.[135] Marius Hautberg complained that the municipal
government issued vouchers to Orléansville residents for tents and blankets
without regard for the ability of the SNPC to fulfill such commitments, and

uncoordinated requests were made to the engineering corps by various authorities, including Debia, at the subprefecture; Freychet, representing the prefecture;
and even the medical service, resulting in wasted time and resources. Meanwhile,
although buildings containing corpses were excavated, no official possessed the
legal authority to order the demolition of the countless other buildings that
stood unusable, damaged by the earthquake.[136] These critiques were analogous
to concerns emanating from within the Algerian government. In an October
7 encrypted telegram marked "secret," Governor-General Roger Léonard expressed alarm that "latent conflicts" between the municipal authorities and the
prefecture prevented effective action, as did the lack of a legal structure permitting the Algerian administration to address the need for repairs to existing
buildings and for permits for new construction. These problems rendered the
administration "paralyzed," according to Leonard, "on the eve of winter."[137]

Kessler argued that the impact of the earthquake was much like that of an aerial bombing campaign, and therefore planners of national defense had much to
learn from Orléansville. Kessler noted that there was one important difference
between the earthquake in Orléansville and the experience of cities destroyed
in war: the Orléansville disaster had occurred when France was otherwise at
peace and had affected only a single region.[138] Given the peacetime abundance
of means in September, the inefficiency of the response was worrisome. For Kessler, the "appalling mediocrity" of the service's own resources and the grossly
"insufficient training" of the local French population would be a wake-up call,
he hoped, for French disaster response.[139] A similar view was made public in
the pages of the newspaper *L'est républicain,* where an editorial titled "Warning" pointed out the growing danger of nuclear destruction of French cities and
cited the inadequacy of the response to the earthquake.[140] Hautberg, too, hoped
that improvements in the organization of disaster response would better prepare
the administration to respond in times of war.[141] Finishing their reports in December 1954, it was not yet evident to Kessler and Hautberg that history would
record the period of relative peace in North Africa as ending within weeks of
the disaster.

Neither Hautberg, the engineer, Kessler, the administrator-in-training, nor
Debia, the subprefect, made any mention of the political agitation of the survivors. Hautberg and Debia, however, addressed the question of deeper causes
of the suffering occasioned by the disaster, and recognized that the problems
revealed by the earthquake went beyond organizational inefficiency. They recognized that the disaster produced disproportionate suffering among Muslim

Algerians living in rural poverty even if they did not accept that this poverty was rooted in the injustices of colonialism, insisting instead that French rule was a force for positive change.

In subprefect Debia's view, the earthquake provoked a "revelation" for outsiders, including "visitors, metropolitan or Algerian [i.e., *colons*], journalists, functionaries—and even very high functionaries." This revelation, for Debia, was not of the iniquities of colonialism but of the harshness of the land, invoking the discourse described by Aït-Saada. For the first time, these outsiders saw beyond the façade of beautiful beaches, impressive dams, and public works usually shown to important visitors and tourists. The disaster brought to the fore "the Algerian reality" of an "ungrateful land" where people toiled in an inhospitable climate, as they had for millennia, but where population growth now exacerbated their poverty. Debia was confident in the French colonizing mission, however: the solution lay in the "mise en valeur" (improvement) of Algeria through economic development.[142] Prior to the earthquake, Debia had dreamed of the Chélif Valley becoming "a new California," and he remained optimistic, although he had recognized, even before the outbreak of war, that the poverty of rural Algerians was "the gravest problem, which risks endangering France's work of civilization."[143]

For Hautberg, too, the alterity of "this land of Africa . . . brutal and savage" was the root cause of Algerian underdevelopment.[144] Like Debia, Hautberg believed that the future of French Algeria depended on economic development. Unlike Debia, however, Hautberg acknowledged the fragility of the ties between Muslim Algerians and France, pointing out that French "penetration" in North Africa was a relatively recent phenomenon. Hautberg argued that poverty was the root cause of unrest in Algeria, inclining Muslim Algerians "to react violently in order to loosen the grip of their misery."[145] For Hautberg, this poverty exacerbated the suffering brought by the disaster and was the primary cause of social disorder. Ignoring the role that the French had played in destroying the rural livelihoods of Muslim Algerians since the nineteenth century, Hautberg assumed that the current underdevelopment reflected the historical status quo ante, perpetuated by a lack of modern agricultural methods and by insufficient French schooling. Echoing a frequent postwar theme in French colonial theory, Hautberg argued that the solution lay in a Keynesian program of state investment in Algerian economic development.[146]

The suffering that followed the earthquake had drawn attention to the inequities of life in Algeria and the need for improvements in the standard of living of the Muslim population. For Debia, French rule was the cure for Algeria's underdevelopment, not its cause, but "two thousand years of backwardness cannot be regained in a century."[147] Algeria's situation, he argued, was not unlike that of America's rural South, where state-led economic initiatives—the Tennessee Valley Authority—had been initiated in response. In contrast, *Alger républicain* asserted that the root cause of Algerian poverty was the state itself, which imposed on poor Algerians "a burdened life, with taxes, caïds, informants, and gendarmes."[148] Nevertheless, there was significant point of agreement between the views of dissidents like the editors of *Alger républicain* and imperialist analysts like Debia, Kessler, and Hautberg. They recognized that disaster response was not enough: reconstruction would be insufficient if it merely returned the Chélif Valley to its pre-earthquake condition. As *Alger républicain* put it,

> The problem posed goes beyond reconstruction, or aid, or even solidarity with the victims of the catastrophe. Because these fundamental problems will not be resolved when everything is put back "in order." When we resume "as before" the neck irons of misery and hunger. A "normal" misery and hunger. A life without school, without doctors, without warmth and without liberty.[149]

Yet even restoring the Chélif to its pre-earthquake condition seemed initially to be beyond the competence of the French authorities; the seismic event had dramatically exacerbated the contradiction between imperialism's promises and the reality of life in Algeria.

Revolution and Reconstruction

On the night of October 31, 1954, a series of attacks were carried out by the FLN (Front de Libération Nationale), across Algeria, mainly in Algiers, Kabylia, and the Aures mountains. On November 1, the FLN issued a proclamation that the "final phase" of the struggle for an independent Algerian state was beginning, a "true revolutionary struggle" that would use "every means" to force the French to negotiate.[150] This was not, however, the sudden start of a conflagration. The insurrection of 1954 started small, with fewer than a thousand armed militants and was part of a history of postwar resistance that included the mass uprisings of 1945 followed by the militant activities of the FLN's predecessor, the MTLD's *Organisation Spéciale*.[151] In 1954, the new insurrection brought no immediate transformation of the situation in the Chélif.

Soon, however, the "events" of the incipient revolution started to compete with the disaster response for headlines in the Algerian press, and voices in the

Chélif Valley began to ask how the earthquake was related to the insurrection. The earthquake, it was assumed, was simply too momentous an event to be irrelevant to the political question. Descriptions of the calm in the Chélif and planning for the reconstruction of Orléansville became elements of the discursive contest to imagine the future of Algeria.

For seventeen months after the FLN declaration, Orléansville remained untouched, as the FLN struggled to gain traction beyond its strongholds in the Aurès and Kabylia in eastern Algeria. The calm in the Chélif, however, was only relative: peace for the colonizer went hand in hand with oppression and violence for the colonized. The nationalists' declaration of war had made all forms of dissidence more dangerous. Survivors' organizations, by operating in the open, gave the administration prime targets for repression: the activist, the disgruntled, the engaged. In November 1954, fourteen people were arrested in Beni Rached, as well as five CGT members organizing in the village of Chouchoua.[152] In January 1955, the CGT's *La voix des sinistrés* reported that homes of its disaster committee organizers were raided by French troops or gendarmes, claiming to search for arms. One local organizer, Ahmed Sameti, was imprisoned on charges that he had stolen the caïd's cow. Two others were imprisoned on charges related to demonstrations. The villages of Yaabouch and Ouled Bendou were also raided. No arms were found, but nineteen people were arrested.[153] The following year, there were more arrests in Beni Rached, and twelve in the village of Taighaout. In August 1955, Kaddhar, the secretary of the Fédération des Comités de Défense des Sinistrés was arrested, as was, a few months later, the secretary for the Comité Intersyndical de solidarité, Dahmane Guessous. In May 1956, the remaining leadership of these committees were rounded up and sent to detention camps.[154]

Yäel Fletcher has argued that the earthquakes of September and the rains of October had given the leftist dissidents writing for *Alger républicain* and organizing the CGT's disaster victims' committees a grand opportunity to promote their narrative of a class-based divide between the workers and the wealthy. The FLN insurrection did not trigger a sudden conversion of these activists to the nationalist cause—at least not overtly.[155] In 1954, however, voices from the Left denounced the repressive measures taken in response to the outbreak of hostilities, and argued that the root causes of the insurrection lay in the oppressive nature of French rule in Algeria. *Alger républicain* declared "the necessity of seeking and finding, QUICKLY, democratic solutions to the Algerian problem."[156] The Fédération des Comités de Défense des Sinistrés (Federation of Committees for the Defense of the Victims) declared that "the deep causes of these events [the

FLN insurrection] reside in the accumulation of the methods of exploitation and oppression, in all domains, of misery and arbitrariness, by a colonial regime which is largely condemned by all humanity."[157] *Alger républicain* also pointed out the contradiction between the French state's inability to supply twenty-five thousand tents in the aftermath of the earthquake and its ability to use aircraft to deploy paratroopers across Algeria in response to the FLN.[158] The Chélif earthquake was once again portrayed as revelatory of the follies and hypocrisy of the colonial regime.

Meanwhile, supporters of French colonialism depicted Orléansville and the French response to the earthquake as a model for Algeria. In September 1955, an article in the *Journal d'Alger* asked, "Do we owe to the [seismic] cataclysm the total absence of political troubles in the region of Orléanville?"[159] Subprefect Debia began his book-length history and memoir of Orléansville, published that year, with a foreword titled "Warning." In the months between his completion of the manuscript's chapters and the publication of the book, Debia acknowledged that "the situation" in Algeria had become more perilous. But Debia remained hopeful, as his book's title indicated: *Orléansville: Naissance et destruction d'une ville: Sa résurrection* (Birth and Destruction of a City: Its Resurrection).[160] Debia portrayed the region as a harbor of political tranquility amidst an Algeria in crisis, and he attributed this to the leveling effect of the earthquake. His memoir is notably silent about the discord of October 1954, when, as the rains intensified, people slept in the open and marched in the streets. Debia elided the entire period of the survivors' protests and their repression in three words: "the months passed."[161] Ignoring these events, Debia focused on the urban housing of Europeans and Muslims of all social classes, first in an improvised "city of wood" and then in barracks constructed, for temporary housing, beginning in December and largely completed by March. There, Europeans and Muslims experienced together the hardships of life after the earthquake.[162] The result was a new solidarity. The final page of Debia's book was blank, except for a photograph of a smiling Muslim Algerian boy.

Given the anger and misery expressed in the Chélif Valley in the fall and winter of 1954, Debia's optimism was Panglossian. In the Chélif the rains continued, as did the survivors' demonstrations, culminating in a demonstration of as many as five thousand people on November 25.[163] Their complaints, as conveyed by *Algèr republicain* and *La voix des sinistrés du Chéliff,* continued to focus on the difficulty of obtaining the twenty thousand francs allotted to rural families rendered homeless, and above all, the lack of housing. The weather had made the need for housing urgent, and, weeks after the disaster, neither the administration nor the survivors saw tents as an adequate solution. On October 11, the administration had promised temporary housing in barracks constructed of prefabricated materials. However, the volume of material ordered was grossly insufficient, having been intended only for the residents of towns and cities. Meanwhile, there was no sign of progress on plans for permanent reconstruction of Orléansville.[164]

Nevertheless, the pace of disaster relief did improve in November 1954. This was partly in response to the political agitation in the Chélif and to negative press coverage about the disaster response in Algeria and metropolitan France. In part it was simply because initiatives begun in late September and October were finally bearing fruit. In October, a meeting of the Algerian Assembly had established a legal basis for funding reconstruction; meanwhile, the mess of heterogeneous tents and poles and canvas piling up in the Orléansville train station was sorted out. By late November, according to official figures, over 6,000 tents had been distributed, including 500 from the Italian Red Cross, 1,474 from the SNPC, and 1,030 from the army. Thirty thousand blankets were handed out: the Red Cross had provided twenty thousand and Secours Catholique, ten thousand.[165] (No mention was made in official counts of the efforts made by the CGT, UDMA, or MTLD.) From Kessler's point of view, the distribution of tents "represented the vastest French housing effort ever achieved in a time of peace."[166] Meanwhile, "Operation Gourbi" was declared, to speed the distribution of funds and supplies to permit rural families to rebuild their homes. Debia claimed that the reconstruction of thirty-eight thousand gourbis was completed by winter (Interior Minister Mitterrand claimed that it was thirty-five thousand), enabled by the aid payments of twenty thousand francs each.[167] Official claims of successes in the distribution of tents and aid for the construction of gourbis are corroborated by a shift in the nature of the critiques leveled by the colonial regime's critics, including not only the CGT and *Alger républicain* but also the metropolitan Comité Chrètien d'entente France-Islam, who now increasingly called for the construction of barracks or more permanent "modern" housing for the rural population.[168]

The question of housing in the Chélif was not just a matter of overcoming logistical, financial, or bureaucratic obstacles, however. In the context of the FLN rebellion, reconstruction took on new urgency. It is important to note that this urgency predated the 1958 Constantine Plan, which is often portrayed as a turning point in the French response to anti-colonial revolt. A massive program of state investment in Algeria intended to undercut the appeal of the FLN by fulfilling some of the promises of colonialism, the Constantine Plan aimed to improve standards of living through investment in infrastructure, industry, education, and particularly the construction of decent housing for swelling urban populations. However, this 1958 initiative was part of an ongoing shift in postwar colonial thinking emphasizing social reconfiguration and economic development through Keynesian investment. French intentions for reconstruction in the Chélif Valley prefigured the Constantine Plan as a response to the threat of nationalism.[169]

Scholars have demonstrated that violent coercion played a central role in this attempt to remake Algeria through an imposed economic transformation.[170] For Debia, the earthquake's violent disruption of traditional patterns of Muslims' lives already represented a helpful "forced step toward assimilation, of which we today see the happy effects."[171] Debia argued that these "happy effects" meant that there was hope that France might "remake the moral, social, economic and administrative conquest of the country."[172] Debia argued that the regime's critics were wrong to focus merely on housing the rural poor without envisioning a wholesale transformation of Algerian life. He saw the inadequacy of the gourbis as a mere symptom of the underdevelopment of rural Algeria; replacing collapsed gourbis with modern housing would not treat the cause of the problem. Roads, he argued, were the key: "It is by road that civilization penetrates and implants itself."[173] Debia's view that the inaccessibility and isolation of remote villages was a major obstacle to the success of the French project was widely shared; and in the context of the war this problem would eventually be addressed through the mass relocation of rural Algerians into dismal centers of *regroupement*.[174] In the Chélif, however, mass relocation had already begun. The population had swelled on the outskirts of Orléansville, as desperate earthquake survivors in rural areas moved closer to the center of aid distribution, many resorting to picking through garbage dumps to survive.[175] The Muslim Algerian population of the city grew to over twenty thousand by 1955.[176]

Reconstruction was slow to manifest, however, and its political purpose was undermined by the inequities of colonial power. In 1955, the Commissariat of Reconstruction rebuilt low-income (HLM) apartment housing in Orléansville destroyed by the earthquake and constructed additional HLM housing, but in some cases European families moved into these buildings. Muslims in the suburbs of La Ferme and Bocca Sahnoune, including many who had migrated from the countryside after the disaster, continued to live in tents.[177]

Owners of European-style buildings in cities and towns were better provided for. Not only were they provided with temporary "barrack" housing, but they had the opportunity to receive substantial compensation for their losses from the state. In October 1954, the Algerian Assembly had authorized assistance from state funds for private property owners (excluding the gourbi dwellers) equal to the value of any property damage valued at more than five thousand francs. This aid included grants of up to one hundred thousand francs per property owner for repairs, and up to the depreciated value of the building for buildings deemed irreparable. Government-backed low-interest loans were offered to cover the remainder of repair or reconstruction costs. This assistance, however, was issued in the form of vouchers, redeemable only when reconstruction was underway, which required obtaining demolition and building permits from the newly created Commissaire de la Reconstruction. The process was slow, and consequently little permanent reconstruction occurred before 1956. When buildings were reconstructed, provisions intended to ensure that renters would be able to return to reconstructed buildings proved ineffective, and many tenants remained displaced.[178]

As one might expect, the well-to-do and the well-connected fared best of all. As the months passed, critics on the Left pointed out that Bisgambiglia and Saïah Abdelkader received state funds to reconstruct their own villas, reportedly costing ten million and sixteen million francs, respectively, while their business enterprises and those of their family members benefited from state reconstruction contracts.[179] The wealthy also benefited from the real estate market created by the process of reconstruction. The rich bought the property and the vouchers of owners left destitute by the earthquake, who could not afford to wait for the Commissariat of Reconstruction to approve reconstruction plans and issue payment for their vouchers. This created a profitable market for those with the means to speculate in a real-estate market propped up by government funds. Meanwhile, in 1955, the municipal council blocked the urban planners' efforts to "construct affordable housing in well-situated locations" such as on the central thoroughfare, the rue d'Isly, and near the train station. In a move paralleling segregationist strategy in the United States, plans for a public swimming pool were thwarted in favor of a privately owned swim club exclusively for Europeans.[180]

In March 1956, 1.3 of the 1.5 billion francs from the national solidarity fund remained unspent. The CGT's Committee for the Defense of Disaster Victims (Comité pour la Défense des Sinistrés) argued that these funds should go directly to the survivors of the earthquake.[181] However, a member of the Saïah family had organized a survivors' group to act as an alternative voice to the CGT, and this group supported the transfer of money from the solidarity fund to the reconstruction budget.[182] The CGT's approach would have ensured that

equal benefits would go, not only to urban property owners, but to urban renters and the rural population in the gourbis, who constituted the majority of
those affected by the disaster. Instead, the funds were transferred to the Algerian Government-General for use by the Commissariat of Reconstruction, in
keeping with the regime's desire to impose centrally directed transformation.[183]

Plans laid out in October of 1954 that had languished for many months were
now put into motion. These included provisions to address the needs of the
Muslim poor. Two hundred fifty thousand francs were allocated for "social improvements in the douars." One hundred nine million francs were allocated for
roads, water supply, and sanitation in Orléansville's Muslim suburbs of La Ferme
and Bocca Sahnoune. In Ténès, apartment housing was to be constructed for 328
families, along with a school, mosque, and bathhouse (a *hammam* or "Moor-
ish bath"). Trade schools for construction were to be built in Orléansville and
Ténès, at a cost of thirty-two million francs; eighty million was allocated for a
cultural center in Orléansville; while only forty million was set aside for a rural
vocational training center in El Attaf.[184] A small portion of these funds were
used to respond to complaints of discrimination against Muslim Algerians. For
example, a supplemental distribution for war veterans of ten thousand francs
from the solidarity fund, originally only distributed in the city of Orléansville,
was extended to veterans in the outlying areas when Muslim veterans outside of
the city complained that geography was being used as a proxy for race in grant-
ing preferential treatment to European veterans.[185] Such measures, however, did
little to counteract the rural catastrophe inflicted, first by more than a century of
settler colonialism and then by the new violence of the earthquake and the war.

The FLN's major operations were largely limited to eastern Algeria in the
first phase of the war, and therefore no clear conclusion can be reached about
the effectiveness of the reconstruction effort as an imperialist countermeasure to
nationalist recruitment. Keynesian effects may have made a contribution to the
relative calm in Orléansville. Certainly, Keynesian stimulus is a more plausible
explanation than Debia's imagined social leveling and post-disaster assimilation
in the tent cities of Orléansville. *Alger républicain* had criticized the state for
earmarking funds for reconstruction and compensation of damages that could
have been directed toward the most urgent material needs of the survivors.[186]
However, after 1954, both the reconstruction of the city and the presence of the
army stimulated the local economy, creating jobs and a demand for goods from
local businesses.[187] Before the earthquake, in an economy long dependent on day
laborers, underemployment had been a major problem. Hundreds of the under-
employed and unemployed had demonstrated in Orléansville in October 1953,

and the administration in 1954 counted 1,026 unemployed workers. The direct, short-term effect of the earthquake was striking: even critics of the colonial state recognized that, in November 1954, more than two and a half times that number were employed in the task of clearing the debris; seven months later, 800 were still working in this capacity.[188] However, this was not a sufficient remedy in the long term for Algerian economic suffering or political discontent.

In 1956, the war came to the Chélif. The Army of National Liberation (ALN) gained a foothold in the mountains north and south of the valley (the Dahra and the Ouarsenis); the Government-General considered the villages in these areas 20 to 50 percent "contaminated."[189] Although the French maintained control of Orléansville, the city experienced attacks and assassinations; meanwhile, the ALN expanded their control of the mountains.[190] In 1957, the Chélif Valley itself was the site of significant fighting, not only between the nationalists and the French and their Muslim allies, but between rival nationalist groups.[191] In some areas affected by the earthquake, disaster reconstruction came to a halt.[192] The army began implementing the massive forced "regrouping" of populations, along with whatever portion of their herds and belongings they could manage to bring with them, out of the mountains and into *regroupement* village centers. By October 1958, in the newly created *département* of Orléansville, which extended north to the coast at Ténès and south beyond the Ouarsenis, over 100 thousand people had been forcibly displaced; two years later, over 260 thousand were housed in 311 regroupement centers, approximately 40 percent of the region's Muslim Algerian population. Many thousands more fled to cities to escape this "regrouping."[193] Although the French organizations Secours Catholique and the Protestant CIMADE (Comité Inter-Mouvements Auprès Des Évacués) provided aid in these camps and distributed donations of food and clothing shipped from the United States, press reports of unhygienic living conditions and malnutrition in the camps scandalized the metropolitan public.[194] Across Algeria, the French state's belated attempts to provide these refugees with food and housing, first in tents and then in barracks, paralleled the belated scramble to provide shelter to earthquake survivors in 1954.

In the Chélif, funds originating in the 1954 fundraising campaign and earmarked for "social improvements in the douars" were now directed toward these "regroupment" centers. As the Commissariat of Reconstruction put it, disaster relief in the countryside was now "tightly associated with the work of pacification."[195] Disaster response in Algeria, in other words, had become a tool used by the colonial state in its efforts to counter the effects of nationalism. Reconstruction efforts were then augmented, beginning in 1958, by the Constantine Plan.

Valentin Pelosse points out that the total sum paid in salaries to Muslim Algerians by the Commissariat of Reconstruction (750 million old francs between 1955 and 1961) amounted to as much as 40 percent of the annual agricultural payroll for Algerian workers in the Chélif area.[196] In addition to providing employment directly, reconstruction had a broader effect on the economy. The Muslim middle class in Orléansville, including business owners, teachers, professionals, and functionaries, grew to perhaps 10 percent of the Muslim population by 1962.[197] However, the fact that Muslims in Orléansville benefitted from post-earthquake reconstruction and the Constantine Plan does not mean that the position of Muslims improved relative to Europeans or that economic power was redistributed. Europeans, already economically and politically better off, tended to benefit the most from the economic stimulus. The Constantine Plan aimed to counteract this by stipulating that contracts for goods and services engage not only the largest firms but also "diverse small and medium-sized local entrepreneurs." However, these small businesses, to a greater degree in Orléansville than in some cities, were often owned by Europeans.[198] Moreover, as Pelosse points out, the combined effects of *regroupement* and public spending on earthquake reconstruction exacerbated the long-standing tendency of the French colonizers to privilege urban areas while carrying out the "devastation" of the rural economy.[199] Construction and reconstruction could do little to address the gross and pervasive inequities of colonialism. Yet it is clear that the response to the earthquake was part and parcel of the French state's response to the political insurrection.

Conclusion

The inseparability of the natural disaster and the war was captured in the interpretation of events presented by the playwright Henri Kréa and in the memoir of Dr. Aït Ouyahia, who each portrayed the earthquake as a harbinger of a nationalist awakening. In Dr. Aït Ouyahia's recollections of his own personal experience, the façade of solidarity in Algeria crumbled within days of the first earthquake. Aït Ouyahia recalled press reports describing how, "during those days, the entire world manifested its compassion and generosity."[200] Yet this talk of universal solidarity did not ring true for him. As a Kabyle-speaking Muslim from a small rural village, Aït Ouyahia had a deep-rooted sympathy for the predicament of the rural Algerians he found suffering in Beni Rached. However, as a French-educated doctor and the son of a French-educated "indigenous schoolteacher," Aït Ouyahia was part of a tiny elite of Muslim professionals who had benefitted from French power and from colonial education.[201] Frantz Fanon wrote that, before the war of liberation, "the doctor always appears as a link in the colonialist network, as a spokesman for the occupying power."[202] Yet, in his memoirs, Aït Ouyahia dated his passage from the realm of the colonizer to the realm of the colonized not to the outbreak of war but rather to the aftermath of the earthquake.

Soon after the disaster, the young doctor observed as crowds of mostly Muslim Algerians gathered to receive aid, and a commotion occurred outside one of the tents where humanitarian aid was being distributed. Soldiers dragged a young Algerian man away from the tent, and a French officer ordered the crowd to disperse, declaring: "All thieves, these Arabs!" This event is not implausible: Hautberg also described incidents of friction between earthquake survivors and French aid workers leading to the intervention of gendarmes during the distribution of aid.[203] For Aït Ouyahia, the angry words of the French officer were an outrage, and a transformative moment for the newly minted, French-trained obstetric surgeon. According to his account, he confronted the French officer and, in front of the crowd, denounced the man's racism. Aït Ouyahia remembered the moment as an epiphany:

> It was as if this insult was addressed to me alone. I decided then to take on, alone, the burden for all the Arabs, and in their name, to respond, alone, to he who had just injured us. I had to do it, me who spoke French. . . . Forgotten was the Muslim intern, all proud of being called "Monsieur," just like his European colleagues in the medical service! Forgotten the young indigene who had been told, more than once, that he was not "an ordinary Arab." . . . To the Devil the privileged Muslim! I was no longer me; I was those, those poor wretches in rags and dirty feet. I felt suddenly strong, all grown up.[204]

At this moment, Aït Ouyahia appears to have experienced a conversion. For this Kabyle-speaking, French-educated doctor, a new ideology of solidarity, that of Arab-Algerian nationalism, had replaced the claims of Franco-Algerian unity and universal brotherhood.[205] Aït Ouyahia would later go on to provide active support to the Algerian Revolution against France.

Although Henri Kréa in 1957 and Aït Ouyahia in 1999 portrayed the earthquake as a definitive trigger event in Algerians' embrace of the FLN cause, this cannot be taken as evidence of a widespread phenomenon. Even regarding his own, personal experience, Aït Ouyahia's story about the French officer seems to fit uneasily with other chapters in his disjointed memoir that treat his wartime support for the FLN without any reference to the earthquake as a formative experience. Clearly, however, Aït Ouyahia's memory of the earthquake itself was strongly tied to his commitment to the nationalist struggle for independence.

The archival record produced in the weeks and months following the earthquake supports the view that disaster and decolonization were linked, as colonizer and colonized interpreted and responded to the seismic disaster in light of the problems of inequity in Algeria and of Algeria's relation to France. By the time the earthquake struck, Algerian nationalism had already been growing for decades, and North Africa already being rent by nationalist violence from Morocco to Tunisia, but the Chélif disaster revealed and exacerbated the very injustices and miseries of colonialism that fueled the nationalist revolution. The bankruptcy of the social contract implied by French promises of "civilization" and economic development was already apparent in Algeria, but by destroying vast amounts of housing, the seismic shocks of 1954 exposed the poorest Algerians to the winter rains and exponentially increased what it would cost the French state to follow through on its promises—at precisely the moment when anti-colonial opposition was gathering strength. Indisputably, when the earthquake struck, questions of decolonization were far from the minds of some—the child Ali Bouzar waking from his bed in Oued Fodda, or those suffocating workers, crushed in the ruins of a high rise in Orléansville. However, the earthquake and the war of independence are not separable objects of inquiry, at least not when the scope of inquiry includes the Chélif Valley. Every action of the French state and of its agents, critics, and rivals in the Chélif was conditioned by the question of whether Algeria was France, and whether and how it would remain so. In the Chélif, disaster relief became a field of struggle over decolonization, as it would in Fréjus and in Morocco in the coming years.

CHAPTER 3

Fréjus 1959, under Water and at War

ON THE NIGHT OF DECEMBER 2, 1959, fourteen-year-old Christian Hughes and his father returned home in Fréjus, on the north shore of the Mediterranean, at the mouths of the Argens and Reyran Rivers, in the French *département* of the Var. They had set out by bicycle in the pouring rain to pull in their fishing nets, but Christian's father had changed his mind, and they returned empty-handed to the house the Hughes family shared with Christian's aunt and uncle. They immediately went to bed. All seemed normal until, as Hughes recalled: "Then, we heard noises, gunshots. My father, my sister came out of their rooms. We were in the midst of the Algerian War. My brother was in the *djebels* [mountains]. We heard rifle shots, the siren, the alarm; it scared us. My sister said, 'It is the FLN attacking. . . .' Everyone shouted, 'It is a revolution!'" There was a loud knock on the door: Christian's father demanded that somebody get him a knife. Then his father and uncle opened the door, and his uncle, "a colossus, two meters tall," fainted. It was not a revolution; the Algerian revolutionaries were not invading the quiet town in the dark of night. It was a flood: there was "water everywhere." A neighbor was at the door to warn them. Christian's uncle revived and shouted: "A tidal wave, we're all going to die!"[1]

The Malpasset Dam north of Fréjus had burst at approximately 9:30 p.m., unleashing a wall of water. Waves, reportedly four to five meters high, swept through the Reyran Valley and into Fréjus. Over 400 people were drowned; 155 buildings were destroyed, and almost 800 were damaged. Over thirteen square kilometers of agricultural land flooded; a thousand sheep drowned, and 471 vehicles were destroyed.[2] Christian Hughes remembered seeing the aftermath the next day: "cadavers, destroyed buildings, hectares of ruined crops, drowned animals."[3]

Hughes's short memoir of the event, published in 2003, illustrates how the environmental catastrophe his family experienced in 1959 was intertwined in his memory with the political violence of the Algerian War. Hughes was not alone in connecting the experience of the flood to that of war and decolonization, and the imbrication of the Fréjus flood with imperialism and the Algerian War was not limited to an ex post facto reconstruction of events. The town of Fréjus had long had a particularly strong connection to France's colonial endeavors, and this connection was evident in 1959. This chapter will make use of sources from 1959 through the 1960s and beyond to demonstrate how the Malpasset disaster was shaped by events in North Africa (including both the Algerian Revolution and the Chélif earthquake), affecting who suffered as a result, how the state responded, and how the disaster in the Var was remembered.

The War at Home

In the grim context of 1959, the Hughes family's initial assumption that the violence outside their door was caused by Algerian revolutionaries was not entirely implausible. The "events" in Algeria, which had begun as a small uprising in November 1954, had metastasized into a major conflict by the time of the Malpasset Dam collapse. The nationalist revolutionaries of the FLN (Front de Libération Nationale) recruited new supporters as their attacks prompted French reprisals against the Muslim population. By February 1955, the revolt had begun to spread beyond FLN strongholds in the Aurès Mountains and Kabylia; in response, the French sent more soldiers to Algeria, including conscripts as well as professional troops battle-hardened by the failed campaign for French Indochina.[4] In 1957 Algiers had been engulfed in urban guerrilla warfare, terrorism, and reprisals, but in Algiers and across Algeria, French forces were able to reverse FLN gains through a brutal campaign that included the "widespread and systematic" use of torture.[5] The "regrouping" of rural populations and then the 1958 Constantine Plan further escalated this French campaign to transform and "reconquer" Algeria.[6]

The population of the metropole was keenly aware of these "events." In the Var, the then-dominant Socialist party lamented "the sacrifices imposed on young Frenchmen, who risk paying—and paying dearly—for the faults of colonialism and the deficiencies of rulers."[7] In Algiers, however, there was no sign of compromise: in 1958 the formation of a government in Paris under a prime minister who had advocated negotiation with the nationalists had precipitated a rebellion of hard-line settlers and French paratroopers in Algeria. This had brought down the Fourth Republic and prompted the return of Charles de Gaulle to power. The war continued.[8]

In the sixteen months preceding the Malpasset Dam collapse, the FLN had taken the war to the hexagon, attacking oil refineries and military and police forces in metropolitan France.[9] In 2011, the German historians Matthias Ritzi and Erich Schmidt-Eenboom uncovered a report by a West German spy, Richard Christmann, indicating that the FLN had been considering attacks on dams in France. Through his informants inside the FLN, Christmann had learned that the revolutionaries were exploring new options for sabotage. They were studying sewer tunnels and public water supply infrastructure in Paris and Algiers, with the idea of planting bombs underneath buildings or using judiciously placed explosions to destroy the public water supply. In addition, the FLN was contemplating the destruction of dams. After the Malpasset collapse, Christmann believed that the catastrophe had in fact resulted from such an attack. In a report for the West German intelligence service, Christmann stated, "After an attack on a small dam in southern France had only a partial success, but took many lives, all other terrorist measures were stopped by order of the group around Ben Bella, then still imprisoned." This was obviously a reference to the collapse of the Malpasset Dam. Ritzi and Schmidt-Eenboom have accepted Christmann's view that the FLN were responsible for the Fréjus disaster.[10] This idea has since been publicized by Schmidt-Eenbaum and by a 2013 Arte television documentary, reviving, in public discourse, the connection between the violence of decolonization and the flood.[11]

This interpretation of the evidence has significant flaws, however. It would be hard to imagine by what standard the destruction of Fréjus would be considered only a "partial success," if the goal of the FLN attacks in France was to spread fear and force France to shift security forces away from Algeria.[12] It is possible that the indiscriminate killing of Algerian Muslims by the disaster caused FLN leaders to think twice about turning dams into weapons. However, it seems likely that Christmann was mistaken or misled about the cause of the disaster, or was speculating based on his knowledge of prior FLN intentions. He may also have misinterpreted later FLN discussion about what lessons the revolutionaries might learn from the Fréjus accident about dam sabotage as a political tactic.

As Benjamin Stora and others have argued, it seems very unlikely that the Malpasset Dam was in fact destroyed by an FLN attack, since no reference to such an attack has ever been made by party leaders or combatants (generally unapologetic about the necessary violence of the war), and no trace of such an attack has been found in the French or Algerian archives.[13] Therefore, Christmann's claim and the Ritzi/Schmidt-Eeboom hypothesis are best interpreted as a parallel to Hughes's recollection of the disaster. Across the decades, some have found it difficult to believe that destruction on such a scale was not somehow connected to the events in Algeria—which it was, even if an FLN attack was not to blame.

No Such Thing as an Accidental Disaster

The flooding of Fréjus, the result of a dam failure, would not generally be considered a "natural" disaster. However, the same analytical questions that underlie environmental historians' efforts to problematize the notion of a natural disaster can be applied to this "accidental" event: how did the unintended movements of the inanimate (earth, concrete, water) interact with historical processes and human power relations? How was the resulting damage and suffering distributed as a result? How did authorities' treatment of the event as an accident obscure the deeper causes of this suffering?

The Malpasset Dam's construction in the 1950s had been a response to drought in the area—that is, to the discrepancy between the availability of fresh water and the demands of the human community that had developed in the region. While the Argens River ran from the west to its mouth at Fréjus, the lands to the north of the town were watered by the Reyran—but the Reyran "river" was a seasonal wadi, often dry except for in the winter months. The population of the coastal region including Fréjus, Saint-Raphaël, Saint-Tropez, and Sainte-Maxime, within the *département* of the Var, had been estimated to be about 45,000 in 1945, and departmental officials expected it to soon reach 150,000 during the summer tourist season. The thirsty summer crowds would require 6.5 million cubic meters of drinking water annually; the agriculture of the area would use over 13 million cubic meters of water for irrigation. Planners concluded that meeting such needs necessitated a reservoir of 22 million cubic meters, to account for evaporative loss. Damming the Reyran produced the necessary accumulation of water for year-round use.[14]

From the start, the dam had served both a political and an economic purpose. As Georges Menant wrote in 1960, "Our lands were thirsty for water like we were thirsty for progress, and progress, that was the dam."[15] Although there had been discussion of various solutions to the region's shortage of water for irrigation, the project that became the Malpasset Dam was proposed in 1946 by a departmental councilman, communist schoolteacher Antoine Foucard. The project was then taken up by the Socialists, who had initially shared power with the Communists and then came to dominate postwar departmental and municipal government for over a decade.[16] The dam's construction began in 1952 and was completed in 1954, although legal disputes with the owner of one of the local mines delayed the complete filling of the planned reservoir basin until 1959.[17] The dam project served not only as a source of water, but also as a symbol of hope after the Second World War, and as proof of the ability of the postwar local leadership, men of the Left and the Resistance, to provide a brighter future for the people of the region. By the time the basin was filled, however, the wars of decolonization had eroded the optimism of the post–World War II decade.

In the weeks immediately following the disaster, connections between French imperialism and the dam collapse were absent in the public discourse. It was quickly accepted that there was no sign of sabotage or saboteurs, and discussions regarding the cause focused on the design and placement of the dam. In the initial absence of details, a failure of state oversight seemed a likely culprit. On December 6, *Nice-Matin* published an article with the headline: "Unbelievable but True! No legislation requires prior geological testing for the construction of dams, which is indispensable."[18] Those who sought to impute blame for the disaster were more likely to point to the Socialists than to the FLN. In 1958, the "events" in Algeria had brought down the Fourth Republic, and the March 1959 municipal elections had brought the French right into power in Fréjus, led by the new mayor André Léotard, a former Vichy official.[19] On December 10, Louis Eugène Joly, a former colonial engineer and the Vichy-era mayor of Fréjus (1941–1944), wrote to Léotard and demanded that justice be brought to bear against those responsible for the dam's creation. Joly was presumably no friend to the Socialist and Communist politicians who had initiated the construction of the dam. Invoking the authority of his experience as an engineer in the colonies, and particularly as head of Public Works for Niger "where had created, in technical matters, the city of Niamey," Joly now stated that he had always had his doubts about the Malpasset project. Joly suggested that those responsible for geographical studies may have been pressured to produce a favorable conclusion, and he claimed that there had been sufficient concerns among engineers to have precluded the decision to construct a "thin" arch dam rather than a traditional "heavy" dam."[20]

Yet official inquiries concluded that no one was to blame for the disaster. In 1960, a Commission of Inquiry instituted by the Ministry of Agriculture considered and ruled out sabotage, judging that some traces of saboteurs operating on such a large scale would have been noticeable before or after the flood. The commission also considered and eliminated as possible causes the use of dynamite in the construction of the nearby highway, an earthquake, or a meteorite strike (the latter possibility suggested by eyewitness reports that lights in the sky preceded the flood). The commission also ruled out human error, finding that the dam failure "must be exclusively attributed to the ground below the foundations," which contained hidden faults or underground weak spots that gave way in November 1959; there were no errors in the design or construction of the dam itself. The final words of the "survey of possible causes" seem to have been directed against the idea that the project itself might have been hubris: "As the result of these investigations, the commission can affirm that the catastrophe of Malpasset should not diminish the confidence of engineers in dams of the arch type [i.e., a "thin" curved dam which bulges in the direction of the reservoir], the safety of which is ensured as long as the entire supporting structure is capable of permanently carrying the loads transmitted by such a structure."[21] In other words, there was nothing wrong with the dam; it was the earth that had failed.

Consequently, the French state acknowledged no responsibility for damages. Laurin, deputy from the Var, objected that the state's response, based on past responses to natural disasters, was inappropriate. This was no natural disaster, argued Laurin: its origins, he argued, should have entitled victims to full restitution from the state for damages which, he estimated, would total 236 million new francs.[22] However, the Senate rejected all such proposals that would have made the state responsible for damages.[23] The state authorized only 40 million new francs for disaster aid (equivalent to 4 billion old francs following a devaluation in January 1960). Of the 40 million, 39.2 million was to be used to rebuild civilian and military infrastructure, including roads, rural engineering, and port repair. Only 800,000 was allocated for urgent, emergency aid to victims.[24] The Department of the Var did provide reconstruction payments for businesses, and the municipal government of Fréjus successfully pursued legal action against the Department of the Var, as the party responsible for the dam, to eventually obtain state compensation for the cost of rebuilding public infrastructure, even in the absence of any finding of fault. However, indemnifying individual victims and their heirs would be left to private donors through a "solidarity" campaign modeled on the one for victims of the Orléansville earthquake.[25] As in the case of Orléansville, an abundance of donations poured in from around metropolitan France, Algeria, and the world.[26]

The initial investigation's finding that no one was to blame for the disaster was upheld by a separate, court-ordered inquiry in 1967, and finally by the Conseil d'État, (France's highest court) in 1971.[27] Some experts did testify that the civil engineers had been negligent in failing to conduct adequate testing of the rock bed, and that therefore ultimate culpability should be attributed to the chief of the Service of Rural Engineering of the Var, who had been in charge of the operation. However, the prevailing argument was that the faults in the rock were not detectable before the disaster by the means then available but became visible only after the flood waters had swept away concealing layers.[28] This conclusion has been upheld by recent scholarship on the topic. Pierre Duffault describes how a lack of regulatory oversight and communication among experts, regulators, and workers meant that there was little awareness of risks and therefore no surveillance for early warning signs. Nevertheless, Duffault argues that would be anachronistic to expect procedures to have been in place that became commonplace and required only as a response to studies conducted after the Malpasset incident.[29]

However, when viewed in a broader historical context, debates about whether the disaster was an unforeseeable accident or the result of human malfeasance or negligence present a false dichotomy. One need not believe that FLN sabotage was involved in order to view the Fréjus disaster as a product of empire. The technical ability to construct such public works, developed to a great extent in the colonies, served to demonstrate a putative European modernity that sought to maintain the distinction between the "modern" metropole and the "backwards" colonial subject. Moreover, the Malpasset Dam was intended to serve not only the basic needs of the Var's population but also the agricultural export market, contributing to the postwar economic recovery that was necessary for France to maintain control over its colonial empire. The Malpasset was also designed to provide for the affluence and leisure of the metropolitan French, who would be vacationing on the Côte d'Azur, and French military personnel preparing for, or recovering from, their imperial duties at the local army post and aero-naval base. These ambitions drove the politicians, engineers, and geographers of the Var to push beyond the contemporaneous limits of geotechnical foresight and bureaucratic oversight and beyond their ability to master the inanimate forces of water, earth, and rock.[30]

However, understanding this disaster is not just a matter of identifying the causes of the construction and dissolution of the Malpasset Dam. The dimensions of the disaster—whom it affected, and how—unfolded over weeks and even years. The cultural and social dynamics of empire shaped how Fréjus's inhabitants of European and Algerian origin were impacted by the disaster: where they lived, where they died, and, for those who survived, how they were imagined and treated afterward by French commentators and French government officials.

Chélif-on-the-Reyran (Algerians in Fréjus)

The history of the city of Fréjus is intertwined with the history of the French empire in Africa, a fact which has been recognized by academics as well as by journalists, amateur historians, and popular writers. The academic historians Gregory Mann and M. Kathryn Edwards have analyzed the connection to empire found in local representations of Fréjus's history. As Mann has argued, Fréjus is not merely a generic French town: "Fréjus has a distinct past, one not shared by the country as a whole ... for Fréjus we cannot read 'France.'"[31] Fréjus's history and locally constructed identity is imbued with what Mann refers to as "residues of empire."[32] Mann notes that the streets and places of Fréjus are named after the heroes of empire, like Gallieni and Lyautey, and that over ten thousand soldiers from the colonies are buried in the area, a byproduct of the Great War and the presence of (segregated) military hospitals that treated the wounded.[33] Later, as local historians and journalists have frequently noted, Fréjus also became the point of departure for many metropolitan troops leaving for the colonies, and the town's population grew by 1959 to include not only over 13,000 civilians but also 6,331 uniformed soldiers.[34] As Edwards explains, Fréjus's connections with the imperial project in Indochina made the town an apparently "natural choice" when, in the 1980s, French associations of veterans and formers settlers sought a site for a national memorial to the Indochina Wars.[35]

As Mann has argued, colonial relationships, networks, and flows connect particular localities in the empire, "and it is not always necessary to pass through Paris."[36] Although Mann has emphasized the movement of troops and Fréjus's unique relation to France's West African empire in the period of the World Wars, after World War II Fréjus also developed a strong connection to Algeria's Chélif valley, and especially to the village of Ouled Fares, near Orléansville.[37] Postwar Fréjus, like other parts of metropolitan France, needed laborers for its mines, factories, and road crews. Fréjus came to depend on immigrant workers, including Spaniards and Italians as well as Algerians, many of whom were housed in barracks. After 1954, the Algerian proportion of the population increased dramatically.

This demographic shift was not in itself unique to Fréjus or to the Chélif: between 1954 and 1962, the Algerian Muslim population in France increased from 211,000 to as many as 436,000, driven largely by migration from rural areas.[38] In the Chélif, however, the geoenvironmental violence of the earthquake was a unique factor exacerbating the suffering of a rural population already impoverished by a century of colonial rule, prompting an additional wave of migration. According to official counts, Fréjus in 1954 was home to "between 300 and 350 souls" from the Chélif; after the earthquake, some immigrant workers in Fréjus brought over additional family members.[39] By 1959, Fréjus was said to be home to as many as 1,100 Muslim Algerians. According to a 1960 account by journalist Gaston Bonheur, earthquake refugees constituted the majority of the North African community there.[40] Some of the Chélif refugees arrived destitute, and were assisted by the Red Cross, using funds from the earthquake solidarity fundraising drive, inciting resentment among the European residents.[41] In Bonheur's telling, one set of twins, Zorah and Nadine Mekki, had been born when their mother went into labor during the terror of the Orléansville disaster; they drowned in the Fréjus flood.[42]

In addition to the earthquake and the economic devastation of wartime rural Algeria, another driver of the migration of Muslim Algerians to France was the effort of the French state to assist Muslims targeted as collaborators by the FLN. These political refugees sought safety in metropolitan France. Archival documents reveal that in 1956 the French government authorized the use of nine million francs remaining in the budget for the reconstruction of Ténès (near Orléansville) for the purpose of resettling "North Africans needing residence in the metropole, with first priority being given, where appropriate, to those from the disaster region of the Chélif."[43] This decision had been urged both in order to resettle Orléansville earthquake survivors with relatives already working in agriculture in several departments, including the Var, and in order to resettle political refugees. The Var was listed among the regions needing immigrant labor. Because of the large proportion of families in Fréjus from the Chélif region and the directive given to prioritize the political refugees who came from the earthquake-stricken zone (a nod to the original authorization of these funds for use in reconstruction), it is reasonable to conclude that the Algerian population of Fréjus in 1959 included these sorts of political refugees as well as other war refugees and earthquake refugees.

When the economic inequalities of empire and the violence of decolonization drove Muslims to migrate from the Chélif Valley to the Reyran Valley, these anthropogenic factors combined with drought in Algeria and "diluvian rainfalls" in France to set the stage for the Malpasset disaster in December 1959. When the water level behind the dam rose to unprecedented levels,[44] the earth beneath the dam gave way, and the residents of the valley below paid the price. The Algerian residents of Fréjus paid a particularly high price.

Official Policies and Local Officials

As Todd Shepard and Amelia Lyons have shown, the French state's official response in the late 1950s to the FLN's nationalist narrative was to proclaim that Muslim Algerians were fully French, while taking ethnic origins into account for the purposes of fighting discrimination and providing social welfare assistance. These policies aimed to improve the economic status of Algerians within France while simultaneously pursuing the goals of the colonial "civilizing mission" among the migrants. Officially, Muslim Algerians were already French citizens, and since the late 1940s the "Muslim French from Algeria" were to possess political equality while in metropolitan France, even if they lacked such equality in Algeria. In 1958, in a belated attempt to undermine support for the FLN, the Constitution of the Fifth Republic had extended this more equal citizenship to Muslims in Algeria as well. Furthermore, the promotion of equality was now extended beyond political and legal equality. The social advancement of Muslims became an official goal of the Republic, and provision of welfare benefits to the "Muslim French" in the metropole became a crucial element of the French effort to mitigate Muslim grievances against French rule.[45]

Nevertheless, as Amelia Lyons has pointed out, resistance to this policy existed both among "well placed government officials" and among the European-born neighbors of migrants, who continued to believe that "Algerians did not fit into France."[46] French responses to the Fréjus disaster demonstrate that the official ideology of integration and equality was not hegemonic in the Var of 1959, as evinced by French accounts of the disaster and by the actions of the local government revealed by the archival record. The municipal government, conditioned by the ingrained colonial habits of discrimination, made use of ethnic categorization to block, rather than assist, the social advancement and well-being of Muslim Algerians in the wake of the Fréjus disaster.

Archival sources reveal the gap between local government actions and the official, legal position of Muslim Algerians within France. After the Malpasset disaster, Algerians in Fréjus, citizenship notwithstanding, were subjected to a discriminatory regime based on suspicion and scrutiny. Payments to "North Africans" or Algerian "français musulmans" (used as interchangeable terms) were set at lower levels than those available to French of European origin, and procedures were put in place to make it difficult for Muslims to obtain disaster relief at all. On December 5, 1959, the municipal council of Fréjus instituted a special service to review the claims made by North Africans who sought aid as victims of the flood. This service would be staffed by the Service of Muslim

Affairs on the first floor of the town hall (*Mairie*) and would authorize North Africans to receive "white cards" entitling disaster survivors to food, clothing, and aid payments—but only if they could provide a signed attestation to their residency from an employer, and another from a "metropolitan Frenchman" or an Algerian accepted as a long-term resident of Fréjus who was also "known to be honorable." This had to be accompanied by an official certificate from the Fréjus police commissariat.[47]

Even for disaster survivors who managed to assemble these documents, the initial approval of disaster survivor status for "les français origins des départements de l'Afrique du Nord," ("the French originating in the departments of North Africa," i.e., Muslim Algerians) was merely provisional, entitling them only to an initial emergency payment of 5,000 francs and a second, "complementary" payment of 10,000 francs per person. Algerians became eligible for the larger amounts received by "metropolitan Frenchmen" (eventually set at 50,000 per household, plus 20,000 per additional person in the family) only "when the state of disaster victim is determined in an [unspecified] formal manner, and unequivocally."[48] The imposition of this special process for Muslim Algerians belied their official status as equal citizens of France.

The distribution of donations from the Fréjus solidarity fund reveals gross discrepancies between the aid for Muslim citizens and for citizens of European descent, despite the official doctrine of the Frenchness of Algeria and the goal of equality of Algerians as French citizens. According to an unpublished report from the *Mairie*, 170 "Muslim" families suffered damages from the disaster, and 1,680 "autochthones" families. The official number of deaths was given as 396, including 27 unidentified dead and 50 missing persons. The report's section on "Categorization by Origin" identified 338 of the dead as "nationals"; 10 were "foreign"; 39 were "Muslim" and 9 were "Africans (soldiers)."[49] Fatalities in the Muslim community may in fact have been more numerous: Oliver Donat has pointed out that the number of dead remained controversial. The official figure of the total dead eventually rose to 423, but some claimed that the real number was over 500.[50] According to the unpublished Mairie report, "North Africans" in Fréjus received 360,212 new francs in aid payments, compared with over 88 million distributed to survivors and victims' heirs not identified as of North African origin. This meant that Muslims in Fréjus constituted at least 9.2 percent of the affected families and at least 9.8 percent of the dead (based on 39 out of 396) but received 0.4 percent of the aid.[51]

The archives offer several clues indicating why Algerian Muslims in Fréjus received relatively little aid in the aftermath of the disaster. The mayor's office's unpublished report listed payments to "North Africans" separately from payments to other residents of Fréjus. These payments were included only after a section listing donations from the city of Fréjus to victims of later disasters elsewhere, including victims of floods in 1960 and a donation of clothing (including transport) to post-earthquake Agadir. The content and organization of this unpublished report is telling: Muslim deaths were not like other deaths, and aid to Fréjus residents of Algerian origin was seen by the Mairie not as a duty of the state to its citizens but as an act of humanitarian generosity to outsiders, who did not merit the same treatment as "autochthon" French in Fréjus. Moreover, resistance to the notion of Algerian French citizenship was not limited to officials at the local level: a December 7 telegram from the Interior Ministry's Service of National Civil Protection to all metropolitan and Algerian departmental prefects included Algerians in an enumeration of "foreign nationals" who had perished in the flood.[52]

This othering of Algeria was also present in Mayor Léotard's 1959 Christmas message to survivors: "We have not been abandoned. From all parts of the world have come evidence of sympathy and solace. A mutilated Algeria has responded with an admirable élan."[53] Léotard's phrasing, placing donations from "mutilated" Algeria after those "from all parts of the world," paralleled the structure of the unpublished report on aid distribution. Algeria was not France, and Algerians were not French, even if French rule over its Algerian subjects and lands was to be defended and maintained.

This, however, was not official policy, of course. When the city government published a brochure on the occasion of the first anniversary of the disaster, the brochure included much of the information from the unpublished report. However, it did not specify payments to North Africans as opposed to other residents and did not divide the affected families into "autochthon families" and "Muslim families." Moreover, the published brochure no longer divided the dead into the categories of "nationals," "foreigners," and "Muslims" (in that order); instead, it recognized Algerian Muslims as a subset of French citizens, listing fatalities as

"*Français métropolitains*"	338
musulman	39
militaires africains	9
Étrangers	10[54]

The difference between the published and unpublished reports illustrates the tension between the status of Algerian Muslims as a category of French citizens and the colonial culture that designated them as "Other," as *indigènes*.

The published brochure served to efface the colonialist thinking, which was still pervasive, in favor of the official doctrines of equality. The unpublished report, however, reveals the colonial culture that was driving decision-making at the local level in Fréjus. Even the published brochure, however, found it necessary to distinguish between the two categories of French.

Being officially French did not help Algerian Muslims in Fréjus: unlike the Italians living there, for example, these "North Africans" had no foreign consulate to advocate for their interests or to raise funds abroad specifically on their behalf.[55] Of course, much of the discrepancy in relief payments was due to the fact that aid distribution was in large part based on property damage, a common post-disaster practice that privileges the owning classes and prevents disasters from having leveling effects. However, the lack of wealth among Fréjus's Muslim inhabitants of Algerian origin was no accident; it was the result of a long story of colonialism dating back to the French invasion of Algeria in 1830 as well as the more recent history of Fréjus and Algeria in the twentieth century.

The relatively meager aid dispensed to North Africans in Fréjus came in the context of a general abundance of donated funds. In the words of Olivier Donat, a local historian of Fréjus:

> The Malpasset disaster was a global tragedy. From all over the world, food, clothing, clothes, money and consolation made their way to Fréjus, a town ravaged by nature. All roads in France were crossed by cars with flags and the placards of "S.O.S. Fréjus" to collect donations for the victims.[56]

In the wake of such generosity, waste was almost inevitable, and there was little oversight of the mayor's office as it distributed the funds. In 1961, a government auditor concluded that "it is not possible to know whether the funds collected were regularly and equitably distributed" but noted that in many cases damage claims seemed to have been inflated. There were, nevertheless, funds left over for use in other catastrophes, and it was suggested that donations might be used for general improvements in the area, such as road construction, although such a diversion of funds, noted the author of the 1961 investigative report, would be a violation of the normal expectations of donors.[57] In the final accounting, in 1968, a total of 100,001,938 francs had been available, including state funds and donations. In the end, the books were balanced and excess funds dispersed by shifting funds to victims of subsequent disasters (over half a million francs) and to expenditures for "Works in the Public Interest" (almost 3.5 million), and "diverse" expenditures, mainly repair of roads and public buildings (almost 5 million).[58] It is notable that the amount diverted into "Works in the Public

Interest" and "diverse expenditures" would have been sufficient to bring relief payments to Muslims in Fréjus in line with that received by the non-Muslim population.

The special surveillance of France's Muslim Algerian citizens was by no means unique to Fréjus but was fundamental to the state's effort to remake these citizens' relationship to the Republic on both sides of the Mediterranean. However, the 1959 flood provided local officials in Fréjus with a particular opportunity, and a choice. They could implement and extend, on a local level, the national policy of promoting the social advancement of Algerians in France in hopes of undermining the case for Algerian independence and bolstering France's image as a universal, democratic Republic. Or the local leadership could reassert the second-class status of Algerians as dangerous, untrustworthy subjects whose alterity made them less, not more, worthy of public benevolence than the "autochthon" French. In 1959 and 1960, the town government chose the latter, even if national policy made them reluctant to advertise this choice.

Memoir and Memory

Far removed from these events, Christian Hughes, the fisherman's son, wrote his memoir of the Fréjus flood "forty years later" from a hospital bed. A new crisis, an unspecified medical crisis, had brought his mind back to 1959: "In this hospital where I have once more had a brush with death, the enormous wave of the Malpasset has torn my memory, leading me, frozen, to the mouth of the Argens."[59] Thus three painful events became intimately connected in his memoir: one that was his alone and two that were shared by many. In Hughes's memoir, the agonies of decolonization, the flood, and the hospital all live on, together.

Hughes's juxtaposition of his personal medical crisis with the paired memories of decolonization and environmental disaster resembles a parallel juxtaposition in the memoir of Belgacem Aït Ouyahia, the Kabyle Algerian doctor who treated the injured in the immediate aftermath of the 1954 Orléansville earthquake (discussed in chapter 2). Dr. Aït Ouyahia linked the disaster to the struggle for independence by describing his sudden shift of identity, days after the earthquake, from an assimilated "privileged Muslim" to a new identification with the downtrodden "Arab" poor, portrayed as a key moment in his conversion to the FLN cause.[60] However, Aït Ouyahia, like Hughes, also associated the environmental and political upheavals that he lived through with his memory of a personal medical crisis. In Aït Ouyahia's case, this was an automobile accident in which he had been injured prior to the earthquake. His account of the scene at the hospital after the earthquake is interrupted by a flashback to this earlier trauma, inserting Aït Ouyahia's own fear and hysteria after his automobile crash into the middle of his account of the bloody scene at the hospital.[61] In Aït Ouyahia's fractured narrative, this digression serves to convey empathy for the emotional suffering experienced by the earthquake survivors. The linkage of personal health crises to the experience of environmental disaster in the memoirs of Hughes and Aït Ouyahia support the idea that suffering, while inseparable from the social or political contexts of history, also functions as a category unto itself; that traumas, however disconnected in origin, tend to become associated in lived experience, memory, and representation, and, therefore, in effects.[62] However, the association of the Fréjus flood with the war in North Africa was not idiosyncratic, or purely psychological, and is repeatedly evinced in historical sources whose dates of origins range from the immediate aftermath of the disaster into the new millennium.

Portrayals of Algerian Muslims in French writers' dramatized accounts of the Fréjus disaster can illuminate the shifting discourses that shaped French attitudes and actions towards Algerians in the aftermath of the flood. Journalist Gaston Bonheur, memorializing the flood just months after the event, saw the history of the Fréjus flood as inseparable from the town's longstanding relation to the French empire. Bonheur—an editor of *Paris-Match* who would later be accused of sympathizing with the French settler terrorist group, the Organisation armée secrète (OAS)[63]—dated this connection to 1916, when general Gallieni had established Fréjus as a military town in order to provide a winter base for "troops of color" from the empire. This event brought the empire to Fréjus in a way that, for Bonheur, seems to have evoked sentimental exoticism: "Senegalese, Malgaches, Pondicherryians, soldiers from the Pacific: they camped here in the antiquated décor of a colonial exposition that lacked neither Buddhist pagoda nor Sudanese mosque."[64] Bonheur's comparison of Fréjus to a festive colonial exposition invoked positive images of the unity of the empire, a comparison that had been used to describe Fréjus during the interwar period.[65] A similarly sunny portrayal of the town's military multiculturalism—concealing the racism faced by African troops in France—was provided by an anonymous memoirist many years later, in a collection published for the fiftieth anniversary of the flood: "The Senegalese riflemen based at Camp de Caïs descended into town, sometimes barefoot. . . . 'There are pretty things!' they said, looking at the shop windows. They called all the women 'Mama,' always smiling; we never failed to salute their passage to honor their bravery on the field of battle."[66] These accounts made it clear that when the Malpasset Dam burst, the flood swept through a crucial part of a benevolent French Empire.

In 1998, local historian Max Prado echoed this sense of Fréjus's imperial role but extended his analysis into the more distant past and combined it with a sense of grief. For Prado, the south of France had for centuries been the site of a cycle of colonialism and rebellion that had begun with the colonization of the Mediterranean coast, first by Cretans, then by Celts, then by Romans. Prado saw the war in Algeria as a tragic continuation of this long historical process. Modern Fréjus was a staging point for French conscripts, "young people in the flower of their life," being deployed to Algeria for a doomed cause: "The barracks in Fréjus in the Robert and Lecocq districts, the aeronaval base and hospital in the military zone, would be the last steps of mobilization on the continent; the origin of many broken lives."[67] Prado made clear that Fréjus, flooded or dry, was bound to the violence of empire.

For many writers, memories and metaphors of war permeated descriptions of the flood, and references to the Second World War accompanied references to the African empire. Martial associations were always close at hand in Fréjus, and the immediate disaster response was led by troops stationed in the area, joined by the gendarmerie and the Fréjus-Saint Raphaël fire-rescue unit (*pompiers*). Assistance was also provided by United States Navy ships in the Mediterranean.[68] In 1968, Régina Wallet imagined the scene at dawn that day as a scene of war: "Jeeps and bulldozers and military trucks ready to attack. Alas, the battle of Fréjus was not won."[69]

Undoubtedly, each person in Fréjus contextualized the disaster in terms of their own experience. Wallet imagined the multiplicity of individual reactions to the deafening roar of the flood wave:

> An earthquake, thought the refugees from Orléansville, who, this time, would not escape their rendezvous with death.—the bombardment of Leipzig. And also of Dresden, specified a former prisoner in Silesia.—no, it was the Deluge, murmured a nun at prayer; God wearied of an evil world, and Fréjus did not repent like Nineva.[70]

In a different era, the religious explanation of Wallet's nun would have been the dominant cultural reaction to a European disaster such as this. But this was the twentieth century, the century of anthropogenic, martial apocalypses, and every adult in Fréjus had experiences related to the wars France had been fighting since 1939 in Europe, in Indochina, and in North Africa.

French accounts of the flood often displayed grim irony in relating the stories of those who survived the horrors of war only to die in the flood. Writing
in 1960, Bonheur described a Madame Legrand, who "during the war, already
had known all the horrors, all the distress, during the terrible bombardments of
Hammanlif (400 dead in a few minutes), as the British 8th Army pushed back
the Afrika Korps in Cap Bon in 1943." The death toll on that day in Tunisia
was similar to that of Fréjus in 1959; in Fréjus, according to Bonheur, Mrs. Legrand lost her seven-year-old daughter.[71] Bonheur's account of Mrs. Legrand's
tragedy is paralleled by Wallet's 1968 portrayal of Sergeant Boul, an Algerian
War veteran:

> On that first of December, Sergeant Boul had rejoined the garrison in
> Fréjus. Two years in Algeria, what an ordeal. Even for a soldier who had
> chosen war. But these guerrillas, these attacks on the sly, these tortures, this
> was not real war, in which one is confronted face to face. Sergeant Boule
> was weary. He aspired to rest, to calm, with his family. He would rejoin,
> that very day, his wife, who was arriving from Thionville with their three
> children.[72]

That night, Boul was awakened by a sudden noise and leaped up. Wallet imagined his thoughts in the last moments of his life. He was startled by the noise,
"but he smiled. He was no longer at war. He was in the Côte d'Azur. Already the
wave, like the enemy, pounced."[73] Boul and his children were killed. For both
Bonheur and Wallet, the tragedy of the flood was amplified by the surprising
irony that the apparently peaceful environment of the Reyran Valley could destroy what the brutality of war had not.

Some representations of the Fréjus disaster reflected the official discourse
claiming the unity and solidarity of Algeria with France, in which the conflict
in Algeria was portrayed as an unnecessary and unfortunate political development. Other accounts, however, embodied the habits of imperial contempt for
a subjugated population. Maurice Croizard's 1960 depiction of Fréjus presented
an image of harmony among the town's metropolitan and African residents.
Croizard's account reconstructs (or imagines), in intimate detail, a reunion between two former military comrades just before the collapse of the dam. Croizard portrays a conversation between a Sergeant Léveillé, just arrived from Algeria, and his friend Mohammed Azzi. Léveillé speaks warmly about the North
African territory: "A beautiful country, really, if it were not for this political
ugliness," and Mohammed Azzi tells his old friend that he has been thriving in
his new home in Fréjus: "I am the happiest man in the North African colony [in

Fréjus]. Now there are eleven hundred of us here, including families. Almost all of Orléansville. . . . And you know, there's no politics here. Just family men. . . . We have social security, we are paid like the French."[74] Minutes later, the dam burst. Azzi survived but had to identify the bodies of many of his friends and their children, drowned when the floodwaters destroyed the Sabagh factory and the nearby homes of "North Africans." Forty-three corpses were found in an area of a few hundred meters. Nearby, according to Croizard, "an Arab . . . beat his chest and repeated: 'I was the one who brought them here after the Orléansville catastrophe. I told them, you will have work, you will be happy. I'm the one who led them to their deaths . . . it was me, it was me.'"[75] Fréjus's memoirists' portrayals of Muslims in Fréjus took on a sympathetic tone as they conveyed the stories of Algerians surviving the earthquake only to die in Fréjus, stories which paralleled narratives about the European French who had lived through war but perished in the flood.

Some French accounts of North Africans in Fréjus mixed sympathy with the paternalism and contempt inherited from over a century of imperialist culture. Régina Wallet repeatedly invoked clichés of Arab fatalism in the aftermath of the disaster: "It was written, declared the Arabs."[76] Wallet seemed to reserve her sympathy for North African children. She enumerated twenty-two children who died in three North African families, identified by name, and ended a chapter with a grisly scene: "A great silence, disturbed by the cawing of crows. Dreadful silence, silence of death. Between two branches, like a bird with wings unfurled, rested a dismembered [écartelé] toddler. One of the little Arabs who, around the barracks, laughed and sang, like the children of the douars."[77] Later, Wallet's book itself ends with the story of an official at the prefecture who received a request from a young French couple from outside Fréjus who hoped to adopt a child orphaned in the disaster. "We would take the most pitiful, even a little Arab," the couple offered, but, according to Wallet, "the little Arabs, they all disappeared in the flood," including those saved from the "bombardment [sic!] of Orléansville." The official wept, having also lost a child in the flood.[78]

Although the writing of history is always an interpretive act, Wallet's "history" of these events, as well as the accounts of the disaster published in Bonheur's 1960 anthology describing the disaster, must be considered, at least in part, to be imaginative rather than strictly historical works. Unlike, for example, the 2003 history of the Malpasset Dam and the Fréjus disaster published by the *Société de l'histoire de Fréjus* written by Vito Valenti and Alfred Bertini (the latter a participant in the municipal response to the disaster), or Max Prado's self-published 1998 history, neither Wallet nor Bonheur explicitly grounded their description of events in documentary evidence. Nor were Wallet and Bon-heur witnesses to the events, as Christian Hughes was. In her preface, Wallet, a novelist by trade and habit, claimed that there was "neither imagination, nor invention on the part of the witness, in this atrocious and authentic story." At the same time, however, she acknowledged that memories change over time, and that "a testimony is only sincere at the exact moment of the event." Conse-quently, Wallet asked her readers for "indulgence, not for myself, who is relating this just as it is told to me, but for the witness, who, time passing, translates his unhappy adventure with an infinite sadness."[79] Nevertheless, Wallet did not limit her account to what could be verified or supported by written sources or oral testimony. Her narrative obviously goes beyond what could be supported by any evidence, documentary or otherwise, in depicting the interior soliloquies of individuals in the moments before their deaths.

Gaston Bonheur, in contrast, was a professional journalist and editor of news magazines (and also a poet in his younger years). His volume on Fréjus, cowrit-ten with Maurice Croizard, Géorge Pernoud, and others, was part of the Map-pemonde series, directed by Bonheur and Pernoud, which included titles such as *The Nazis since Nuremberg* and *Anastasia, if It Is True?* The back cover of Bonheur's Fréjus volume claimed that "the authors published in this collection do not write. They transcribe. The real author is the march of time, creator of tragedies, comedies, catastrophes, which will draw out literature and which History will put in order." Nevertheless, Bonheur's and Wallet's accounts might be classified as pseudo-histories, claiming creative license in works aimed at a mass audience.

As Claire Eldridge has argued, memories are "socially-framed, present-orientated, relational, and driven by specific agents."[80] Like the memoirs written by survivors of disasters, these pseudo-historical works demonstrate how the constructed narratives of the event intertwined the memory of environmental disaster with memories of war and empire. These sources also reflect and reveal how decolonization impacted representations of disasters.

Unlike the accounts by Bonheur and Croizard, Wallet's 1968 work was penned after Algerian independence. As Todd Shepard has demonstrated, after 1962 the Frenchness of Muslim Algerians was erased from French law and from public memory, consecrated through a mass genuflection to "the tide of history."[81] Before 1962, there had been a political incentive for French writers to deny belief in the otherness of Muslim Algerians in hopes of undermining the nationalists' narrative and advancing the notion that Algeria was France. Decolonization brought an end to this incentive. It also brought thousands of "repatriated" Europeans from Algeria to the Var, a demographic event that would arguably play a role in making the area a stronghold of the nationalist, anti-immigrant Front National, culminating in the FN's 2014 victory in the Fréjus municipal elections. Wallet's portrayal of sympathy for "little Arabs" and solidarity among the bereaved drew on notions of childhood as a universal category. However, it also reflected post-1962 fears that France would be "overrun" by the children of an alarmingly fertile Algerian migrant population, an attitude which contrasted sharply with the natalism of French policy in the 1950s, when women and children were seen as a moderating (or "civilizing") element among Algerian migrants in France.[82]

The sympathy Wallet expressed toward children was less evident in Wallet's portrayals of North African adults. Wallet conveyed the suspicion and contempt, replete with the cultural legacy of imperialism and orientalism, that some European French held toward the North Africans in Fréjus:

> Eleven Arab families were lodged at the Sabac [sic] factory. All were swept away by the wave. But one could never count the victims, and the Arab mystery was never resolved. How many Algerians or Moroccans worked at the factory, or on the nearby farms? Many were employed as casual hires and their foremen had never declared them. Some, in transit, had no fixed domicile. They lodged with a comrade, in a ruined shed, or even under the light of the moon, under an olive tree. At the whim of friends and acquaintances, they took turns with the employers, who never recognized them. An Arab greatly resembles another Arab. The same bronzed face, the same look, evasive or timid, the same childish language. An Ali is always an Ali. . . . None of them possessed a work permit. Thus, how can they be counted? Especially those who were lost. Impossible to identify disfigured faces. And how, in Algeria, to find their families? All their wives were illiterate and incapable of making a claim. . . . At the same time, some crafty Arabs, who were working very tranquilly outside of Fréjus, attempted to get indemnities as victims. In the chaos, all regulation became impossible and there are always some vultures who profit from tragic circumstances.[83]

Eldridge has shown, in her scholarly examination of the construction of memory and historical self-understandings among the communities of Europeans who left Algeria for France circa 1962, that the pieds noirs' communal construction of memory denied that racial injustice was the basis of colonial Algerian society and that these constructions described positive interactions with Muslims in Algeria (mainly those in servile positions). However, as Eldridge argues, these positive portrayals were also accompanied by "a category of unknown or threatening 'Arabs' or 'Muslims,' who, although mentioned much less frequently, underlined the continuing need for a 'civilizing' French presence and the challenges associated with that endeavor."[84] This pattern is evident in Wallet's account of "crafty" and dishonest Arabs in Fréjus.

Wallet's 1968 portrayal of Arabs exploiting French generosity also resembles a 2008 description of Orléansville by Jacques Torres, a "repatriated" pied noir. Torres describes a joke that circulated after the 1954 earthquake: "When you asked an Arab what his profession was, he responded, laughing: 'Me? *Sinistri ou labbès* (a disaster victim, and all is well—that is enough)'. . . . The status of *'sinistré'* [disaster victim] gave the right to significant aid, which was enough for certain people to live on."[85] When Wallet and Torres wrote their stories of disaster, Algeria was independent, and it was no longer necessary or plausible to affirm the solidarity of Algerians with France, or vice versa.

However, Wallet's portrayal of Arabs as "crafty" or "vultures," unrecognizable and transient, defying French attempts to name and regulate them, were not merely a product of decolonization, but was also shaped by old habits of racist, orientalist colonial discourse. Hostility to Muslim North Africans in Fréjus did not begin in 1962. Ingrained colonialist habits of thought were evident in Fréjus in 1959, and such thinking had demonstrable consequences for Muslim Algerians, both immediately after the 1959 disaster and for many decades to come.

Fréjus and Le Pen's Front National

In 1987, local historian and Var schoolteacher Louis Robion described anxiety over "cultural identity" in Fréjus which prompted people to ask, "Is Fréjus still Fréjus?" Robion attributed this "millenarian disquiet" to rapid population growth and urbanization, as well as the transformations brought about by the expansion of tourism, globalization, and the cultural influences brought by new migrants. As the tourism industry grew, some residents feared that Fréjus might be on the verge of becoming a generic beach town: as Robion put it, residents were "haunted by the image of the American '*sun-belt* [*sic*, in English].'" Meanwhile, cultural change meant that Provençal was disappearing, and "anchovies have given way to pizza and *méchoui* [North African roasted lamb]."[86] The cultural anxieties described by Robion resemble those that might be found among longtime residents of many a European (or American) small town, reflecting the rapid changes of the twentieth century. Yet, despite the commonalities that cultural anxiety in Fréjus shared with the angst of globalization elsewhere, the anxieties in Fréjus were shaped by the particularities of its history.

Robion argued that the changes Fréjus was experiencing were not to be condemned because the only "tradition" in Fréjus was one of cultural transformation and successive new influences, from the Romans to the tenth-century Bourguinons, seventeenth-century Moriscos, and eighteenth-century Germans, to the "Portuguese and Maghrebians of today."[87] It was no shame and no novelty, argued Robion, that Fréjus had in the twentieth century become home to a pagoda and a mosque as well as a cathedral. But Robion's analysis ignored the particularly strong historical connection that Fréjus had with the French colonial project and omitted the fact that Fréjus's first mosque and first pagoda were built by and for colonial soldiers of color forcibly conscripted to serve their French conquerors.

Many Fréjus residents would not join in Robion's embrace of cultural fluidity. Fréjus was one of several French localities where Marine Le Pen's anti-immigrant National Front party emerged victorious in the elections of 2014; National Front mayor David Rachline, elected in 2014 at age 26, had made opposition to a large new mosque complex central to his campaign. The mosque, approved by the previous mayor, opened in January 2016, but the Rachline government unsuccessfully sought a court ruling requiring its demolition on the grounds that the building permits had been illegally obtained.[88] The reasons for the rise of the nationalist, anti-immigrant right in Fréjus, as elsewhere in France, are complex, and include disenchantment with the established parties as well as the popularity of the National Front's nativist, anti–European Union rhetoric. Explanations often include the influence of the *pieds noirs* in France after 1962, who formed an important source of support for the party and for Jean-Marie Le Pen, the ex-paratrooper and Algerian War veteran who led the party from its founding in 1972 until 2011. After the 2014 elections, nostalgic support for the French colonial project in Algeria figured prominently in the Rachline administration's construction of nationalist patriotism and was embodied in a new monument "to all those who died so that France could live in Algeria."[89] Unlike the Indochina war memorial built in Fréjus in the early 1990s, this new monument made no attempt to balance competing views and experiences of the war and of French colonialism.[90]

Despite the immanence of the Algerian question in the distribution of aid after the 1959 Malpasset Dam collapse and in the work of memoirists and local historians of the disaster through the turn of the millennium, today the

Malpasset disaster seems to be absent from the rhetoric of Fréjus's National Front and its critics and from the identity anxieties described by Robion. Nevertheless, the history of the flood of 1959 offers important insights into a strain of political culture within this town that has been so receptive to the message of the National Front, and to the Rachline administration's support for an *Algérie française* that had not existed for fifty years. No town is a monolith, of course, and there were contradicting voices, both in 1959 and in 2014. But in 1959 and 1960, Fréjus's first right-of-center municipal government in decades spoke, wrote, and acted in ways that contravened official French policies and that cast Algerians—French Muslims from Algeria, citizens of the French Republic—as foreigners and outsiders and denied these disaster victims and survivors their fair share of disaster aid donations sent, not only by white Europeans, but by Muslims in Algeria and by Moroccans and Tunisians. This suggests that the appeal of the anti-immigrant Front National in Fréjus, insofar as it relates to nostalgia for the French colonial project and hostility to immigrants of North African descent, cannot be attributed simply to the post-1962 influence of the pieds noirs settler-refugees, or to more recent events such as terrorist attacks of the 1990s and 2000s. The influence of deep-seated colonialist ideas about the venality and alterity of North Africans was evident in Fréjus after the 1959 flood and cannot be attributed to the influence of the pieds noirs. If one was to point to a turning point in the political history of Fréjus, it would not be the arrival of the "repatriated" refugees of 1962 but rather the end of Socialist hegemony in early 1959 with the election of André Leotard to the mayor's office (a local consequence of the collapse of the Fourth Republic and the return of De Gaulle in 1958). Such a conclusion would be oversimplistic, however. The attitudes of the local authorities in Fréjus, revealed in the aftermath of the flood, were rooted in the town's long and intimate association with the French imperial project.[91]

Conclusion

The Malpasset disaster may have been "a global tragedy," as Olivier Donat has written, but it was also an imperial drama. When the inanimate intervened in the lives of Fréjus's residents by behaving in unexpected ways, the gradually unfolding results of the sudden collapse of concrete and the rush of water were shaped by the history of France and its empire, as the local leadership acted to reinforce the distinction between colonizer and colonized, regardless of the strategies and policies that emanated from Paris before being overturned by the independence of Algeria. The solidarity of Algeria with France celebrated by

Fréjus's mayor in 1959 did not extend in return from the mayor's office to Fréjus's most vulnerable population.

"Solidarity" campaigns for local catastrophes such as Orléansville and Fréjus were meant to function much like the war memorials to colonial troops discussed by Mann and Edwards. These fundraising drives were "simultaneously 'local' and 'national,'" serving to foster a sense of republican unity between the suffering, sacrificing locale and the benevolent, unifying empire, as Mann has argued.[92] For the Algerians living in Fréjus, however, there was more suffering than benevolence. In the months following the collapse of the Malpasset Dam, an optimistic vision of imperial unity in the form of a new brotherhood of equals was advanced by the official proclamations of the French state, as well as by chroniclers of the disaster like Bonheur and Croizard. However, the long history of colonial distinctions between citizen and colonial subject could not be overcome. For the Muslim Algerians of Fréjus and for the municipal government that had the ability to help them recover from the 1959 disaster, the subordinate status of Muslim Algerians remained.

The French appeal for donations as a show of imperial solidarity did little to strengthen the affections of the colonized for the colonizer. Indeed, post-disaster solidarity could be turned to serve the interests of anticolonialism. Donations were also sent from Morocco, already independent but still struggling to negotiate its post-independence relationship with France. Three months after the Fréjus disaster, when earthquake-stricken Agadir was in need, and French aid seemed to come with neocolonial strings attached, Moroccan nationalists would angrily recall the unconditional generosity Morocco had offered to Fréjus.[93] Although the disaster in Fréjus was a much smaller disaster than those that afflicted Algeria and Morocco during these years, Fréjus thus became part of the game of "disaster diplomacy" played by great and small powers in 1959 and 1960.

CHAPTER 4

Poison, Paralysis, and the United States in Morocco, 1959

O N SEPTEMBER 10, 1959, three years after Moroccan independence, a previously healthy twenty-five-year-old man was brought to the Mohammed V hospital in Meknes, unable to walk. He had been suddenly afflicted with cramps in his calves the day before and had awoken to find he had lost control of the muscles in his lower legs and feet. He had no fever, but was subjected a spinal tap, which revealed no signs of polio or other pathogens. Ten more patients with similar symptoms arrived at the hospital later that day. Thirty more came the next day. Seven hundred more cases were identified by the end of the week. The epidemic of partial paralysis continued to spread, and by November, approximately ten thousand cases had been identified in cities, towns, and rural areas throughout Morocco.[1]

The origins of the epidemic remained a mystery for weeks, but investigators eventually identified contaminated cooking oil as the cause of paralysis. On April 13, 1959, Mohammed Bennani, an automobile supply wholesaler, had purchased a large quantity of surplus jet engine lubricating oil from the US Air Force base at Nouasseur, near Casablanca. This oil contained triaryl phosphates and cresol phosphates. The lot purchased by Bennani was just one lot out of over fifty lots of various sorts of obsolete or excess oils sold from US airbases in Morocco that year. Bennani then sold portions of the oil, still in its original drums, to another merchant, Ahmed ben Hadj Abdallah. The oil was then resold to approximately two dozen cooking oil wholesalers, in Casablanca, Fez, and Meknes. The Meknes wholesaler later explained that he had been seeking a way to increase his profit margin from fourteen to twenty-four francs per liter. He had purchased several tons of the engine oil, mixed it with vegetable and olive oil, and bottled the concoction as cooking oil, labeling the bottles "Le Cerf" and "El Hilal." As this oil went to market, he purchased another six tons of the engine oil, in expectation of increased demand from pilgrims arriving in Meknes for the coming celebration of Mouloud, the birth of the Prophet.[2]

Soon after consuming the adulterated oil, the victims typically suffered a brief period of gastrointestinal distress. This quickly passed, and all seemed to be well for just over two weeks. Then, however, the motor neurons of the spinal cord were damaged by the toxins.[3] The victims began to experience fever accompanied by numbness in the hands and feet. This also passed, but was replaced, within about forty-eight hours, by pain and weakness in the feet and legs, eventually leading to paralysis of all muscles below the knee, followed by spasticity and muscle atrophy. Most victims' hands were also affected, and impotence afflicted the men. For some, paralysis of the pelvic and abdominal muscles rendered them non-ambulatory, but in most cases the paralysis was limited to the muscles of the extremities, so that the victims could still walk, with the "ungainly high-stepping gait" that soon became the identifying characteristic of the "Meknes disease."[4]

The 1959 oil poisoning was an event produced neither by freak chance nor by malevolent intent. This disaster, like the Malpasset Dam collapse, might be understood as the product of what Jane Bennett calls the "distributive agency" of an "agentic assemblage" consisting of a "confederation of human and non-human elements."[5] Neither the former French colonizers, nor the US Air Force, nor the profiteering oil wholesalers intended to poison thousands of people in Morocco. But declining colonial powers needed superpower allies, and, to be competitive, twentieth-century superpowers needed overseas bases and jet aircraft. Twentieth-century industrialism produced synthetic cresyl compounds, providing lubrication for those jets and a means for grocery merchants competing in a capitalist economy to cut their production costs. Just as French colonialism and the neo-imperialism of America's Cold War influenced human thought and action to produce certain patterns of behavior and cultures of imperialism, they also produced particular arrangements of inanimate matter. Together, these arrangements of the animate and inanimate constituted the military bases, schools, hospitals, and embassies that were meant to serve colonial and neo-imperial goals. However, neither humans nor the inanimate behaved quite as the colonizers desired—and the behavior of manufactured chemicals could prove as intractable as those of colonial subjects or tectonic plates.

The very large scale and rather sudden onset of the Morocco oil poisoning produced effects that were in some ways akin to a seismic disaster or a flood. Much like the 1954 earthquake in Algeria, the paralysis epidemic overwhelmed the hospitals, tested the state's ability to respond to the needs of the people, and opened the state, for its failures, to criticisms by the political opposition. However, unlike the earthquake in Algeria's Chélif Valley or the anthropogenic flooding of Fréjus, the oil poisoning did not destroy the built environment and was not confined to a specific region or locality. The silent destruction and bloodless suffering brought by the oil poisoning did not invite comparisons to the bombing campaigns of the Second World War, and when the disasters of Orléansville, Fréjus, and Agadir were discussed as kindred catastrophes, the oil poisoning was often left out. Nevertheless, the Morocco oil poisoning was connected to the other disasters, especially the 1960 earthquake in Agadir. The Moroccan political opposition made use of comparisons between the oil poisoning and the Fréjus flood to critique both the continued French presence in Morocco and the Moroccan state's response to the oil poisoning. The Moroccan monarchy used both the oil poisoning and the 1960 earthquake to enhance royal authority and prestige, and international responses to these two disasters were interrelated. For the diplomacy of the United States, in particular, the oil poisoning and the Agadir earthquake were of profound mutual relevance.

The origins of the Morocco oil poisoning were inseparable from the social and political contexts of imperialism, the Cold War, and the still-incomplete process of decolonization. So, too, were human responses to this disaster. In 1956, after several years of escalating violence, France had permitted its Moroccan protectorate formal independence. The Alaouite monarch, Mohammed ben Yusef, was retrieved from a two-year exile and became King Mohammed V. Morocco owed its independence not only to the courage and ruthlessness of its freedom fighters, but also to France's determination to keep Algeria at all costs. Morocco's independence meant that France could concentrate its resources on the fight against the FLN. It remained to be seen whether independence would lead to the French ending their military presence in Morocco, and whether independence would mean that the French would cede their political and cultural influence over Morocco to the Americans, or, less plausibly, to the Soviets, or whether the Moroccan state would be able to assert full sovereignty.

In particular, responses to the oil poisoning by the French, American, and Moroccan states, and by the Moroccan political opposition, were tied to one of the legacies of colonialism that had not been undone by formal Moroccan independence: the presence, on Moroccan soil, of military bases occupied by the former colonizers and their American allies. Just as the question of whether or not Algerians were to be French shaped responses to the disasters in the Chélif and Reyran Valleys, responses to the oil poisoning by the Moroccan, American, and French states, and by the Moroccan political opposition, revolved around the question of whether these remaining plots of foreign-occupied land would be decolonized.

Epidemiology and Crime

The mass poisoning in Morocco was distinctive in its relation to imperialism and the Cold War, but there had been cases of mass cresyl phosphate poisoning before. Tasteless, odorless, and soluble in vegetable oil, triorthocresyl phosphate has a particular tendency to find its way into the food supply through a variety of means. In 1930 and 1931, as many as sixteen thousand people were poisoned in the American Midwest through the adulteration of Jamaican ginger extract, or "jake," an alcoholic patent medicine. In that incident, an unscrupulous entrepreneur had used triorthocresyl phosphate to circumvent government regulations stipulating the minimum solid content of alcoholic "medicinal" extracts. Another case affected hundreds of women in the 1930s who had consumed a parsley extract abortifacient that included the toxic chemical. There were also several cases in Germany in the 1940s in which food shortages had led to the use of engine oils in cooking as well as the accidental poisoning in 1940 of ninety-two Swiss soldiers after machine-gun cleaning oil was mistaken for cooking oil. In 1942, forty-one people in Verona, Italy, were afflicted with paralysis, which has been traced to ground contamination caused by the use of discarded military engine oil containers in the handling of farm compost and manure.[6] In its scale, the Moroccan case was most like the American jake poisoning; in its causes, it also resembled the German incidents, in which poverty incentivized the substitution of machine oil for vegetable oil, and the Verona case, in which the outbreak was caused by improper repurposing of military surplus material.

In 1959 Morocco, it took some time for investigators to determine the chemical and human vectors by which imperialism, industrial capitalism, and the Cold War had led to mass paralysis in Meknes. Initially, polio or other viral infection was suspected. The appearance of a few cases between August 31 and September 2 followed by an explosion of cases between September 18 and 24 suggested a pattern of contagion. Indeed, the delayed onset of paralysis following consumption of the poison mimicked the incubation period of a contagious disease, making the true origin of the crisis difficult to identify. In addition, all of the afflicted lived in poor neighborhoods, suggesting disease vectors related to housing patterns and unsanitary living conditions, such as sewage or insects.[7] Fearing the outbreak of some new virus, King Mohammed V ordered the cancellations of all festivities for the September 19 celebration of Mouloud, which had already brought pilgrims to Meknes from the surrounding region. The King also banned all travel and public meetings; public pools and athletic facilities were closed.[8]

Yet the symptoms did not correspond to polio or any other known illness, and blood and spinal fluid showed no infection.[9] By the end of September, the hospitals in Meknes were overwhelmed, and the sick were being housed in ad hoc locations. The French military hospital in Meknes opened beds to the afflicted, and French military doctors arrived from Casablanca to help treat the influx of patients.[10]

In late September, Dr. Youssef Ben Abbes, the Minister of Public Health, met with doctors from the Avicenna Hospital in Rabat, and the staff of the National Institute of Health. The doctors and investigators were puzzled by the odd distribution of the malady. While all of the victims were poor (typically the families of day laborers), the very poorest members of the society (who had been too poor to buy oil that month) were spared, as were infants. Furthermore, the "absolute immunity" of Europeans (with one exception) and of Jewish Moroccans seemed to defy explanation simply in terms of superior sanitation, and very few of the fifty thousand people who traveled to Meknes from other regions for the celebration of the Mouloud in mid-September had become afflicted. Of the hundred Moroccan soldiers stationed in Meknes, only two suffered from paralysis.[11]

Several patients had pointed to the strange, dark, reddish-colored oil as the probable cause of their affliction. One family, described by a doctor Baillé, had noticed that a bottle of cooking oil that they had recently bought was unusually dark in color—in hindsight, "as dark as old motor oil."[12] After using the oil to cook a meal, they had been concerned enough to offer a portion to their dog. After the dog seemed to suffer no ill effects, they went ahead and ate, but remembered the strange oil after they later fell ill.

The experts, however, were not to be convinced by anecdotal evidence alone. Albert Tuyns, a Belgian epidemiologist working for the World Health Organization (WHO), began to investigate the outbreak, applying the methods of the "new epidemiology" of the period, which included expanding epidemiological analysis to include maladies other than infectious disease. Tuyns initiated a survey of 250 patients on September 21 and found that cooking oil was the common factor in all the responses. This conclusion was also supported by the correlation of paralysis cases, noticed by the Ministry of Health, with the areas frequented by street vendors who sold fried pastries and by the suspicions of the patients themselves.[13]

When the cooking oil was identified as the cause of the paralysis, the investigators suspected a neurotoxin was involved and sent a sample of the oil to the Institute of Hygiene in Rabat. The initial chemical analysis on September 30 indicated the presence of "phosphates, phenols, and cresols." World Health Organization investigators identified triorthocresyl phosphate as the culprit, although other triaryl phosphates and cresol phosphates also present were later seen as having a significant role.[14] These chemicals were binding with acetylcholinesterase in the neural synapses, causing a buildup of the neurotransmitter acetylcholine and a breakdown in the functioning of the motor neurons. This diagnosis was not good news; there was no pharmaceutical cure—the only course of treatment was physical therapy, and in certain cases of severe spasticity, orthopedic surgery.

Once the chemical cause of the affliction had been determined, the focus of the investigation shifted from the epidemiologists to the police. By October 9, the Moroccan government's Criminal Investigation Division in Meknes tracked the tainted oil to a warehouse in Meknes, where they discovered a second shipment of "six tons of oil in 31 drums ready for bottling and delivery to consumers."[15] Initially, the investigation focused on Meknes, and warnings to the public identified only the local brands Le Cerf and El Hilal as potentially dangerous. Soon, however, it became evident that this was not a localized disaster; nor was it contained. The number of diagnoses continued to rise sharply around the country, reaching ten thousand within a few weeks. On October 28, the Moroccan government declared a "national calamity" and provided emergency funds (one hundred million francs) to the Ministry of Public Health for rehabilitation and treatment of those afflicted.[16] The government also began a public information campaign, warning the public against the purchase of any oil that was not bottled in a factory, and identifying twenty of the more inexpensive brands of cooking oil sold throughout the country as potentially dangerous. Orders were issued that all persons should turn over bottles of the suspected brands to the police, and the government announced that that house-to-house searches would be conducted to confiscate any undeclared household stocks. Following the initial arrests of alleged culprits in Meknes, interrogations led the investigators to other wholesale centers in Fez and Casablanca, from which the toxic oil had been sold across Morocco, and thirty-one people were arrested. The authorities seized 190 metric tons of suspected oil, including 600 kilograms of machine oil.[17] As in the Chélif region of Algeria in 1954, however, the mountainous territory in parts of Morocco and the dispersal of the rural population inhibited the disaster response, and authorities feared that the inhabitants of villages in areas "with neither radio nor roads" might continue to possess and consume adulterated supplies of oil.[18]

The impact of the disaster extended beyond the individuals afflicted and was economically devastating to entire families. Minister of Health Youssef Ben Abbes estimated that "there are at least 30,000 made destitute" by the disaster.[19] Abdelmalek Faraj, head of the National Institute of Health, noted that 80 percent of the afflicted were unskilled workers, who lived by means of their physical labor. Such workers had already been struggling to sell their labor in an economy characterized by high unemployment; for the newly disabled, finding work would be impossible. Moreover, as Faraj noted, afflicted families often lost both wage labor and the unpaid labor of women and children, producing a desperate situation.[20] Medically, the long-term prognosis was equally grim. WHO physicians offered little hope of recovery, stating that "a probable 10,000 cases, including small children, will be completely paralyzed for the rest of their lives, and their upkeep will depend on the good will of their neighbors or the State."[21]

Treatment and the Habits of Imperialism

In November, the World Health Organization sent Denis Leroy, head of the anti-polio center at the University of Rennes, to Morocco. Dr. Leroy stressed that physical therapy needed to begin immediately for the many thousands of people afflicted, and Leroy developed a plan for a massive, long-term rehabilitation effort, led by the international Red Cross.[22] Departing from their usual policy of providing only immediate, urgent disaster relief, the Red Cross launched an eighteen-month effort that began in January 1960 and that would involve 167 foreign medical professionals (typically in rotations lasting several months) as well as 150 Moroccan nurse's aides.[23]

Five treatment centers were established around the country. In the most severe cases of spasticity, medical and surgical treatments were attempted, ranging from alcohol injections and casts to the surgical severing of tendons.[24] For the most part, however, the program was based on physical therapy, while also addressing the social impact of the catastrophe described by Faraj. In Khemiset, fifty kilometers west of Meknes, a reportedly successful program of occupational therapy was implemented, thanks to the presence of a Swiss occupational therapist, the only occupational therapist present at the start of the program. Vocational rehabilitation programs were also later initiated in Meknes in May 1960 and in early 1961 in Fez, Sidi Slimane, and Sidi Kacem. In Khemiset, the Swiss therapist developed a program providing groups of sixty patients at a time with practice in "weaving, tapestry-work, sewing, leather-work, knitting, production of strings of *perlas*, embroidery, cord-knotting, basket-work, mat-plaiting, writing, painting, drawing, type-writing, wood-work and gardening."[25] These activities were chosen partly for their value as physical therapy: they "favored the rhythmic change between contractions of agonists and antagonists," and therefore, it was claimed, were not fatiguing.[26] Vocationally oriented tasks were also included in hopes that they would provide paths to employment for those who could no longer work in their prior occupation or who needed occupational therapy to relearn skills. The choice of tasks obviously was influenced by the predominance of women among the afflicted, although the medical professionals who reported on the program were almost silent on the role of gender; the significance of the fact that the Swiss occupational therapist was female went unremarked, despite the relevance that her gender must have had when treating female patients. No occupational therapy program was implemented at the treatment center in Al-hucemas; it is unclear whether this was related to the predominance of men among the patients in the northern region.

In addressing the "social aspects" of this disaster in independent Morocco, the international assistance teams went beyond the treatment of affliction and assumed a role once played by colonial schoolteachers and doctors. Medicine had long been an instrument of colonialism in Morocco. As historian Ellen Amster has written, "Histories of colonial medicine illuminate how native bodies were invented as objects of scientific knowledge, racisms were naturalized, and health dictatorships were designed to sanitize, rationalize, and control native bodies."[27] The occupational therapy program operated on the assumption that the patients needed not only to recover the use of their extremities but also to adopt Western, industrial attitudes toward work and time. Like the French doctors and educators of the colonial ("protectorate") era, the international rehabilitation staff made Moroccans into objects of observation, engaging in what, given the brief period of time spent in residence, must have been a very superficial study of "native arts and crafts."[28] However, they also asserted knowledge of their Moroccan patients' mental characteristics, which they aimed to improve through the bodily discipline of physical therapy. Dr. W. M. Zinn, of Switzerland, wrote that "the majority of patients were illiterate, owned no watches, and often had no concept of time," traits that complicated the scheduling of outpatient treatment.[29] Consequently, the goals of the rehabilitation program came to include the remediation of these perceived failings. Zinn reported that, in Khemiset, "the patients were gradually accustomed to working increasingly long hours, in the end six hours a day, and to turn up for work regularly and punctually."[30]

Amster has argued that the colonized were often successful in resisting such discipline, refusing to adopt the norms and standards of their Western doctors and teachers.[31] Whether this was the case in the physical therapy centers is difficult to assess. In terms of the more measurable goal of vocational rehabilitation leading to employment, only the Khemiset treatment center had success in finding jobs for its patients upon completion of treatment. In part, this was attributed to the delayed start of occupational therapy in other locations, leading to less successful recovery of patients' hands; in Khemiset, the early presence of an occupational therapist and the use of therapeutic hand splints prevented irreversible hand deformities from developing. Largely, however, the success of the Khemiset program in the economic reintegration of victims was attributed to the area's relatively low unemployment rate; elsewhere, finding jobs for patients with impaired mobility was virtually impossible in an economy in which, as in Meknes, 80 percent of the adult male population lacked permanent employment.[32] The state of the local economy was the determining factor rather than the content of the curriculum or the transformation of the Moroccans' mindsets—a reality experienced by colonial schoolteachers in earlier decades.[33]

Latent colonialist habits notwithstanding, there is little doubt that the international physical therapy effort made a positive difference in rehabilitating the bodies of the patients. In January 1961, half of the ten thousand victims of the oil poisoning were still receiving physical therapy, but five thousand had been judged fit to discontinue the treatment; of these, three thousand were kept under supervision, and two thousand were considered "cured to all extents and purposes, and able to resume their former occupation."[34] In June 1961, a Red Cross "follow-up" review judged 6,695 to be functionally rehabilitated; another 732 were still using orthopedic devices such as splints, but were no longer in need of physical therapy. Only 399 were still receiving treatment. This does not mean that the "recovered" patients suffered no permanent loss of motor function (and the persistence of impotence was not addressed by the review). Furthermore, over three thousand patients, probably in rural areas, were not included in the 1961 review, and the closure of most of the treatment centers meant that outpatient care only remained available in Meknes, creating an incentive to classify patients elsewhere as sufficiently recovered. Some initial gains would be reversed over time, as mobility impairments exacerbated the effects of aging, and vice versa. Nevertheless, the outcome for the victims was better than had been feared, and the physical therapy program was believed to have played a significant role in this outcome, although patients' recoveries were also attributed to reinnervation.[35]

The Opposition, Fréjus, and the King

For the Moroccan political opposition, the salient connection between neoco-lonialism and the oil poisoning lay not in the paternalism of the occupational therapists but in the toxic oil's origins on one of the many foreign military bases—French, American, and Spanish—that remained on Moroccan soil three years after independence. Most American forces had been evacuated from North Africa soon after the end of the Second World War, including those at a wartime base at Agadir, although the United States had maintained a naval presence at Port Lyautey (Kenitra) since 1942 (albeit under a French flag since 1948).[36] How-ever, a 1950 agreement with France—made without consulting the Sultan—had permitted the United States to once again expand its military presence in Mo-rocco with the construction of three Strategic Air Command bases, as well as naval and Air Force communications and radar installations, making Morocco an important part of the American nuclear deterrent against a Soviet assault on Western Europe.[37] Moroccan independence had thrown the status of these bases into doubt, as Moroccan nationalists of various political parties challenged the validity of previous US agreements with France, the former protectorate power. The ongoing Algerian Revolution undermined America's relationships with nationalists across North Africa, as the United States played what Mat-thew Connelly has described as a "double game," providing food aid and refugee relief to Algerians while providing France with arms to fight the FLN, includ-ing the retaliatory bombing of Algerian refugees by American-made planes at the Tunisian village of Sakiet Sidi Youssef on February 17, 1958.[38] Initially, the nationalists' demands had focused on the evacuation of the bases of France, the former colonial power. However, according to I. William Zartman, the nation-alist press's discourse on the bases had begun to shift in 1957 from "regularizing" the American presence, in light of Moroccan sovereignty, to demanding evacu-ation—a goal that was adopted by the Moroccan government and monarchy in 1958. The American intervention in Lebanon in July 1958 further associated the American military with Western imperialism in Arab lands, and demands for the evacuation of bases intensified.[39]

In this context, it was inevitable that the oil poisoning would have implica-tions for the question of bases and Morocco's relationship with foreign powers. The government desperately needed US medical, economic, and military aid, and refrained from openly pointing fingers. The press, however, was quick to raise the issue, demonstrating the growing divergence between the nationalist movement that had led the independence struggle and the newly independent state. *Avant Garde*, the newspaper of the newly created Union Nationale des Forces Populaire (UNFP), the party of the royally appointed prime minister, printed an article on November 15, 1959, suggesting that "foreign military authorities ... may have been negligent," in the sale of the oil.[40] This did not mean that Prime Minister Abdallah Ibrahim agreed. The party was fractured, and Ibrahim served at the pleasure of the monarch: a few weeks later Ibrahim would ban his party's paper and have the editor arrested (for questioning the authority of the king to appoint the government).[41] But *Avant Garde* was not alone in pointing out the causal role of the American military presence in the oil poisoning. In early December, the photo magazine *Al Machahid* (described as a Moroccan version of *Life* magazine) had run the headline "American Military Surplus are at the Root of This Disaster: Those who Exploit the Weakness and Poverty of the People: 10,000 Moroccans, one Frenchman, zero Jews Poisoned."[42] While the text of the article was more nuanced, the inflammatory headline reflected the role of the disaster in exacerbating nationalist concerns about the American bases in Morocco.

The dominant nationalist party throughout the postwar period had been Istiqlal ("Independence"), but the main Istiqlal leadership had been forced out of the cabinet by King Mohammed V in December 1958 in favor of the Ibrahim government, and the left wing of Istiqlal, including Ibrahim, split off to form the UNFP in 1959.[43] The remaining conservative body of the Istiqlal party found itself the party of opposition and was subjected to a ban on its publications in early October 1959. On November 29, the party's National Council passed a resolution denouncing the "negligence" of the authorities with regard to the oil poisoning and demanded aid and compensation for the victims; a separate resolution demanded the evacuation of all foreign troops, "considered an assault on the sovereignty of Morocco, a humiliation, and a permanent provocation."[44]

When the weekly newspaper *Al-Istiqlal* resumed publication in December, it addressed the crisis of the oil poisoning in an article titled: "From Meknes to Fréjus: The Responsibilities in the Oil Affair." *Al-Istiqlal* accused the government of failing to fulfill its responsibilities on two fronts: it had failed to protect the health of the people and had failed to bring an end to the colonial legacy of foreign "enclaves" on Moroccan soil. The Americans, the author asserted, knew full well that the oil being sold was toxic. Although the article stopped short of stating that the Americans knew the engine lubricant would be added to cooking oil, the author faulted them for their willful ignorance of Moroccan laws and regulations. The article then went on to implicitly connect the American presence not only to the paralysis epidemic but also to food shortages and famine by comparing this contamination of the Moroccan food supply to the recent introduction of crop parasites from foreign sources. Yet the Americans were not the only target of *Al-Istiqlal*'s wrath. The author asked, "Could they [the unscrupulous merchants] have done it if the minister of Agriculture had exercised vigilant control over foodstuffs, as is his responsibility? Could they have continued their deadly traffic if the minister of Health had not, through his delayed declarations, failed to sound the alarm?" Once the epidemic had begun, the authorities had compounded the crisis through incompetence and indifference—they had been too slow to identify the cause of the paralysis and too slow to realize that the problem extended beyond Meknes and beyond the two brands of oil initially identified as contaminated. According to the article, the ultimate blame lay with the Moroccan government for its lack of "vigilance." *Al-Istiqlal* contrasted this portrayal of Moroccan state incompetence to a rosy view of the national solidarity shown by the French state and people for the victims of the Fréjus flood. While the Moroccan authorities responded with an "indifference" which stemmed from a "convenient fatalism" and an overreliance on foreign aid, French authorities had immediately begun to investigate the causes of the Fréjus dam collapse and to search for those responsible.[45]

In fact, the French inquiries into the dam collapse would, in the end, identify no wrongdoing (as discussed in chapter 3), while Moroccan government had not only conducted inquiries but made arrests. King Mohammed V had issued an edict, or *dahir*, on October 29, retroactively legislating the death penalty and creating a special court for "those who have consciously manufactured, stored with intent to sell, distributed, put up for sale or sold products or foodstuffs destined for human consumption, which are a danger to public health."[46] Far from demonstrating passivity or indifference, the king's response was sufficiently decisive to alarm American officials, who correctly sensed a portent of the monarchy's tendency toward authoritarianism; the French ambassador called the edict "very severe."[47] The Moroccan state cast a wide net in identifying those responsible and erred on the side of arresting marginal suspects, including grocers who sold the bottled oil directly to customers and three men who were unconnected to the adulterated cooking oil but were using the American machine oil as an ingredient in "brilliantine," or hair oil. When defense lawyers argued that the king's edict specified only those who sold harmful substances for the purpose of consumption, judges ruled that one could consume a substance through the hair follicles as well as by mouth.[48] In pretrial hearings, defense attorneys also challenged the arbitrariness of the king's ex post facto decree, but without success. *Al-Istiqlal,* however, though quick to point fingers at the Ibrahim government and the bureaucracy, would not go so far as to challenge the king. Bowing to royal authority in a time of crisis, the paper would later present the king's edict as an exception to normal rules justified by the gravity of the disaster.[49] Catastrophe thus bolstered the king's authority.

Disaster Diplomacy

An examination of the Morocco oil poisoning offers insight into the public health consequences of Cold War militarism, the culture of international humanitarian activities, and the politics of Moroccan nationalism soon after independence. In addition, the diplomatic archives concerning the epidemic also permit a historical case study in what has become known as "disaster diplomacy." Most such studies have focused on very recent disasters, related to weather and volcanic or seismic activity. The 1959 oil poisoning provides an opportunity to extend such investigations to the diplomacy of the Cold War and decolonization and to an overtly anthropogenic disaster.[50]

As Gaillard, Kelman, and Orillos have written, "scholars across the disciplines have recently shown an increasing interest in 'disaster diplomacy,' which focuses on how and why disaster-related activities do and do not yield diplomatic gains, looking mainly at disaster-related activities affecting diplomacy rather than the reverse."[51] This literature has aimed to illuminate the potentially positive effects of disaster response on relations between otherwise antagonistic states.[52] The Morocco oil poisoning, however, suggests that a broader approach to the study of diplomacy and disaster is warranted. The oil poisoning involved semi-adversarial relations between allies and potential allies (the United States, France, and Morocco), and also involved non-state actors, including the Red Cross and the Istiqlal party. Furthermore, this case involves a complex reciprocity between the effects of diplomatic strategies on disaster relief and the effects of disaster relief on diplomatic activities. In post-independence Morocco, diplomatic concerns not only incentivized but also distorted and inhibited disaster response, as American fears of acknowledging culpability overshadowed the desire to make a show of American generosity to an emerging Cold War ally. Furthermore, the examination of this tragedy also allows the historian to explore how responses to one disaster can be intertwined with the experience and diplomacy of other disasters—in this case, the earthquake of 1960.

In 1959 and 1960, Moroccan, French, and American diplomats, as well as the Moroccan political opposition, were keenly aware of the potential impact of disaster responses on diplomacy, for good or for ill. For the Moroccan state, the goal was to maximize aid from both the United States and France without offering concessions that would undermine the government's nationalist credentials or add credibility to the arguments of its domestic critics, most notably the Istiqlal party. For the United States, the explicit goals of disaster diplomacy were to promote the tenure of the American airbases in Morocco and to promote a positive image of the United States in comparison to the Soviet Union, its Cold War rival.[53] French diplomats were likewise concerned about the future of French military bases and about international communism but also had broader goals of maintaining France's influence in Morocco and preventing the United States from usurping this role. The Moroccan state proved adept at exploiting the anxieties and rivalries of its international benefactors.

American disaster assistance, like other forms of US humanitarian aid, constituted part of what Kenneth Osgood has called Eisenhower's policy of "Total Cold War," a means by which the "United States would wage the Cold War assertively through nonmilitary means in the political and psychological arenas."[54] Since 1958, when US military personnel had responded to floods and fires in Morocco with search and rescue teams, the State Department had self-consciously sought to use disaster relief as a form of diplomacy to promote a positive image of the United States in Morocco. In June 1958, US airmen from Nousasser had responded to a major fire in the Derb J'did *bidonville* (shantytown) in Casablanca. In December 1958, US Navy and Air Force squadrons had provided emergency relief when floods struck the Gharb plain north of Rabat, distributing food and airlifting people to safety, feats the Americans would reprise when the Gharb was inundated again in January 1960. Among American officials, however, views were mixed as to whether such efforts produced satisfactory coverage of American heroism in the Moroccan press.[55]

If the positive publicity generated by disaster response was sometimes underwhelming, the oil poisoning presented American officials with a looming public relations catastrophe: the specter of ten thousand disabled Moroccans living out their lives as permanent, visible symbols of American imperial harm. Would the Moroccan public, or the Moroccan political leadership, blame the Americans for that damage? By November 5, 1959, the American embassy received word that the US airbase was suspected to be the original source of the adulterating substance. This suspicion was soon confirmed after the Moroccan authorities requested samples of machine oils from the airbase, which the Americans provided.[56] At the request of Moroccan ambassador Ben Aboud in Washington, the State Department called upon the US Surgeon General to consult with American experts on rehabilitation.[57] As the weeks passed, the Moroccan government made no official complaint but anxiously requested American assistance in responding to the disaster.[58]

The Belgian WHO epidemiologist Alfred Tuyns told American diplomats that the Moroccan government was plagued by fatalism, a complaint also expressed by the writers of *Al-Istiqlal*. According to Tuyns, the government shared the attitude of the victims that the affliction was an "Act of God": "'They are waiting to die,' said Tuyns, 'and one gets the feeling that the authorities share this attitude.'" Tuyns complained that, in the early weeks of the epidemic, the Moroccan government, although active enough in pursuing the perpetrators and confiscating bottled oil, was guilty of "criminal" sluggishness in requesting aid from international agencies. This orientalist description of Moroccan passivity and fatalism was potentially reassuring to the Americans, who feared accusations of culpability but also suggested, to Embassy counselor David Nes, grave failings of the Moroccan state: "this account well illustrates the Moroccan Government's deficiencies in any decision-making process."[59] Tuyns's description of Moroccan government passivity was contradicted, however, by Red Cross officials who stated that Moroccan officials were not blaming the United States for the incident but were "desperately anxious to hear what the US might be able to contribute" and were frustrated by the unwillingness of American organizations to do more than send survey teams.[60] Moroccan hopes for American unilateral assistance might explain the delay in requesting international aid that had frustrated Tuyns.[61]

The magnitude of Moroccan need for US assistance limited the adverse diplomatic impact of the tragedy for the United States. In addition to needing aid for the victims of the disaster, Morocco was heavily dependent on American economic aid, which totaled approximately $50 million in 1959, and the US had just begun to supply military assistance, valued at $30 million that year.[62] The Moroccan government was also concerned that fears of adulterated oil in Morocco would harm the country's canned fish exports, and Moroccan officials hoped that the State Department would help to reassure "all interested agencies" (presumably the US Food and Drug Administration and the Department of Agriculture) that Moroccan canneries had not been affected.[63] Hostility to the American presence in Morocco was a public force to which the Moroccan government had to be sensitive, but the government had no interest in inflaming anti-Americanism. As one US diplomatic dispatch stated, "Whether or not there was any exploitation of the situation from hostile sources, the Moroccan authorities would certainly do nothing to foster unfriendly feelings. As far as the US was concerned, they were entirely preoccupied with the hope that they would be receiving some favorable notification regarding the prospects of equipment and personnel contributions very soon, in view of the desperate problem which the government faces."[64]

The Moroccan government's attitude was a relief for the American officials, who were also relieved that radio appeals for donations for the victims made by President Bourguiba of Tunisia made no mention of America's role in supplying the adulterated oil. The Americans, however, never felt confident of Morocco's loyalty to the United States or of Moroccan dependence on American aid. The fear that Morocco might turn to the Soviets as an alternative source of aid, including military aid, prevented US officials from taking Moroccan dependence for granted. Normally, this insecurity would have led to a generous American aid response.[65]

However, for American diplomats, the origin of the toxic oil in an American airbase made the oil poisoning unlike other humanitarian crises in North Africa. The American response to the poisoning was restrained by the fear that American assistance might encourage the Moroccan public to associate the tragedy with the US military presence. The Nouasseur airbase willingly provided oil samples to the Moroccan authorities for chemical analysis and donated twenty-five thousand dollars' worth of surplus food supplies to the Red Crescent to assist with relief efforts, but otherwise, the US Air Force undertook to "respond only to specific Moroccan requests for assistance with a view to avoiding additional publicity and appearance of culpability."[66] The American Embassy did not declare an official disaster and hoped that the WHO and the Red Cross would be able to respond adequately to the crisis.

American disaster diplomacy was also shaped by tensions between funding rules, which emphasized visibility, and publicity guidelines, which stressed subtlety. Ham-handed publicity about aid provided by the US might seem transparently political and calculated, drawing attention to the bases while undermining the goodwill generated by the relief efforts themselves. Since the work of Edward Bernays in the 1920s, "public relations" had emerged as an endeavor which, to be successful, had to conceal its own deliberate efforts, and its authors.[67] Eisenhower had adopted this approach wholeheartedly: in the realm of Cold War psychological warfare, "the hand of government must be carefully concealed, and, in some cases I should say, wholly eliminated."[68] This approach was not universally accepted. For example, Leon Borden Blair, the Navy political liaison, ardently favored more explicit publicity about the actions of US servicemen in relief efforts. Looking back in his 1970 history/memoir of postwar Morocco, Blair scorned the civilian diplomats' approach; Blair believed that the key to American relations

with Morocco had been the open presence of American servicemen (especially
the Navy men at Port Lyautey Naval Air Station in Kenitra), not only in disaster
relief but also in day-to-day interactions with Moroccans.[69]

Nevertheless, the policy of public relations subtlety and restraint, enunciated
in Washington, had been reinforced in Morocco during the summer of 1959,
when an American payment of $15 million dollars to the Moroccan government
was attacked as a bribe in the Istiqlal press: "Does this mean that this amount
is the price paid for the American bases in Morocco? . . . American dollars have
sealed the lips and appeased the Moroccan Government, may God forgive it."[70]
The lesson learned from this blowback was reflected in the instructions that
USIS reporting on disaster aid in Morocco stress "mutual cooperation" rather
than the heroism of Americans.[71] American reluctance to draw attention to the
US role in the oil crisis through overt generosity was clearly rooted in concerns
about culpability, but such concerns were reinforced by the public relations principle that publicity (and disaster aid was in itself a kind of publicity) should not
be obviously related to its political purpose. US policymakers believed that the
provision of supplies or personnel directly from the US bases would seem like
too transparent a diplomatic ploy in the aftermath of the oil poisoning.

The French Position

The French situation was different. The politics of decolonization gave the
French state every reason to respond vigorously to this disaster. On November
5, 1959, the French ambassador in Rabat, Alexandre Parodi, learned from Dr.
Leroy that the American airbases had been identified as the origin of the epidemic. Parodi's initial response was that this should be kept quiet, presumably
either out of concern for France's American ally or for fear that French bases
might be maligned with guilt by association. However, the Foreign Ministry in
Paris noted that Leroy had credited two French doctors, Hugonot and Geoffroy,
and a French social worker, Ms. Barbet, as playing a crucial role in uncovering
the cause of the epidemic. This offered an opportunity to create positive press
for France at a time when France was subject to much criticism by official news
outlets in Morocco. In addition, France had recently announced plans to test
atomic bombs in the Algerian Sahara, tests which would take place on February
13 and April 1, 1960. The foreign ministry feared that if the true cause of the paralysis was not widely publicized by the time the tests took place, rumors might
spread that fallout from the atomic explosions was the cause of the paralysis.[72]
Consequently, Parodi asked Mohammed Taïbi Benhima, then the Secretary

General for the Moroccan health minister (later the governor of Agadir), to publicly credit the French doctors for their work; by November 12, the names of the French doctors and their contributions had been mentioned by Radio Marocaine.[73] Moreover, the French state had rivals in its pursuit of publicity. When an FLN chapter of the Red Crescent donated 500,000 francs to its Moroccan counterpart in December 1959, and another million francs in March 1960, a new front opened up in the struggle for public opinion in Morocco.[74] Faced with this competition, the French state, unlike the Americans, had every reason to take vigorous action.

The problem for French disaster diplomacy was a lack of budgetary resources. The French Red Cross was already overwhelmed by the needs of a massive program in Morocco and Tunisia to address the needs of more than 250,000 refugees from the Algerian War.[75] French diplomacy suffered a setback in late November when Dr. Leroy mistakenly told Moroccan officials that the French government would supply the buildings ("caserns") needed to house the physical therapy program. When the French embassy informed the Moroccan government that this had not in fact been approved, Ibrahim and officials at the Palace were furious. The French consulate in Meknes expressed alarm at "paradoxical" statements made by Moroccan officials comparing the generosity of the Americans and the WHO to the "ill will of the French." It was hoped, however, that this unfortunate turn of events would be smoothed over by a decision to offer two hundred beds at the French hospital in Meknes to victims of the poisoning, and that any ill will would be mitigated by the "good sense" and influence of Dr. Benhima of the Moroccan health service.[76] Within days, the French government announced its plans to send a medical team consisting of two doctors and eight to ten physical therapists to participate in the international effort for a period of four to six months, and granted the Moroccan government the use of a casern in Fez, recently evacuated by French troops.[77] Nevertheless, officials at the French foreign ministry expressed concern that French aid would fall short of expectations. The French Red Cross could not respond adequately to Moroccan aid requests, and its contribution to the international response was smaller than that of other countries' Red Cross contributions. Meanwhile, the Moroccan government had pledged fifteen million francs to aid the victims of the Fréjus disaster, and French officials feared that "this gesture may be quite adroitly exploited by it [the Moroccan government] to prove that, in proportion, Morocco is more generous than France."[78] This was a prescient concern: *Al-Istiqlal* would make a very similar argument about Fréjus after the earthquake struck Agadir several weeks later.[79]

Charlie Yost's Cold War

In contrast to the French, who sought to maximize publicity so that their response to the disaster would be both "important and known,"[80] American officials endeavored to keep a low profile, hoping to minimize public associations of the United States with the toxic oil. Soon, however, US Ambassador Charles Yost began to press the State Department to commit to a more robust American response. Yost's position on this matter may have been due to his regular contact with Moroccan officials, who pleaded for additional aid. It may also have been due to Yost's view of Morocco's strategic role in the Cold War.

At the strategic level, the American government was divided as to whether the priority in Morocco ought to be to preserve the bases, in order to fight the Soviets in World War III, or to mitigate anti-Americanism, in order to fight communist propaganda. The Truman administration had recognized that, in Europe, "the primary threat was not that the Soviet Union would seize territory through direct military intervention but that it would capitalize on economic and social unrest, expanding its power through subversion and manipulation."[81] For North Africa in the Eisenhower years, however, this remained a point of contention. The construction of new bases in Spain and plans for long-range nuclear bomber routes from the US diminished the need for the North African bases,[82] but the resulting policy change was gradual and fraught. Base tenure officially remained the top priority for US policy in Morocco into the early 1960s, although friction developed over this question both in Washington, between State and Defense, and in Rabat, between the Navy and the Embassy. Yost was an early advocate of the view that preserving the bases was less important than preserving a positive image of America among Moroccans.[83] Yost's prioritization of public opinion rather than base tenure seems to have affected his approach to the oil poisoning. Yost saw base tenure as expendable in the larger, "total" Cold War. From this perspective, a vigorous American contribution to relief efforts might draw attention to the origin of the toxic oil at Nouasser, but if it promoted favorable views toward Americans in general, then that might do more to thwart Soviet ambitions in North Africa than the Strategic Air Command's bombers ever would.

Yost requested that the State Department approve the release of "Cold War" (Mutual Security Act) contingency funds, or, failing that, a smaller $50,000 emergency fund, to assist with relief efforts. The Moroccan government, having learned of the jake poisoning epidemic in the 1930s, hoped that American expertise could offer something beyond the grim prognosis provided by the European doctors.[84] Indeed, experts at the Bellevue Medical Center suggested that successful treatment was possible and recommended that the United States send an "expedition team" comprised of six specialists representing various branches of medicine and therapy.[85] Yost argued that the United States would not make itself conspicuous by providing such aid: the International Red Cross had issued a global call for assistance, which had been publicized in the Moroccan press. He noted that the Austrian government had already pledged to provide a hundred-bed hospital,[86] and Denmark, Norway, Sweden, Britain, and Switzerland were sending physiotherapists.[87] When, on December 3, 1959, Moroccan health minister Ben Abbes announced the details of the Leroy plan for the rehabilitation of the victims, the enormity of the Moroccan need for assistance became more evident.[88]

In this light, the American response started to seem stingy. The Institute of Physical Medicine and the American Red Cross were willing to send a medical survey team for two weeks, and the United States agreed to pay for transportation, but the World Health Organization had proceeded beyond mere evaluation and needed a "semi-permanent treatment team," which the American organizations were unable or unwilling to provide.[89] US Air Force C-124s airlifted the twelve-ton Austrian field hospital to Morocco, and American Red Cross did provide a donation of $5,000, but in the critical area of medical personnel, the Americans came up short.[90] The State Department took care, in its contacts with the international Red Cross, to stress American reluctance to provide further aid and "to avoid any impression that the United States is prepared to contribute."[91] The Moroccan government complied with the American desire to keep a low profile, and no public or official request for aid was made to the Embassy or the State Department. However, the international League of Red Cross Societies appealed to the American Red Cross for two complete field hospitals, which American Red Cross president Alfred Gruenther urged the State Department to provide, pointing out that thus far, the Europeans and Canadians had been providing all the personnel. The international Red Cross suggested that President Eisenhower's upcoming visit to Morocco on December 22 might be the perfect occasion to announce such an American donation.[92]

Ambassador Yost repeatedly made the case for a more generous American response, pointing to the more generous responses of European countries and to the threat that the Soviet Union might seek to gain an advantage from the crisis.[93] Yost also argued that, while the Moroccan government had been silent on the issue of American culpability, the Moroccan press had not, and he nervously anticipated the return of *Al-Istiqlal* to the news shops: Mohamed Lyazidi, editor of *Al-Istiqlal*, had indicated that his paper would demand an investigation into American responsibility when it resumed publication.[94]

Yost's appeal for additional aid was complicated, however, by Eisenhower's scheduled visit. Eisenhower was coming to Morocco as part of an "eleven-nation goodwill tour" intended to counter the publicity impact of Khrushchev's visit to the US in September. The point of the tour was to counter Soviet propaganda and generate a positive image of the United States, and in Morocco, the US Information Agency had plans to make the most of Eisenhower's visit using film, radio, pamphlets, photographs, and window displays.[95] The oil poisoning issue threatened to subvert the narrative that Eisenhower wanted to create: if Eisenhower announced the donation of American field hospitals for Meknes, then the oil poisoning would dominate the headlines. The expected coincidence of the resumption of *Al-Istiqlal* and the president's tour was also unfortunate, for it assured that the focus would be on American culpability and not just American generosity. For these reasons, Yost recommended a "short postponement" of major American aid.[96]

Yost continued to be troubled, however, by the urgency of the human crisis and the inadequacy of the American response. Winter was setting in in Meknes, and on December 17, representatives of the health ministry contacted the American embassy and described the suffering of the paralysis patients being housed and treated in tents. Yost urged the State Department to arrange to meet the Moroccan request for ten thousand each of long underwear, sweaters, wool blankets, pajamas, and wool socks, ideally through the Red Cross.[97] The American Red Cross soon arranged a shipment of sweatshirts and union suits from stores in Switzerland, but only a fraction of the number requested.[98]

What the international Red Cross rehabilitation program most desperately needed was personnel. However, the Americans were participating in the rehabilitation program only minimally. The American Red Cross did arrange to send two nurses to Morocco but the fact that the Americans were sending no doctors or physiotherapists was noticeable, since the Red Cross societies of other Western countries had provided altogether a dozen doctors and thirty physiotherapists, as well as ten nurses. In the early months of the rehabilitation program, the American staffing contribution was on par with that of Austria, Australia, Denmark, Finland, Greece, and Norway, but was outstripped by the efforts of Canada, France, West Germany, Britain, the Netherlands, Sweden, Switzerland, Turkey, and Iraq.[99] The American Red Cross cash donation of $6,000 was more impressive, making up over half the total cash donations at the time, but it was less conspicuous, by design.[100]

The return of *Al-Istiqlal* to publication on December 19 increased the political pressure on the Americans, as expected, explicitly linking the oil poisoning to the American bases and to a general American disregard for Moroccan rights and sovereignty. *Al-Istiqlal*'s December 19 article on the oil poisoning exemplified the nationalist view that the American bases were colonial "enclaves" (implicitly comparable to the Spanish enclaves at Septa, Melilla, and Ifni) although the article placed equal blame on the Moroccan state, as befitted Istiqlal's role as the political opposition.[101]

The Tide Turns

Four events soon transformed the landscape of American disaster diplomacy in Morocco: a treaty, a debarkation, an earthquake, and a trial. Together, these events brought an end to American reluctance to participate in the international medical response to the oil poisoning. First, at the end of December 1959, the Ibrahim government achieved a major success, reaching an agreement with America's president Eisenhower for the evacuation of the US bases by 1963. This facilitated Moroccan relations with the United States and also solidified the nationalist credentials of both the monarchy and the Ibrahim government. The agreement did not end the need for US disaster diplomacy in Morocco, however, because the Americans still desired to negotiate a continuing military presence in Morocco under a different guise and needed to insure against possible demands for an earlier evacuation.[102] Nevertheless, the agreement meant that the nationalist opposition would largely shift its focus to the French bases and away from the role of the US. The editors of *Al-Istiqlal* announced the advent of a "new phase" of relations between Morocco and the United States and hoped that the example of the Americans would push the French and Spanish to make similar agreements.[103] Protests in February 1960 focused on demands for French base evacuation, objections to the French atomic bomb tests, and Moroccan claims to the Algerian Sahara; the Americans were no longer in the crosshairs of nationalist politics.[104] The US-Morocco agreement maintained the incentive for the United States to provide disaster assistance, while lessening the chance that the nationalist press or the government would attack the American military presence by invoking the origin of the toxic oil.

The second major event that altered American calculations was the materialization of responses to the disaster from the West's Cold War enemies. Whereas Istiqlal had been pointing out the link between the health crisis and American bases, the international communist response had been muted. On January 4,

1960, David Nes at the American embassy expressed relief at the "puzzling" silence of Soviet and Chinese officials in Morocco, who seemed to be passing up a grand propaganda opportunity. Nevertheless, Nes argued that the potential remained that the oil poisoning incident "could be used to channel the resentment and desperation of the great number of Moroccans affected—either directly as victims of paralysis or indirectly as indigents deprived of the support of breadwinners—against the American military forces and installations in Morocco."[105]

That same day, a Soviet ship, the *Volkhovgen*, sailed into the port of Casablanca with nine tons of blankets, milk, and sugar for donation to the relief effort. The Soviets had entered the fray of disaster diplomacy in Morocco, giving credence to Yost's warnings. The ship's arrival provided an occasion for the Soviet captain and the Soviet ambassador to meet with Moroccan representatives of the local Red Crescent chapter.[106] The danger to American interests was compounded on January 5, 1960, when the minister of health in Qasim's Iraq announced that Iraq was sending a "medical mission" to express the "solidarity of the Republic of Iraq . . . with the Arab peoples."[107] NATO had already lost their friends on the thrones of Egypt and Iraq, and American fears of Qasim's ties to the Soviet Union were at their peak. Although neither the Soviets nor the Iraqis seemed to be publicly emphasizing America's causal role in the tragedy, the combination of Soviet and Iraqi activity and American inactivity concerned Yost. However, it also presented him with an opportunity: Yost now had significant new evidence to bolster his argument for increased disaster assistance.[108]

However, although the US-Morocco agreement on base evacuation and the arrival of Soviet and Iraqi aid incentivized and facilitated a more generous US response to the oil poisoning, US policy directives continued to inhibit that response. Disaster assistance funds from the State Department's Cold War "Mutual Assistance" program account were only authorized for "meeting immediate needs." This short-term assistance was meant to "obtain maximum political and psychological impact in meeting the immediate disaster situation and to forestall possible need for greater expenditures for other, slower forms of assistance."[109] Long-term assistance to the Red Cross physiotherapy program for paralysis victims was a poor fit for this State Department policy of concentrating aid where it was most noticeable and most temporary. When, in January 1960, the International Red Cross rehabilitation program faced a critical shortage of vehicles to transport therapists and outpatients in rural areas, Yost hoped that this might be an opportunity for the United States to more actively curry the favor of the Moroccan public. The embassy once again pled with the State Department for a generous response, proposing that the State Department arrange for the Department of Defense to not only provide vehicles but also maintenance for them as well. Yost reminded the State Department that the oil poisoning was still being covered in the Moroccan press.[110]

Yost was informed, however that the "Cold War" fund was "currently being reviewed," and hence "unavailable," and that disaster relief funds could not be authorized for uses other than "immediate and temporary relief."[111] Although this decision was influenced by the looming threat of Congressional budget cuts, it was also part of an overall strategy of concentrating aid where it would be most visible, to the press as well as the public. Moreover, the principle that publicity should not be obviously related to its political purpose made the provision of vehicles directly from the US bases particularly problematic.

On the afternoon of February 29, 1960, Yost once again sent a telegram to the State Department, reporting that the Moroccan cabinet director for Public Health, Abdel Hamid Ben Yaklef, had again reached out to the Embassy, stressing the "urgent need" for vehicles, and Yost lamented the view of Moroccan officials that "the US is uncharacteristically at the tail end of the procession providing help for the oil victims."[112] That night, Agadir collapsed.

Agadir's Loss, Meknes's Gain

The earthquake struck Agadir shortly before midnight. News of the disaster reached Rabat, Paris, and Washington within hours, based on reports issued by radio from the French military base on the outskirts of the city. It was immediately clear that the devastation to the city and the surrounding area was enormous: estimates of the death toll soon rose to twelve thousand and continued to climb. When the quake hit, French airmen from the naval base a few miles away arrived quickly. On March 2, a French fleet arrived, as a well as a Dutch naval cruiser, along with American sailors from Port Lyautey, and airmen from the Strategic Air Command bases. Late at night on March 3, the USS *Newport News* arrived from Italy. The Americans brought with them heavy machinery for excavation.[113] In sharp contrast to the oil disaster, the American response to the earthquake was immediate and vigorous.

Even more than the US-Morocco base agreement or the arrival of the Soviet and Iraqi aid, the Agadir earthquake transformed the politics of disaster relief in Morocco. The Agadir disaster, similar in scale to the oil poisoning but fresh and with greater lethality, created competition for aid resources, and the Red Cross immediately diverted personnel, equipment, and supplies from the paralysis rehabilitation project to Agadir.[114] However, the earthquake provided US officials in Morocco an opportunity to overcome any ill will created by the American sale of toxic oil as well as any perceptions of stinginess in the American response. The earthquake, unlike the oil poisoning, allowed the Americans an opportunity to be innocent, helpful, and brave. It also altered the culture, habits, and budget of American aid in Morocco, changing the American response to the oil tragedy.

Analysts have been divided on the role that American disaster relief efforts played in the diplomacy of the time. In 1964, political scientist I. William Zartman argued that American disaster aid had been ineffective as diplomacy aimed at preserving base tenure because aid benefitted the masses, but decisions were made by an isolated elite. Yet the diplomatic archives now available to historians reveal that Moroccan government ministers were putting pressure on Yost with their repeated requests for American aid. This suggests that the ministerial class was not as indifferent to the well-being of the disaster victims as Zartman believed. However, there is little to contradict Zartman's assertion that good relations with ministers with portfolios such as Public Health and Public Works had little impact on the Moroccan Foreign Ministry.[115] As Zartman noted, the nationalist attack on the bases was ideological, in that it was conceived within the context of anti-imperialism and based on questions of sovereignty and legality. The American bases had to be evacuated as part of the logic of decolonization: they were Moroccan territory that had to be reclaimed. Consequently, accidental harm or humanitarian benefit were largely beside the point. This made the oil incident less damaging to the Americans than they might have feared, but it also made humanitarian aid less effective in winning over either the relevant government ministers or the political opposition.

However, as Leon Borden Blair has argued, there was a parallel channel of diplomacy between the Navy and Crown Prince Hassan, not least Blair's own personal diplomacy. Blair's prominent position in the earthquake response as the embassy's representative in Agadir, "with instructions to coordinate and direct the American effort,"[116] combined with his close relationship with Hassan, suggests that American earthquake relief was not entirely disconnected from the American diplomatic successes concerning the bases. These successes consisted not only of securing a continued role for the Navy at Port Lyautey, under the cover of a "Training Command,"[117] but also the ability of the Air Force to remain at the bases until the agreed-upon deadline in 1963, which was by no means certain at the beginning of 1960.

Yet base tenure may not be an appropriate rubric for measuring the success of American disaster diplomacy in Cold War Morocco. Even if the United States had won over the hearts and minds of the Moroccan government ministers (or, as Blair believed he had, the Palace), American disaster response efforts were insufficient to deflate public criticism of the American military presence. As early as March 10, Istiqlal was once again protesting "sequels of colonialism," arguing that American relief efforts, however welcome, were undermined by American unwillingness to evacuate the bases before the agreed deadline of 1963.[118] The Embassy was uninterested in pushing for the preservation of the Air Force bases in the face of public hostility, and saw base tenure as secondary to the goals of rescuing America's image in Morocco and bolstering the legitimacy of the pro-American monarchy.[119]

Of Fish and Angels

Whether the earthquake served as a "tipping point" in the diplomacy surrounding the American bases is debatable, but the earthquake certainly served as a critical juncture or tipping point in the American politics regarding aid for the toxic oil victims.[120] Unlike the poisoning incident, the needs of the earthquake victims were for immediate, short-term aid, and the images that earthquake relief offered were purely positive: American soldiers bravely performing rescue work, in contrast to the feared images of Moroccans crippled by American poisons undergoing painful physical rehabilitation.

The Agadir disaster also differed from the oil poisoning in that it was suitable for that other aspect of the American "total Cold War"—the cultivation of US domestic public opinion and congressional support. Although, unlike France, the United States was not (yet) bogged down in a war of decolonization, funding for foreign aid (both military and non-military) through the Mutual Security program was under attack by a group of congressmen led by Louisiana representative Otto Passman, whose fierce opposition to foreign aid since 1953 was approaching fruition. The Eisenhower administration feared that Congress would impose disastrous budget cuts.[121] The earthquake struck just as Congress was about to begin hearings on the Mutual Security program.

At the hearings, both supporters and critics of the program noted their constituents' distaste for foreign aid spending. But the Agadir earthquake provided an opportunity for supporters of the program to argue that more aid was needed. The ongoing rescue efforts in Agadir were brought up during the testimony of Undersecretary of State C. Douglas Dillon on March 3. Rep. James Fulton of Pennsylvania suggested that the $260,000 released by the State Department to fund rescue efforts was inadequate for a disaster of such magnitude and also suggested that more publicity was needed for such efforts in the US, to create public support for foreign aid. Fulton also argued that the US would win more loyal friendship abroad in situations where it provided visible aid directly to the people. In Fulton's words, "If an angel hands you some food you are certainly surprised and gratified and you thank heaven. But if you are walking along the seashore and you pick up a fish, it is you who has found it, because there is no particular source of it and you don't thank anyone." On the subject of US domestic publicity, Fulton suggested that more needed to be done: the "top angels" needed to get involved.[122] These concerns were echoed in the Senate hearings, where supporters of the program stated that there was a need to inform the public of the magnitude of need in the developing countries, and also that aid might be better channeled directly to the people rather than through often distasteful regimes.[123] The Agadir catastrophe was perfect for both Moroccan and American consumption: an opportunity for dramatic, short-term aid provided directly by the Americans to the people of Morocco, with plenty of opportunities to involve the domestic American public as well. After Dillon's exchange with Fulton at the House hearings, the State Department immediately doubled the earthquake relief funds to $500,000, to reimburse Defense for goods and services in order to assure that military relief efforts could continue.[124] In March, $1.5 million was designated for Agadir in a "Foreign Disaster Emergency Relief Account."[125] Meanwhile, charity drives were organized by private and public agencies across America (including a Navy clothing drive named "Angels for Agadir"). Within days, donations of basic relief goods such as tents and blankets (from Europe and Tunisia as well as from the US) were in excess of what was needed.

As supplies and funds flowed into Morocco, Yost saw an opportunity to reverse the American reticence in supporting the rehabilitation of toxic oil victims. Yost complained to the Secretary of State that American inaction on the Red Cross request for vehicles was becoming "more and more embarrassing."[126] Although the State Department was unwilling to approve additional funds for the oil poisoning response, the Air Force expected that its earthquake relief expenses would fall short of the $1.5 million account designated for the purpose. Consequently, $16,000 was authorized on April 6 for use for the toxic oil response. This was a small sum, but its impact was magnified by the USAF's cooperation in designating vehicles and hospital supplies as surplus, allowing the US International Cooperation Agency to obtain them at 10 percent of their normal value.[127] On March 8 the US International Cooperation Administration office agreed to fund the transportation cost of a team of American therapists out of its "technical support" budget.[128] The gates of US aid were starting to open.

The Trial

The trial of twenty-four men accused in the oil poisoning case began on April 11, 1960, and constituted another major turning point in the international diplomacy surrounding the disaster. In its coverage of the trial, *Al-Istiqlal* again targeted the failures of the Moroccan state, arguing that, whatever penalty the accused might receive, real justice would demand that the government officials who had failed to prevent the disaster should also be held responsible. In contrast to its earlier reporting, however, *Al-Istiqlal* now shifted blame away from the American airbases. *Al-Istiqlal*'s article on the trial stated that the oil, when originally sold, had been clearly labeled as jet engine lubricant: "in white letters, on a green background, one could read: 'Lubricating Oil Ancraf Gaz Turbine Engine' [*sic*, in English], *huile lubrifiant d'avion à la reaction*. There was, therefore, no ambiguity about its purpose."[129] This provided context and corroboration for the testimony of Mohammed Bennani, the original purchaser of the American oil and a dealer in automobile parts and engine oil. Bennani, quoted at length in *Al-Istiqlal*, stated that the sign on his warehouse proclaimed, in words a meter high, "oil and grease for automobiles." Bennani argued that Ahmed ben Hadj Abdallah, the customer who had repurposed the toxic oil as a comestible, was "only one of hundreds" of customers who bought oil from him, that no merchant could be expected to verify the uses to which his customers put his wares. Furthermore, Bennani argued that he had sold the oil

> in its original packaging: U.S. Army. I bought this oil by lots, according to the rules of sale by auction by submitting a bid to the seller, in this case the USAF. I paid customs on this merchandise. If by having sold this oil that I had bought, which I sold in complete ignorance of what my customers might do with it, is a crime or infraction, then I think that the commander of the American base who sold it to me is in the same position as I am, and logically speaking he should be at my side on the bench of the accused.[130]

The court was not interested in pursuing the defense's rhetorical attempt to put the US Air Force on trial, as Bennani undoubtedly knew. Not only had the Americans agreed to evacuate the bases, they had also become an important supplier of various forms of aid to Morocco, lessening Morocco's dependence on France. When a defense attorney asked, "By virtue of what convention did the Americans have the right to sell in our country dangerous materials such as the oil that has caused this great catastrophe?" the advocate general intervened, saying, "Address this question to the Foreign ministry. This question is outside of the authority of this tribunal."[131] The court agreed. The court also dismissed the defense's objections to irregularities in the trial procedure ("It is not the oil that is adulterated, it is the trial!" one defense attorney reportedly protested), and refused to allow the defense to call expert witnesses or to present as witnesses victims of the oil poisoning who had recovered from its effects.[132]

Five of the accused, wholesalers in Meknes, Casablanca, and Fez, were sentenced to death; they were held responsible for using the jet engine oil bought from Bennani to adulterate vegetable oil in large quantities for sale to grocers. Three others were sentenced to life in prison. Judgment on the sellers of hair oil was deferred pending further inquiry. Bennani, however, was acquitted of all charges.[133] For the victims, the *Affaire des Huiles* was far from over. For the Americans, however, the outcome of the trial was a tremendous source of relief, removing lingering concerns about culpability.[134]

By the time of the verdict, the Red Cross had grown impatient with the US and purchased from private sources a number of three-ton trucks to transport patients; the French had also donated ambulances. However, there was still a shortage of medical supplies, since the US had, at that point, declined to donate field hospitals and the idea of supplying goods *à la carte* from base surpluses had not come to fruition. After the trial, however, the procurement of USAF medical supplies by the International Cooperation Association Mission was soon approved, and 75 tons of hospital equipment were donated and delivered on June 13. American officials were pleased to note that the event was covered in the press.[135]

Conclusion

The conundrums of American disaster diplomacy in Morocco revolved around American and Moroccan attempts to assess the political implications of the 1959 oil poisoning's complex origins. This disaster could not be easily attributed to the culpability of a single human agent, but was rather the consequence of Bennett's "distributive agency" or what epidemiologists call a "web of causation."[136] American responses to the disaster were shaped by American uncertainties about how Moroccans would interpret this murky causality. The acquittal of Mohamed Bennani helped to alleviate this uncertainty, producing an authoritative Moroccan rejection of the chain of events that could be traced back to American culpability. Even before the trial, however, the earthquake at Agadir had transformed the calculus and the culture of American disaster assistance in Morocco, liberating American officials to respond more effectively to the needs of the toxic oil victims. It was not just the budget for disaster relief that was liberated: after February 29, 1960, there were no more expressions of concern that American aid might invite thought of American culpability. In the minds of American policymakers, American aid no longer needed to be explained or excused: the images of American soldiers rescuing earthquake victims (displayed prominently in US publicity in Morocco)[137] had changed expectations.

French and American disaster aid in Morocco was both motivated and distorted by disaster diplomacy concerns, as decisions were made based on publicity value rather than on the needs of the victims. In the end, thanks to the transformation brought about by the US-Morocco base agreement, the arrival of Soviet and Iraqi aid, the Agadir earthquake, and the outcome of the trial, the Americans made a respectable contribution to the international rehabilitation effort for oil victims. By the end of the rehabilitation program, which lasted until June 1961, the US had contributed not only a substantial amount of critically needed equipment but also the expertise of two American nurses, five doctors, twelve physiotherapists, and one secretary, typically for six months, or more, for the non-doctors.

France, in contrast, contributed only two doctors, four therapists, and a nurse to the Red Cross effort, who were in residence only one or two months at a time.[138] Although French staff at the military hospital in Meknes also treated patients, the Moroccan government was billed for a portion of the costs.[139] Although the Eisenhower administration's agreement to evacuate US bases by 1963 placed France at a public relations disadvantage vis-à-vis the United States and created an additional incentive for French aid, the war in Algeria was consuming French resources, making it difficult for France to compete with the United States in the realm of disaster diplomacy.

As with the Algerian earthquakes of September 1954 and the Malpasset Dam collapse of 1959, responses to the Morocco oil poisoning were inseparable from the politics of decolonization. Just as the political struggles over Algeria's relation to France shaped responses to the Chélif and Fréjus catastrophes, responses to the oil poisoning by the Moroccan, American, and French states, and by the Moroccan political opposition, revolved around the question of whether Morocco was truly sovereign over the land occupied by foreign military bases, the most obvious vestige of colonial rule. The disaster politics surrounding the aftermath of the Agadir earthquake and the reconstruction of Agadir were, meanwhile, opening new fields of political contestation in Morocco and in France.

CHAPTER 5

Death, Diplomacy, and Reconstruction in Agadir, 1960

T HE EARTHQUAKE THAT STRUCK Agadir late at night on February 28, 1960, led to the deaths of twelve thousand to twenty thousand people. The densely populated Kasbah, high on a hill overlooking the south-facing bay and known to the Tashelhit-speaking population as Agadir Oufella, was almost totally obliterated. The Founti quarter adjacent to the beach below also sustained very heavy damage, as did the Talborj, an ethnically mixed commercial and residential district situated on a plateau to the east of the Kasbah, as well as Ihchach, a village on the northern outskirts of the city. The Anza district to the west and the *Ville Nouvelle* to the east were damaged to a lesser degree; farther east, the industrial district and the French aero-naval base were largely unscathed. To many observers, however, there was an impression of almost total destruction. Mohamed Taïba Benhima, who would soon be appointed governor of Agadir, later described the scene: "At first sight, everything in Agadir built of reinforced concrete, buildings, mighty superstructure, was on the ground. I will go further: everything that was on the ground was twisted, tormented; from the pier of the port, however well-anchored, to the road network that the seismic shock had literally twisted."[1]

The impact of this seismic event on human history was shaped by the legacy of French imperialism. In February 1960, Agadir, like the rest of Morocco, was still in the midst of decolonization, four years after formal independence. The demographics of the city had been shifting as French colonists "repatriated" to metropolitan France and to Algeria, but the French presence remained unmistakable. The 1960 earthquake was a cataclysmic environmental intervention in the decolonization process. The disaster precipitated a fresh exodus of the French population and offered a test of the French state's commitment to its overseas citizens. Simultaneously, the earthquake challenged the Moroccan state's ability to provide for the needs of the people. The increasingly authoritarian Moroccan monarchy took this opportunity to increase its power and prestige within the

Moroccan political field. The monarchy also made use of American aid to lessen the Moroccan state's dependence on the French, another example of "disaster diplomacy" that does not fit the paradigm of hostile states growing closer through disaster response.[2] Because of the prominent role of the United States in the disaster response and in planning for the reconstruction of the city, the Agadir disaster seemed to strengthen American relations with Morocco while weakening French influence. Consequently, the disaster became a focus of anxiety for French officials and politicians worried about France's declining role in the former protectorate, the future of French colonists in post-imperial spaces, and the post-disaster reconstruction of those spaces. The political contestation that followed the earthquake did not just concern the future of those who survived, however: the enormous numbers of the dead posed particular problems. After the earthquake, the treatment and disposal of the festering dead was intimately tied to controversies over power and boundaries in the postcolonial city. Even after the immediate disposal of corpses, French attempts to seek the repatriation of remains created new points of friction as the meanings of these burials and exhumations were created and negotiated. In both the disposal of the dead and the reconstruction of a new city for the living, the disaster became connected to the struggle to work out the meaning of decolonization.

In the hours of chaos and grief immediately following the disaster, survivors struggled to free their family members and neighbors from the ruins and to treat the injured as best they could. French and American sources (diplomatic correspondence, memoirs, and journalism) emphasize the role of French and American troops in the disaster response and tended to portray Moroccans as passive and helpless. However, Moroccan memoirs describe survivors engaging in organized rescue efforts. According to Tariq Kabbage, who was twelve years old in 1960, his father, local landowner Abbès Kabbage, organized nearby farmers to bring workers and tractors to assist in the immediate rescue efforts; in the Kasbah, survivors reportedly organized themselves and began rescue efforts during the night of the disaster.[3] As in Orléansville, however, the local authorities seem to have been unable to respond effectively. The gendarmerie and police barracks had collapsed in the earthquake, killing many; other would-be first responders had died in their homes with their families or were searching for family members. The Royal Moroccan Army troops in the city had taken heavy losses. The Moroccan governor, Bouamrani, had lost several family members. In contrast, the French aero-naval base on the eastern edge of the city was largely untouched, and the commander immediately initiated operations for the rescue of the living and the disposal of the dead.[4]

Soon, however, the Moroccan state took charge. In the morning, King Mohammed V and Crown Prince Hassan arrived by plane from Rabat, along with Colonel Mohammed Oufkir. The king put Hassan in command of rescue operations as Moroccan troops began to arrive from throughout the kingdom, followed by American, Dutch, and Spanish soldiers and sailors, and French reinforcements. Oufkir was responsible for the refugee camps; he would later, as minister of the Interior, command the Moroccan state's repressive security apparatus during Hassan's reign as king.[5]

In a statement to the nation, the king presented the royal family as leaders of the disaster response:

> A great and terrible catastrophe has struck our country. A horrible cataclysm has destroyed the city of Agadir, made its inhabitants victims, and left it in ruins. Language is incapable of describing this calamity. It is not the hour for words, for those whom God has saved await your acts of solidarity, not your tears and words. We have charged our crown prince Hassan with directing the rescue and emergency operations. Likewise, we have charged Princess Aïcha with organizing a campaign of solidarity throughout the entire kingdom and to collect donations for the victims. We have also allocated the funds necessary for the immediate response. Human, religious, and national duty demands of each person to come to the aid of our brothers, survivors of the martyred city, and bring them all forms of assistance, cash and otherwise, thus manifesting his solidarity and accomplishing at the same time his obligations, both religious and national.[6]

Patriotic, moral, and religious duties were thus to be united in service of the king. During the years of the struggle for independence, the monarchy and the nationalist political parties, particularly Istiqlal, had depended on each other. In contrast, the years of disaster, 1959 to 1960, were pivotal in the development of the authoritarian monarchy, and both the oil poisoning and the earthquake facilitated this development, as Ibrahim's ministerial government was sidelined.[7] Less than three months later, the king would take charge of the government directly, appointing himself prime minister, with Hassan as deputy-premier and minister of defense.[8]

For three days after the earthquake, under the direction of Hassan, Moroccan soldiers worked alongside foreign troops to rescue the survivors, and the troops commanded by Colonel Oufkir established refugee tent cities for at least fourteen thousand Moroccans outside the city limits.[9] European survivors had different options: the French military base became a makeshift refugee camp for over three thousand Europeans. The base also provided medical treatment to injured Moroccans, but the guards turned away uninjured Moroccan refugees.[10] Conditions in the Moroccan camps was less than ideal: despite the distribution of aid by the Red Cross, Ahmed Bouskous, who survived the disaster as a teenager, would recall the "inhumane conditions" of these camps.[11] Later, temporary prefabricated housing for the displaced Moroccan population was established in the areas around the existing workforce housing of the two industrial zones in Agadir.[12]

Meanwhile, thousands lay dead, buried in the rubble. In the immediate aftermath of the earthquake, the handling of the dead was inseparable from the rescue of survivors, as bodies were dug out of the ruins. There was an effort to segregate bodies by nationality and religion, but the difficulty of identifying many corpses sometimes made this impossible, with the result that some mass graves included both Muslims and non-Muslims, Moroccans and French. Most bodies identified as French, or of unidentified European origin, were initially brought to the French military base. A tent was erected to shelter hundreds of arriving cadavers, which were then buried in communal graves in what had been the athletic field, wrapped in sheets or placed in crates or armoires scavenged from ruined homes.[13] Moroccan dead were brought primarily to Ihchach, where French marines dug a large mass grave; soon hundreds of Moroccan bodies awaited burial there.[14] Thousands more were never recovered. Seventy-eight Moroccan Muslims were buried at the French base, most likely after arriving there alive, brought for medical treatment that proved to be in vain.[15]

Soon, the bodies of the dead began to be seen as a threat to public health. Crown Prince Hassan feared that the putrefying dead would soon produce epidemics of cholera and typhoid, and the authorities cordoned off the disaster area, evacuating survivors from what became known as the "dead city," which was blanketed with quicklime. (The possibility of dropping napalm on the ruins was reportedly discussed, provoking some alarm, but came to nothing.) The fear of pestilence meant that the treatment of the dead was as great a priority as the rescue of survivors. Hassan ordered a halt to rescue operations after three days, a decision which outraged both Moroccan and foreign residents and observers, and which was soon reversed, allowing for the rescue of a few more victims over the course of the following seven days.[16] After that, thoughts turned toward assisting the survivors, recovering the remaining bodies of the dead, and rebuilding the city. None of these tasks could be accomplished without grappling with the meaning of decolonization.

Diplomacy's Discontents

The fears, frustrations, and resentments of decolonization and the Cold War permeated international public responses to the disaster. For advocates of the French empire, the earthquake unleashed resentments and fears related to growing American influence and the decline of French hegemony in Morocco. Meanwhile, critics of French imperialism suspected malfeasance by the French state. A month after the disaster, the president of Liberia, William Tubman, accused the French of causing the earthquake by conducting a nuclear arms test at Reggane in the Algerian Sahara on February 13.[17] This idea was also reflected in a memoir by a French officer at the base outside Agadir, who described the arrival of Moroccans at the base immediately after the earthquake who had come to express anger at the French, which the officer attributed to a Moroccan belief that the atomic tests had caused the earthquake.[18]

Both French and American diplomats hoped that disaster aid would bring "political benefit,"[19] particularly regarding the issue of base tenure. The international character of the disaster response was shaped not only by the legacy of French colonialism but also by the consequence of the Cold War geopolitical situation: the American military presence at Port Lyautey and three Strategic Air Command airbases meant that American as well as French forces had played a prominent role in the immediate response, alongside Royal Moroccan Army troops. As in the case of the floods of December 1958 and January 1960, those French and American officials who hoped to use disaster response to foster goodwill were disappointed when the resulting publicity fell short of expectations. Competing to gain public relations capital in the Moroccan political market, both French and American diplomats complained that their NATO ally was suppressing information. The French embassy in Rabat complained that the Moroccan press gave full treatment only to American earthquake relief, a fact which officials attributed to the more accommodating policy of the United States regarding base evacuation.[20] In addition, a daily paper in Tangier, *España*, had run photographs of military rescue efforts provided by the US Consul General there; one of these photos depicted French sailors and airmen, but they were not identified, and the credit caption "U.S. Navy" seemed misleading. An American diplomat meanwhile accused the Agence France-Presse and the French-language papers of Casablanca of giving short shrift to American contributions.[21] The French also hoped that their efforts in Agadir would earn good publicity elsewhere in Africa, but expressed disappointment when the press in

Accra, forged in the anti-colonial struggle, failed to mention French disaster assistance when covering Agadir.[22]

Both the Moroccan opposition press and the Palace were quick to capitalize on the apparent crassness of French hopes that gratitude for disaster aid would translate into the extension of French base tenure. Information Minister Ahmed el Alaoui stated, "Aid from a foreign country in such a catastrophe does not mean the foreign country has a right to bases there."[23] The Istiqlal opposition party's Arabic-language daily, *Al Alam*, was more acerbic, stating that, if disaster aid were to result in permission to maintain bases, then it was the Americans and Spanish who should keep their bases, since, according to the paper, these countries had played the largest role in rescue efforts. Moreover, the paper asserted sarcastically that Moroccans might as well invite Italy and West Germany to establish bases, since they, too, had provided aid.[24] The weekly *Al-Istiqlal* called upon Moroccans to remain focused on nationalist priorities despite the earthquake, citing the importance of not only base evacuation but also Moroccan support for the cause of Algerian independence and the pursuit of Moroccan control of the Sahara—the latter cause made more urgent by the French atomic testing "in our territory."[25]

To the dismay of would-be disaster diplomats, voices of discord soon emerged among the French. In Orléansville and Fréjus in the 1950s, advocates of empire had stressed solidarity between the European French and the Algerians, who had recently been declared fully equal citizens of France. Journalists and commentators such as Gaston Bonheur had often added their voices to the official chorus of solidarity. Because Morocco was independent, there was less French motivation to present a public face of solidarity, resulting in hostile polemics in the press.

Tangier, the most international of Moroccan cities, became a focal point of French anxieties. Far from the carnage, resentments about France's new relationship with Morocco erupted, resentments which centered on the apparent lack of a sense of dependence on the part of the Moroccan leadership, and on fears of France's declining influence vis-à-vis other foreign powers in the kingdom. On March 3, Pierre Bouffanais, the French minister plenipotentiary at the Tangier consulate, accused Radiodiffusion Marocaine of "disloyalty," for "systematically minimizing the contributions of the French armed forces in organizing relief."[26] Bouffanais also resented the lack of coverage given by the Agence-France Press, which, in his view, should have devoted less front-page space to Sekou Touré's visit with Mohammed V, and more to the Catholic Mass for the dead celebrated at the French church in Tangier. Bouffanais blamed this on the "Moroccaniza-tion" of the agency's staff.[27]

Bouffanais's complaints went beyond the frustrations of underpublicized di-saster diplomacy, however. A speech by Crown Prince Hassan declared that a new Agadir would be inaugurated on March 2, 1961, and connected this inau-guration with the five-year anniversary of Morocco's independence. Bouffanais alleged that French colonists interpreted this as a continuation of Moroccan "anti-French excitation campaigns."[28] According to Bouffanais, the French of Tangier had "reacted forcefully" against this alleged ingratitude, with the result that their "initial grand élan of solidarity with all the victims" became more "nu-anced," and the French community very quickly shifted their generosity toward the goal of assisting only the French disaster victims, making it difficult to coor-dinate relief collection efforts with the Moroccan authorities. Bouffanais linked the purportedly new "cleavage" between Moroccans and foreigners to anxieties about the new French relationship with the whole Arab world, "where Islam reigns, where the forces of pan-Arabism are unleashed." Bouffanais also stated that suspicions about religious discrimination in the distribution of donations had been expressed by Moroccan Jews and were shared by the French as well.[29]

French colonists, according to Bouffanais, deplored what they saw as the hos-tility and incompetence of the Moroccan state, characterized by "panic and inef-ficiency" as well as by publicity-seeking egotism. (How colonists in Tangier were well-placed to judge the emergency response in the Moroccan south remains unclear.) The French minister concluded that the root of the problem was "the power of the word in Arab countries," and the "illusion" that words could sub-stitute for effective government.[30] The ideas expressed by Bouffanais were not new: they echoed old colonial discourses about threats to French dominance. Accusations of "verbalism" had been a common disparagement of dissidents who challenged French power in the colonies.[31]

By March 12 the uproar in Tangier had reached Paris. Senator Bernard Lafay of the center-left *Gauche démocratique* formally asked whether the Ministry of Foreign Affairs might request that the Red Cross conduct an inquiry into the "hesitations" and "counter-orders" that had resulted in the deaths of individuals, buried in the ruins, who might have been saved by quick and resolute action.[32] Going beyond Bouffanais's vague assertions of state "incompetence," and La-fay's thinly veiled implication, *Paris Jour* explicitly accused Prince Hassan of misconduct for his decision to halt rescue operations on the third day after the earthquake.[33] *Al-Istiqlal* responded by accusing Lafay and the French press not only of insensitivity toward Moroccan suffering but also of violating Moroccan sovereignty by questioning the handling of internal affairs. When catastrophe had struck Fréjus, the paper noted, Moroccans had sent donations without meddling in French domestic matters.[34]

On March 30 *Europe-Magazine* in Brussels put forward still more acerbic accusations of Moroccan incompetence, alleging that Moroccan troops had arrived tardily on the scene and had "accomplished practically nothing except issue lamentations and implore Allah" while French, American, and Dutch troops engaged in rescue efforts. The article went on to mock Moroccan ambitions for base evacuation, despite the role played by the bases in rescue efforts, and ridiculed plans for the reconstruction of Agadir ("With what money? Undoubtedly with ours, and that of the other European powers, and America"). Hassan was accused of self-aggrandizement in pursuit of personal popularity in a country where he was "unanimously detested."[35] Similar arguments were expressed in still stronger terms in *Le Figaro* by André Figueres, who stated that more lives would have been saved "if the rescue work had been directed by someone serious." Figueres singled out the thirty-year-old Hassan, stating that the "panic and nonchalance of a ridiculous adolescent had condemned people to death, undoubtedly including French families."[36] Figueres blamed the French left for handing over power to such incompetents: the problem, for him, was inherent in decolonization, which spelled doom for the accomplishments of the colonial period. For French diplomats, however, the anti-Moroccan sentiment exhibited in the wake of the disaster threatened French interests. The minister-counsellor at the French embassy, Le Roy, pointed out that Crown Prince Hassan was one of France's most important allies within the Moroccan government, and that by attacking him, the French press was playing into the hands of the Palace's opponents, presumably meaning the Istiqlal party that was demanding the immediate evacuation of the French military bases. Moreover, Le Roy feared that these attacks might undermine any possibility that the positive role played by the French base in Agadir would lessen the king's support for their evacuation.[37]

French Survivors and the Decline of Empire

Soon, criticisms emanating from French politicians and journalists shifted their targets from the Moroccan monarchy to the French state, and from unnecessary deaths to the alleged neglect of the living. Two years before the independence of Algeria in 1962, when the stream of repatriated colonists would become a flood, hostility toward an independent North Africa was already linked to the concerns of repatriated French refugees, as it would be in the politics of the far right in France for the next four decades. Lafay became a supporter of victims' organizations in Agadir,[38] and an article by General Bethouart titled "Le scandale d'Agadir" appeared in *Le Figaro*, contrasting the "magnificent" French rescue effort immediately after the earthquake to inadequate aid for survivors who hoped to reestablish their lives and livelihoods. Bethouart acknowledged that funds had been dispersed by the embassy, but the general pointed out that these met only the most immediate needs. Existing procedures for the disbursement of unsecured loans were being accelerated, but Bethouart argued that these existing programs failed to meet the unique needs of Agadir survivors. The general supported proposals providing indemnities for "reconstitution de foyer," the reestablishment of residences, but so far the government had failed to approve such measures.[39]

Bethouart's editorial drew attention to three key distinctions. First, by comparing the "magnificent" immediate response with the inadequate post-disaster aid, the general was implicitly drawing a contrast between the heroism of the military and the ineffectiveness of the civilian government. Second, Bethouart noted that "there are two sorts of victims: those who desire to remain in Morocco and those returning to France." For the former group, Bethouart allowed that it might be necessary to wait until the Moroccan government had announced its policies before France could act. But for the latter, the decolonized, such delay constituted neglect.[40]

Third, Bethouart accused the French government of a double standard with regard to French citizens. According to Bethouart, the French state privileged the survivors of the Orléansville and Fréjus catastrophes while neglecting French citizens in Agadir: "For the administration, the French victims of cataclysm occurring abroad have rights to nothing . . . There are thus two categories of French, treated differently: those of the metropole or overseas departments and those in foreign lands [*de l'étranger*]." After 1962, the politics of decolonization and repatriation centered around the fate of Algeria's *pieds noirs,* who migrated to France by the hundreds of thousands at the end of the Algerian War of Independence.[41] Since the mid-1950s, however, increasing numbers of French colonists had been repatriating from Morocco and Tunisia. Bethouart felt that these colonists were being betrayed: "It was they [the French abroad] who, spread out across all the continents, made our country a world power. It was they who created the markets supplied by French factories. It was they who taught and propagated our language. But, in exchange, they have rights to nothing, and the survivors of Agadir, who, in other times would have been indemnified, no longer interest the administration."[42] Bethouart's anxieties about the commitment of the French state to Agadir's French refugees were intertwined with concerns that, with the end of empire, the government was turning its back on colonists. Notably, however, Bethouart, like many French in 1960, still believed that the Fifth Republic would win its fight to keep Algeria.[43] Hence, Bethouart still counted the French survivors of Orléansville among the privileged, French on French soil, though two years later they would become refugees and migrants themselves.

As the press continued to draw attention to the hardships of the French survivors of Agadir,[44] the French government maintained that, since the disaster had occurred on foreign soil, in an independent country, "the indemnification of the victims of the Agadir earthquake is incumbent on the Moroccan government."[45] (The French state would take a similar position toward the recovery of bodies and the reconstruction of the city.) The Moroccan government offered 1,000 dirhams, or 975 new francs (about $200 in 1960), to each family that remained in Agadir, whether Moroccan or foreign, for the purchase of household goods, and would later offer a combination of grants and subsidized loans to cover the cost of reconstruction.[46] As in Orléansville, Fréjus, and most other disasters, this meant that the vast majority of public aid would go to property owners, with minimal provision for the renting poor. The majority of the French property owners, however, were relocating to Casablanca, Algeria, or France, and were therefore ineligible for reconstruction assistance from the Moroccan government. The combination of political independence, exodus, and natural disaster left many French survivors, accustomed to being part of a privileged elite in Morocco, in an unusually weak position.

However, French officials pointed out that French aid was already being given, and more had been approved. Each French survivor remaining in Morocco was eligible for an emergency payment of 300 new francs (100 for children) from embassy funds.[47] In order to allow survivors the opportunity to return to Agadir to retrieve belongings, a daily stipend was available until May 31.[48] Initially, for those who chose to resettle in France or Algeria, it was proposed that heads of household would receive 500 new French francs (equivalent to 102 US dollars in 1960), and 300 new francs per dependent, renewable once, in addition to transport and temporary lodging. However, apparently in response to pressure from senators Lafay and Tomsamini, larger subsidies were soon approved to assist in the reestablishment of households. Beginning May 7, both repatriates and those remaining in Morocco would receive compensation based on damages, capped at 2,500 new francs per head of household plus 500 new francs per additional surviving resident, although the French consul in Agadir argued that this was not terribly generous.[49] By the end of 1960, 1.9 million new francs had been distributed by the French government for this purpose.[50] In addition, a solidarity campaign for French victims of the disaster, collected in Morocco and France, eventually raised over 700,000 new francs.[51] Most of this solidarity fund was distributed by the French embassy to "survivors whose situation merited particular interest."[52] The remaining 70,000 was later spent on the Agadir cemetery for Europeans and on the exhumation, transport, and reburial of French corpses, both within Morocco and to France.[53]

In striking contrast to Orléansville, there was no expectation in any quarter that the French state should provide equal assistance to European and North African Muslim victims of the disaster. This was not just because of formal Moroccan independence in 1956. Moroccans, unlike Algerian Muslims, had never been depicted as French citizens, or even as future French citizens. The protectorate arrangement had meant that, after 1912, Moroccans had remained subjects of the sultan, producing a politics of "two weights and two measures," with overtly separate systems of justice, education, and rights applying to the French and Moroccan populations. Nationalists during the protectorate period had denounced this "politics of racial privilege."[54] Independence, however, meant that the French state no longer faced criticism for its preferential treatment of French citizens in Morocco. French expatriates in Agadir had what Algerians in Fréjus had lacked: an embassy to look after their interests. However, the Moroccan state—in many respects a continuation of the colonial state—now had to take care not to perpetuate the protectorate practice of special treatment for Europeans. In Agadir, this was as much an issue regarding the dead as regarding the living.

The Decolonization of the Dead

As French politicians, journalists, and diplomats in Tangier, Paris, and Rabat wrangled over the purported failings of the rescue efforts and the compensation of French survivors, another drama unfolded in Agadir concerning the bodies of the dead. As the chaotic initial phase of disaster response passed, a new question confronted both the Moroccan and French states. How much effort and expense would be devoted to extricating the thousands of bodies that lay buried deep under tons of rubble? In a modern state, the management of the dead is a critical function and demonstration of state authority. Since the 1930s, the colonial state—which in Morocco meant the sultan's "Makhzen" state in Fez operating under the control of the French Residency in Rabat—had provided the land necessary to accommodate the funereal needs of Agadir's growing population, both Muslim and non-Muslim.[55]

The families of French colonists had not always wanted their loved ones to be buried where they perished, however, and it had been the French state that had regulated the transport of colonists' bodies within Morocco and to France. Although the authorization of the French secretary general of the protectorate had been required for transfer of bodies out of Morocco, such requests seem to have been routinely granted. Some rather malodorous problems developed, however, involving the inevitable results of transporting bodies in hot weather. In the summer of 1950, there were unspecified "incidents," in which bodies arriving at their destination seem to have been rejected and returned to Morocco. This led, in May 1951, to a moratorium on such international transfers from June until the end of September each year.[56] This moratorium remained in effect after Moroccan independence, exemplifying the often-noted continuity between the French and Moroccan regulatory state. In this case, the continuity is unsurprising— the management of the dead is a required duty of the state, and burials had not been at issue during the struggle leading up to Moroccan independence.[57] New implications of the political independence of Morocco surfaced, however, following the Agadir earthquake.

Shana Minkin has recently explored the role that bodies and burials played in the implantation of European power in North Africa.[58] In the era of decolonization, mass migration of European colonists also produced burials and reburials that were fraught with cultural and political tension. In 1960, excavations and burials became central to the negotiation of boundaries between the French state and the newly independent Moroccan state and hence to the very meaning of the political independence of Morocco that had been recognized in 1956.

The Agadir earthquake temporarily produced a transnational space where the boundaries of sovereignty and even citizenship were unclear, as the French and Moroccan states were overwhelmed by the scale of the disaster. For several days, French troops moved freely through a Moroccan city once again, while Moroccan police made irrevocable, crucial decisions about the fate of European bodies. Nameless French and Moroccan corpses were buried together in mass graves, an irreversible mingling of the dead. Of the French population, a total of 404 bodies were eventually identified. Another 131 were missing and presumed dead.[59] Immediately following the disaster, after the burial of several hundred bodies at the French base,[60] the base commander had declared that further burials posed a health risk. European bodies were then sent to Ihchach, on the northern outskirts of the city, where the Muslim dead were already being buried. In the European cemetery there, Moroccan police oversaw the burial of Europeans in mass graves, generally without coffins; a French municipal engineer reportedly assisted.[61] For the French dead, death certificates were filled out by military first-responders and later ratified by the French consular authorities. About thirty victims, either dead or dying, made the journey to other cities in Morocco.[62]

Once the initial phase of emergency response had passed, French authorities showed no desire to shoulder the staggering financial and moral burden of responsibility for the retrieval of corpses that remained buried under the ruins. The official French policy was "that the search for bodies is incontestably incumbent upon the Moroccan authorities, and that the French state is not to substitute itself for the Moroccan state in this matter."[63] Nevertheless, Moroccan authorities granted the French base commander access to the burial sites of French military dead and involved the consulates of various countries in the process of identifying bodies. Many French survivors, however, expected their government to do more for French citizens.

The first organized attempt at the excavation of ruins in order to retrieve corpses began at the Hotel Saada, an upscale establishment that had suffered a catastrophic collapse. The first four bodies were retrieved on March 12, 1960, and were identified as a Frenchman from Casablanca and a Frenchwoman from Paris, a German tourist, and a young Austrian woman. Although during the first few days after the disaster, the French base commander had made unilateral decisions about the disposal of bodies, the Moroccan authorities were now firmly in charge. The Moroccan state delivered coffins from the French base to the hotel, which were numbered to correspond with new graves that the Royal Moroccan Army, under Colonel Driss, had dug at the Ihchach cemetery.[64]

But the excavation of the Hotel Saada was an exceptional effort, led by the Moroccan army, an effort that was no doubt motivated by the diplomatic importance of retrieving the elite and foreign clientele entombed there. Otherwise, however, the retrieval of bodies came more or less to a halt. Thousands of unretrieved corpses remained, buried deep within treacherous ruins. International donations poured in, but the Moroccan state had thousands of refugees to feed and house, and resources were finite. In the two months after the earthquake, the excavation of the Hotel Saada, which had cost approximately twenty million francs (the dirham was not yet in use), was the "only [such] operation that had been methodically carried out and completed."[65]

By early May, a transnational "Association of Disaster Victims" (L'Association des sinistrés d'Agadir) had been formed, led by landowner Abbès Kabbage, who was a local leader of the UNFP (Union nationale des forces populaires), joined by representatives from the Red Cross and from French companies such as the Compagnie du Souss. This association planned to present a variety of complaints, including the need for action on the retrieval of bodies, directly to King Mohammed V.[66] By this time, however, some French survivors (some of whom had just returned to the city to claim their moveable property), had also begun to pressure the French consulate and the French Red Cross to step in and take action to retrieve the buried bodies; these French survivors were frustrated by what they perceived as the "prolonged inaction" of the Moroccan state, "which they saw as incapable, due to lack of means and money, of carrying out the clearing of ruins."[67] The frustration of these survivors was sufficient to worry the French ambassador that this issue might cause further scandal in the press. Far more alarming for the new French consul in Agadir, Jestin, were indications that some of the French survivors were prepared to take matters into their own hands and enter the cordoned-off area to excavate their own dead. Jestin feared that, with armed Moroccan soldiers guarding the perimeter, under orders to shoot looters, new tragedies with international implications might result.[68] The Moroccan government had agreed to study the possibility of a massive excavation project, but in the meantime, Jestin urged the governor of Agadir, Mohamed Benhima, to approve the excavation of one or two of the most suitable buildings, in hopes that this "first swing of the pickaxe" would calm the nerves of the disgruntled French survivors.[69]

While awaiting Benhima's response, French tempers smoldered. Rumors spread that in one building seven bodies lay on the surface, easily retrieved but for the inertia of the French and Moroccan authorities. An inspection of the site in question by embassy personnel revealed, however, that the bodies were visible but inaccessible, blocked by "enormous masses of concrete." One French official had heard mutterings from his own disgruntled colleagues that some might storm the ruins of the old consulate building, where their fallen coworkers still lay. He added that the horrific, nauseating odors emerging from the ruins did little to calm the nerves of the bereaved. The French Embassy hoped that the Red Cross might serve to fill the vacuum between the French and Moroccan states and take the lead in the retrieval of bodies. However, the Red Cross indicated to French officials that "the Red Cross' mission is to offer relief to the living; they were not in the habit of intervening for the dead."[70]

The case of the Hotel Saada as a successful example of excavation became a liability, rather than a model, for foreigners seeking more aggressive Moroccan action. The Saada excavation had been extremely expensive, and it had clearly been a case of special treatment for foreigners. This made excavation a sensitive issue for the Moroccan authorities. Governor Benhima accompanied the French consul, Jestin, on a trip to meet with survivors outside of Agadir proper, and Benhima reportedly explained, both in public and in private, "the impossibility, for him, of undertaking systematic excavations of the modern city to recover the bodies of 300–400 Europeans, when, at the same time, nothing was being done in the quartiers where Moroccans were the majority."[71] The French Embassy recognized that "it would be hardly imaginable to envisage that only the Ville Nouvelle (where most of the French resided) should be excavated, while classifying the Talborj, Founti, and Yachech [Ihchach] as zones destined to be necropolises in their current state."[72]

To placate the French survivors, Benhima authorized individual excavations, at private expense, reportedly to allow the public to see how difficult it was. Technicians employed by the Moroccan government (many of whom were French) estimated that excavating the collapsed consulate building, a project comparable to the effort made for the Saada, would cost eighteen million francs; to excavate the whole city would cost 1.2 billion. This, in their view, was "materially impossible."[73] Several private excavations did follow, (one large building, "Immeuble Le Nord," and some small homes in the Talborj). Jestin argued that, rather than prove the unfeasibility of such efforts, as Benhima had predicted, they demonstrated the contrary, and, moreover, they had been better organized and consequently much less expensive than the excavation of the Saada. Jestin concluded that the Moroccan government would have to give in and engage in a costly and time-consuming excavation before beginning the planned reconstruction of the city.[74]

Mass excavations did indeed begin on July 5, 1960,[75] in both the Ville Nouvelle and the Talborj but not in the Kasbah.[76] In May, Benhima had subtly introduced the formula which would determine the fate of Agadir's fallen. Preferential treatment for Europeans was unacceptable to the Moroccan government and the Moroccan public, but a functional distinction approximated the national one: the site where the new city would be constructed would be excavated for the purpose of recovering bodies. The areas where seismologists had proscribed construction—most notably the Kasbah, with its exclusively Moroccan population and its towering elevation—could be left to lie in their present state. By October, it was reported that 1,400 bodies had been recovered from the Ville Nouvelle and the Talborj. The latter, with its ethnically mixed and disproportionately bourgeois population, was just outside the area zoned for reconstruction but was excavated nonetheless. This suggests that Benhima's formula provided political

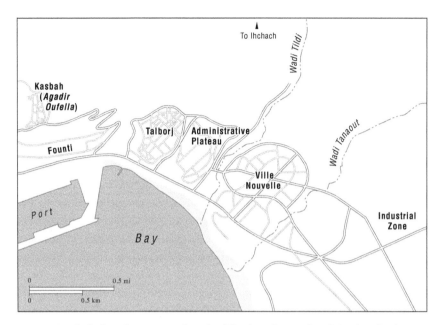

MAP 2. Agadir before the 1960 earthquake. The densely populated, Saadian-built Kasbah, atop Agadir Oufella (Agadir Heights), at an elevation of over 230 meters, towered high above the city's other districts. (Erin Greb Cartography.)

cover for the excavation of areas housing Europeans and more privileged Moroccans, while leaving the Kasbah untouched.[77]

Benhima remained sensitive to charges that the French were getting preferential treatment. When the French base commander asked that a representative from the base be present whenever excavations occurred involving the bodies of French military personnel or their family members, Benhima reportedly protested that the French were asking him for the "creation of a system of exception in favor of a category, the French, upon whom all the Moroccans had their eyes fixed."[78] Although Benhima ultimately accommodated this French request, his concerns demonstrated the new politics surrounding the French presence in Morocco.

French expectations about the retrieval of loved ones were often accompanied by a desire to repatriate the bodies for burial in France. The expense of shipping bodies out of Agadir to Europe, estimated to cost 4,000 new francs,[79] remained prohibitive for many European families. Even burial in Morocco was a burdensome expense for those left homeless by the disaster. Survivors were outraged in June 1960 when they learned that the missing and fallen were to be excluded in the calculation of the stipend authorized for the reconstitution of residences ("aide à la reconstitution des foyers") for survivors. Many families had been counting on the extra 500 per deceased family member "either for a tomb or for reducing the cost of repatriating the bodies."[80] For those French survivors whose loved ones were buried at the French airbase, their burial in the soil of a French military cemetery may have been of some comfort. Some, however, hoped for a mass transfer of French corpses to the metropole. The first suggestion that the French state undertake (and fund) the repatriation of the corpses of French earthquake victims came in April 1960, proposed by the former French consul in Agadir, who had been reassigned soon after the death of his son in the earthquake. The French foreign ministry's *Direction* for Morocco and Tunisia was supportive, and noted that a collective repatriation would be more cost-efficient than individual shipments.[81] Nevertheless, it would not be until the evacuation of the base that such a mass expatriation of the dead would occur, organized and paid for by the French state.

Four Decolonizations of Agadir

The destruction of Agadir in 1960 produced a far more dramatic demographic transformation than had occurred with the decolonization that accompanied Moroccan independence in 1956. In the 1950s, amidst the violent struggle for independence, the city's French population shrank from 15,000 to 5,200. An estimated 535 French citizens lost their lives in the earthquake, but beyond this, the disaster precipitated sudden new exodus. After the earthquake, Agadir's French civilian population dropped precipitously, to 200, by June 1960, with another 300 or 400 displaced to the nearby towns.[82] The departure of many French survivors after the earthquake constituted a second decolonization of Agadir, a demographic revolution comparable to that of Algeria, two years later, at the end of the Algerian War.[83]

After years of polemics and negotiations, a third decolonization—the evacuation of French military bases—would bring formal resolution to the nationalist struggle to restore Moroccan sovereignty over what nationalists considered Moroccan soil (at least insofar as Franco-Moroccan relations were concerned; the Spanish and the FLN were another matter entirely). An agreement between France and Morocco was finally reached on September 1, 1960, providing for the "progressive evacuation of operational bases by the fifth anniversary of independence [i.e., by 1961] and of training bases by the end of 1963."[84] Disaster diplomacy had failed to save the bases, but the evacuation agreement saved disaster diplomacy, removing a major political grievance against France that had frustrated French efforts to use disaster aid to win goodwill.

For Agadir, the base evacuation agreement meant that plans had to be made immediately, not only for the withdrawal of French troops but also for the future of the bodies, French, Moroccan, and others, that lay buried in the base cemetery. For French families seeking the repatriation of loved ones buried at the French military base, the September 1 agreement was a boon, for the evacuation of the base included the exhumation and repatriation of those buried at the base cemetery—the beginning of a fourth decolonization of Agadir. A French naval vessel was used to bring bodies to Marseille, where they were shipped on to their final destinations at private expense. Bodies that were unidentified or whose shipment to France was not requested by their families were exhumed and reburied at the European cemetery in Ihchach. Meanwhile, the families of some victims already buried at Ihchach seized the opportunity of the base evacuation and requested the exhumation from Ihchach of their loved ones and their transfer to France. This was accommodated, but only in the case of bodies buried in individual graves; the exhumation of communal graves was interdicted by the Moroccan government. In February of 1961, the French Navy transported 191 bodies of French citizens to Marseille.[85]

Grimm's Tale

The evacuation of the French base in Agadir did not close the book on the decolonization of the French dead, however. Following the evacuation, a survivors' organization was founded in Paris ("l'Association française des sinistrés et rescapés d'Agadir"), and politicians and private individuals lobbied for more exhumations. Some families also sought the repatriation of bodies that had been buried outside of the Agadir area and thus had not been eligible for the group expatriation in 1961. Others sought loved ones whose bodies had never been identified.[86]

In August 1962, the director of the Morocco office of the mortgage bank Crédit Foncier d'Algérie et de Tunisie, Mr. V. A. Munier, wrote the French authorities about a particularly complex case of missing bodies. The remains in question were those of a Mr. Jacques Bordeaux, also an administrator at the Crédit Foncier, and his wife, Monique. Mr. and Mrs. Bordeaux had died in the earthquake when the roof of their apartment building caved in, crushing them.[87]

The bodies of Jacques and Monique Bordeaux were missing, and thus they were presumed to be among the thousands of victims of this earthquake who were buried anonymously immediately after the disaster or the following summer as the terrain was prepared for the construction of a new city. It was also possible that their unretrieved remains had been ploughed under with bulldozers, along with the wreckage of their homes. Although most bodies from the Ville Nouvelle had been recovered and identified, Mr. and Mrs. Bordeaux were not among them.

However, Mr. Munier, the author of the 1962 memo, believed that he knew where the Bordeauxes had been buried, and he wanted their bodies to be exhumed and repatriated to France. The evacuation of the base cemetery had inspired a flurry of French requests for exhumation and transport of the deceased from other burial sites in Agadir. Munier's request for the exhumation of the Bordeauxes was problematic, however, not because the whereabouts of the bodies was in question but rather because Munier traced their postmortem itinerary to a grave shared with other disaster victims, some of them unidentified—and therefore of unknown nationality and religion. Even more problematic was Mr. Munier's request that this grave be opened to retrieve the bodies of French citizens when many hundreds of Moroccans remained interred in communal graves. Munier's petition again raised this basic question of decolonization: would Europeans still be treated as a privileged class in independent Morocco?

Central to Mr. Munier's petition was the story of a certain Ms. Grimm, who lived across the landing from the Bordeauxes with her sister and her brother-in-law, the Macans. Both Mr. and Mrs. Macan died in the disaster. Ms. Grimm survived ("woke up on top of the body of her sister"), managed to extricate herself from the ruins, and then left the building.

When Ms. Grimm returned to her home (reportedly after "a few moments" but probably at least an hour later), she discovered that the Macans' bodies were gone. In order to find them, she ran to the French base on the outskirts of the city, where many of the dead were being buried. Not finding the remains of her loved ones at the French military base, however, Mrs. Grimm hurried to the civilian cemetery for Europeans in Ihchach, north of the city, where burials were also underway. She arrived in Ihchach in time to witness the burial of her sister. Her sister's husband Mr. Macan was not there, however.

Mr. and Mrs. Macan, who had died in the same bed, ended up buried miles apart. Apparently, the deceased couple had become separated from each other after a truck had delivered them to the base, just as the base commander was ordering a halt to the burials there, declaring that the cemetery was over capacity. Some of the bodies from that truck were among the last buried at the base, but others, including Mrs. Macan's, were again placed on trucks and sent to the Ihchach cemetery. Mr. Macan's body had remained at the military base and had been identified only later, during the 1961 exhumations that accompanied the base evacuation.

Where, then, were the remains of the Bordeauxes, those neighbors of the Macans' sought by Mr. Munier of the Crédit Foncier? Munier believed he knew: they had been transported by truck, along with the Macans, to the French military base. Another surviving neighbor, the one who had rescued their daughter, had reported that he had identified both bodies there. There was no record of Jacques Bordeaux's burial there, but the base's burial records listed a grave containing a Mrs. Bordeaux and a second where a woman was semi-identified as possibly a Mrs. Bordeaux. However, the exhumations that took place in 1961 during the evacuation had revealed that neither of these two graves contained Monique Bordeaux—one turned out to be an unidentified man, and another was identified positively as a different woman. No other bodies exhumed there seemed to match the Bordeauxes. So where were they? Munier concluded that, like the Macans, their bodies must have been unloaded at the base, but then, with Mrs. Macan, they were shipped to the European cemetery at Ihchach.

At the Ihchach cemetery, five communal graves were known to contain European victims of the earthquake, but only two contained unidentified corpses, including one in which the Bordeauxes's neighbor, Mrs. Macan, had been buried. So, Munier argued, the French authorities needed to arrange the exhumation of these two communal graves.[88]

Munier, however, was stonewalled by the French government, which emphasized the technical impossibility of exhuming and identifying entangled bodies in communal graves. He responded by citing expert opinion to the contrary. But the situation was not quite as simple as Munier claimed. After conducting an investigation, the French Embassy concluded that, after the transfer of bodies to Ihchach from the French military base in 1961, up to 156 French bodies were interred at Ihchach, including 14 unidentified corpses. Any of these, officials argued, could have belonged to the Bordeauxes. The Embassy also estimated that 158 of their compatriots had been buried in unknown locations, without a "decent burial," presumably in the large mass graves near the French base and in the razed Talborj quartier.[89]

Nevertheless, Munier was right to sense that the French authorities were offering him flimsy or phony excuses. It would have been a relatively small task to give it a try and excavate the two small communal graves requested by Munier, and in 1963 the French consul in Agadir conceded that the Bourdeauxes were "presumably" in grave number two.[90] Ultimately, however, the real issue was not technical or budgetary, but political, and concerned the Moroccan state.

As then-governor Benhima had explained, the Moroccan state was obliged to treat European bodies and Moroccan bodies in an egalitarian fashion. It was politically impossible to permit the exhumation even of small communal graves in the European cemetery while denying a proper burial to thousands of Moroccan Muslims.

Although there was obviously a concern about offending religious sensibilities by handling Muslim remains in reopened graves, the denial of Munier's petition for exhumation, and others like it, was based on the distinction between communal graves and individual graves and not between identified or unidentified corpses or even between Muslim and non-Muslim graves. One of the two graves identified by Munier (number five) was determined to include only European bodies: but opening even this grave, it was feared, would open, in the memories and politics of a traumatized Agadir, undesirable repercussions. In the words of Consul René Cader, "The question of opening the communal graves at the Yachech [Ihchach] cemetery cannot be raised without evoking the memory of the thousands of dead who rest in the immense communal grave in the Talborj quartier (population very mixed) and in the no less immense communal grave located across from the military base."[91]

Although in many respects the more significant decolonizations of Agadir were those of 1960 and 1961, the decision to leave French dead in the communal graves at Agadir and not to decolonize them was predicated on the political and legal decolonization of 1956. The Moroccan state was sovereign over the Ihchach cemetery for Europeans, and Moroccan public opinion would not tolerate a protectorate-style system of "two weights and two measures" when it came to exhumation. In this matter, the French and Moroccan authorities were in agreement, despite the objections of some French citizens. In matters concerning the retrieval and exhumation of bodies, we see here a case where—as in the decolonization of Algeria in 1962—the desires of the French residents of North Africa conflicted with what French officials saw as state interest, the French state being much more willing to prioritize relations with the post-colony.

Reconstruction

Another field of contested decolonization concerned planning for the permanent reconstruction of Agadir. Urbanism and architecture became fields of struggle for the final horizon of decolonization—ending the hegemony of French culture in Morocco—and would be fought out for many decades. In the short term, reconstruction was a desperate necessity for the Moroccan population of Agadir, who did not have the option of "repatriating" to France. For elites in Rabat, Paris, and Washington, however, other issues were at stake. Reconstruction became a test for the ability of the Moroccan sovereign to respond to the needs of the nation, for the French to maintain their influence, and for the Americans to demonstrate their resolve as Morocco's new benefactors.

As rescue efforts wound down, Crown Prince Hassan declared that a new city would be constructed in a year's time, and an *Al-Istiqlal* editorial called upon Moroccans to rebuild "not only a new Agadir, but also, and above all, a new Morocco."[92] French critics of decolonization responded with hostility. André Figueres proclaimed in *Figaro* that "Lyautey [Morocco's first French Resident General] had conjured Agadir out of the marvelous but deserted sands of the Moroccan south." Now that the French had handed the country over to an archaic "feudal regime," Lyautey's legacy would be squandered. The reconstruction of Agadir, which the king had set for the 1961 anniversary of independence, would be a test to see whether "Morocco did not still need Lyautey."[93]

While eager to show its ability to fulfill its obligations as an independent state, the Moroccan leadership recognized that Morocco could not rise to this challenge unilaterally. But while French and American officials in Morocco were quick to recognize the possible benefits of disaster diplomacy, neither the French nor the American government wished to accrue major financial obligations for the reconstruction of the city. The French repeatedly stressed their need to prioritize relief for the French survivors of the disaster.[94] Both nations stated that, in their massive contributions to the rescue effort and in providing immediate relief to survivors, they had done enough. Enthusiasm for disaster diplomacy was diminished by concern over the lack of initial publicity for the foreign role in the emergency response phase and by a recognition that the Moroccan government had in the past been resistant to publicizing foreign aid and to projects with an obviously foreign origin.[95]

Yet international power politics provided considerable motivation. French reluctance was mitigated by the fear of losing, to the Americans, its role as the primary provider of technical assistance, while American budgetary concerns were offset by the fear, voiced also by the British, that the Soviet Union might step in with "a spectacular offer of aid which Morocco would be unable to refuse."[96] The US State Department hoped that an appeal to a United Nations agency such as the UN Technical Assistance Board might reduce pressure for Franco-American support, while diluting the impact of possible Soviet aid. This approach, however, also entailed risks that donations might be expected from member nations and "the risk that Afro-Asian enthusiasm could lead to a proposal of unmanageable proportions."[97]

For the French, anxieties about growing American influence in Morocco dated back to the early years of the Second World War. Even before the landing of a North American army on the shores of northwest Africa, the war had prompted a new intensity of transatlantic contacts. After the fall of France in June 1940, French authorities in Rabat had sought American aid and trade to alleviate the economic hardships caused by the collapse of the French metropole. The British had grudgingly consented, and a modest American aid program operated, with some interruptions, until November 1942.[98] As rumors of the impending American invasion spread, French prestige faltered, and there were reports of Moroccan troops "refusing to obey their French officers because they knew the Americans were coming."[99] In November 1942, they came, and by the end of the month there were sixty-five thousand US soldiers in Morocco.[100] Along with these troops came American lend-lease and an end to the partial British blockade: Morocco was "now open again to the markets of the world."[101] After the war, American imports increased, including cars and durable goods for the benefit of the more well-to-do *colons* and the Moroccan elite.[102] French anxiety about the growth of American influence—political, economic, and cultural—became a tool that the Moroccan state could use for diplomatic leverage.

Within days of the 1960 disaster, the Moroccan government approached the Republic of France to request assistance. The boundaries of the two states were still porous and blurred, four years after independence. Morocco's Ministry of Public Works and its Service of Urbanism were still dependent, even at the highest levels, on French *coopérants*, thirteen thousand French professionals who worked for the government of Morocco under the terms of a convention signed between the two countries. The presence of these coopérants could serve as a backdoor diplomatic channel. It was two such high-level coopérants, the secretary-general to the minister of Public Works and the engineer-in-chief of the Service of Urbanism, who were dispatched to Paris on March 7 to request reconstruction aid from the former colonizer.[103]

These two Frenchmen requested that France send a team of urbanists to develop a plan for the new Agadir, reinforce the staff of the Ministry of Public

Works with an additional ten technicians, and commission three companies that had worked on the reconstruction of Orléansville to analyze and inventory damage to standing buildings and roads, and to define anti-seismic building standards for reconstruction. The total cost was estimated at 1.6 million new francs, but there was a clear French interest in maintaining their influence in the realm of culture and technical advising and thus in taking part in urban planning for the new Agadir. The extent of the proposed assistance went beyond city planning and represented more than the French wanted to spend, but, as one French official noted, if France did not rise to the occasion, other states would.[104]

The assessment that independent Morocco would not rely on France alone was correct. On March 4, Hassan had already outlined a plan to the American ambassador, Charles Yost, in which the US, France, and one other unnamed country would each contribute to the design and reconstruction of a different part of the city. Hassan requested that an "imaginative, modern" American planner be dispatched to Morocco as soon as possible.[105] Over the next several days, it became increasingly clear that Hassan sought as much American support as possible. A formal request was made for an American city planner, preferably (Spanish-born) Josep Lluis Sert, dean of the Harvard Graduate School of Design. The minister of Public Works, Abderrahmane ben Abdelâli, also requested the services of an American housing expert as well as a geologist, a seismologist, and an architect.[106] This was a clear statement about distancing Morocco from its former colonial "protector" and signaled a Moroccan strategy of provoking competition between the two NATO allies.

From the beginning, French and American officials were skeptical of Moroccan ambitions about the future of Agadir. On March 12, Le Roy had voiced his suspicion that Hassan's decision to halt rescue operations on March 4 was motivated, not only by epidemiological fears, but also by excess "haste to pass from the phase of disaster to that of reconstruction."[107] Americans officials felt that Hassan's ambitions for reconstruction were "grandiose."[108] The Americans pressed the Moroccan government to include the French in a joint planning commission,[109] and initially expressed reluctance to displace France in the realm of civil technical assistance. Neither foreign government, however, wanted to be left out of the project altogether. As events developed and the Moroccans presented United States diplomats with opportunities to play a prominent role in the "rebirth" of Agadir, the Americans felt obliged to be responsive to Moroccan needs.

Alarmed by the growing American involvement, the French government also tried to accommodate Moroccan requests. France agreed to a Moroccan request to send a "grand urbanist" to Morocco and suggested several names.[110] Abdelâli, minister of Public Works, wanted a bigger name, however, and requested the great planner Le Corbusier.[111] For the French ambassador in Rabat, this was an important opportunity, "at a moment when international interventions are multiplying regarding Agadir."[112]

Le Corbusier, however, did not turn out to be the solution for the French in Agadir, although his visit to Morocco seemed to go well. When Le Corbusier arrived in Casablanca on March 25, he made a brief statement to the press indicating that his experience working in Japan had prepared him well for the challenges in Agadir.[113] In Rabat, he met with Abdelâli and the professionals of the Public Works ministry's Service of Urbanism and with Crown Prince Hassan. However, bad weather prevented him from traveling to Agadir prior to his return to France on the 27th. This may have been an influential environmental turn of events; perhaps scenes of disaster might have given Le Corbusier more motivation. Although by all accounts Le Corbusier made a fine impression, he did not reach an agreement with his hosts and seems to have declined to participate in the project. According to one account, Le Corbusier chafed at certain requirements set forth by Crown Prince Hassan, who stated "I do not see Agadir without a mosque, and I do not see a mosque without green tiles."[114] French Ambassador Parodi noted that Corbusier "constantly expressed, with great frankness, and did not conceal that he could not accept responsibility for the reconstruction of Agadir unless he was given the liberty and the means necessary."[115]

Although Le Corbusier's visit ultimately failed to produce an agreement, his trip to Morocco provided leverage for Abdelâli to press the United States to send an "expert of similar caliber."[116] The State Department had approved the Moroccan request for a housing expert to advise Public Works, but the Moroccans clearly wanted someone more prestigious for the urban plan. For US ambassador Yost in Rabat, the disaster diplomacy possibilities were clear: "American association with the rebirth of Agadir is so obviously desirable that special efforts appear more than warranted." Yost argued that this would not imply additional commitments to the actual reconstruction, but added, however, that a prompt "no" would be better than an "embarrassing" delay.[117] The implied priority was that the US be seen as a reliable partner. However, Yost stressed, as he had throughout the oil poisoning crisis, that American stinginess might provide the Soviets with an opportunity. Yost also noted that communist China's ambassador had already made a substantial cash donation to the king.[118] The next day, in Moscow, Morocco's ambassador reassured his American counterpart that American aid had in fact been much more forthcoming than Soviet aid and "showed Moroccans who their true friends were."[119] Despite the anti-Soviet character of the remark, the Moroccan diplomat nevertheless reinforced the connection between disaster aid and Morocco's position in the Cold War.

The US State Department, however, preferred to "let the French take the lead, as they are doing," and suggested sending, rather than a "top planner," a "working-level planner," who would participate in planning during a year-long assignment, but take no leadership role.[120] Abdelâli, however, insisted that a top-level planner was necessary to build a city that would fulfill the king's desire to create an "expression of modern Morocco"; he noted that the Ministry of Public Works was already well-staffed with rank-and-file planners quite capable of "following up and executing plans."[121] The State Department relented.[122] Josep Lluis Sert was apparently unavailable, uninterested or deemed unsuitable, and so the State Department selected the prominent American planner Harland Bartholomew, chairman of the Washington, DC, National Capitol Planning Commission.

The United States authorized $50,000 to pay Bartholomew for the planning. As American officials put it, "Although the project has materialized somewhat differently than originally envisioned, withdrawal of US assistance at this time would not be easy to explain without embarrassment."[123] The Americans still wished to avoid paying for actual reconstruction, although a contingency policy was outlined in the event that political pressures required it. This would entail American involvement in the reconstruction of a single "model" residential area; "necessary safeguards would have to be taken to assure proper recognition of US sponsorship." In this case, the "project would specifically emphasize simple construction technique to permit maximum use of locally available labor."[124] This backup policy never came into play but demonstrates the gap between the expectations of Washington and Rabat. The king and crown prince wanted an impressive, modern city that would be a flagship for the new Morocco; American officials expressed anxiety about these "grandiose" ideas.[125] In Agadir, however, a different dynamic would develop, in which Governor Benhima's desire for a practical, swift reconstruction plan conflicted with the more ambitious scheme proposed by Bartholomew and endorsed by Rabat.

In Rabat, Moroccan expectations were high. Abdelâli hoped that preliminary plans could be produced by late April.[126] Bartholomew arrived in Rabat on April 9. However, the American planner produced only a basic report, recommending that the city be rebuilt on much of its current site (a major question among the seismologists in the preceding weeks), but that construction be prohibited in the Kasbah, Talborj, Founti, and Ihchach districts. Bartholomew stated that a first draft of the comprehensive city plan would take two months, followed by meetings with Moroccan authorities and the staff of the Ministry of Public Works and then would require three months for revisions.[127] The Moroccan state accepted Bartholomew's terms.

In May 1960, the king dismissed the government headed by Abdallah Ibrahim, and appointed himself head of government, with Crown Prince Hassan as deputy-premier and defense minister. This was a momentous event in Moroccan politics, but because the monarchy had always been the main representative of the Moroccan state in the response to the earthquake, the impact of the change in government on the question of reconstruction was minimal.[128] In a speech in Agadir on June 30, Mohammed V declared:

> Here we see today Agadir at the hour of its resurrection, through the execution of the plan, the preparation of which we have ensured: this plan, the essential objective of which is to reconstruct the city on a secure location, chosen by the architects, Moroccan and foreign, after a detailed scientific study, and by modern methods which are applied in cities affected by earthquakes.
>
> This program will make of Agadir a modern and active city, endowed with all the equipment necessary for life today: broad avenues, pleasant gardens, abundant light, mosque, schools, administration, etc. . . . Forward for the reconstruction of Agadir! Forward for the renaissance of the Sous! Forward for the new Morocco![129]

The king thus associated the monarchy and future of Agadir with the "modern methods" of a transnational community of seismologists, engineers, and architects. It appeared that the United States would play a prominent role in this new Morocco. In contrast, French influence seemed to be at a low point in the summer of 1960, eroded by the unpopularity of France's ongoing war in Algeria, atomic testing in the Sahara, and France's apparent intransigence (prior to the September agreement) on base tenure in Morocco. The United States, in contrast, was enjoying the benefits of Eisenhower's base evacuation treaty.[130] Bartholomew's contract for Agadir seemed to indicate a trend. French fears were compounded by the appointment of another American, George Schobinger, to serve as an advisor to Abdelâli, the Minister of Public Works. French officials worried that they were losing "the traditional influence of France in the domain of public works." This did not please American officials, however. According to Marcus Gordon, the State Department's director for Europe and Africa, "the worst eventuality would be that US assistance caused the French to become so angered as to make a complete withdrawal, leaving us with a void we could not possibly fill."[131] The State Department did not wish to be stuck with the bill for developing Morocco.

France's role in civil engineering had not vanished: private French companies were contracted for the clearing of debris. However, the French consul in Agadir became concerned that even this role might be lost, due to the high prices charged.[132] Moreover, in addition to the new role played by Americans in high-level technical assistance at Public Works, the ministry was engaging in a "Moroccanization" of its staff. The leadership of the ministry's Service of Urbanism had already been Moroccanized. In June, the Public Works engineer-in-chief in Agadir, a Frenchman, was replaced with a Moroccan engineer, Mohammed Faris. The consul recognized that, for the Moroccan state, "The reconstruction of Agadir constitutes, in effect, a political act; this gesture will only have full effect if the Moroccan authorities place technicians of Moroccan nationality as sector chiefs."[133] If the reconstruction of Agadir was going to be a national and royal act, an act of decolonization, it had to be done by Moroccans.

French hopes for a leading role in the redesign of the city were revived, however, when serious concerns emerged about Bartholomew's commitment to the project. The Moroccan leadership expressed dissatisfaction with the American planner's long absence through the summer of 1960; Bartholomew's dispatch of a junior employee to Morocco in late July failed to assuage these concerns. Bartholomew's representative presented plans which the high commissioner for the reconstruction described as "childishly superficial." Even American officials recognized that Bartholomew's company representative was "obviously not of a caliber to deal as an equal with Ministry technicians."[134]

For the US embassy, this was no longer just about Agadir. Minister of Public Works Abdelâli had survived the change of government, but he and the Americans had invested much political capital in each other, and they would stand or fall together, it appeared. As Ambassador Yost put it, "If Abdelâli fails on Agadir due to our promised support and is made to suffer for it as he doubtless would, the future of our entire technical assistance program will be jeopardized."[135] Yost's alarmism was apparently influenced by Abdelâli, who described the shortcomings of American assistance in the project as "shocking" and who emphasized the strength of internal opposition to the government's increasing turn toward American aid.[136] Bartholomew's lackadaisical efforts had provided an opportunity for those opposed to American leadership in the reconstruction, including "jealous" French coopérants within the Moroccan Ministry of Public Works. Soon, Abdelâli's frustration was compounded by the failure of the Americans to provide an advisor to Public Works in Rabat to replace George Schobinger, who departed in August.[137]

Thus began what became known as the "Battle of the Plans." This was both a competition between France and the United States for influence and prestige, and between Minister of Public Works Abdelâli, who reportedly aspired to become the Moroccan Ambassador to the United States, and Benhima, the governor of Agadir, who had good relations with the French in Agadir and who was skeptical about the Bartholomew plan, finding it overly ambitious. Benhima wanted a plan that could be implemented more readily to house the displaced population of the city. Benhima and Abdelâli also differed in their attitudes toward Crown Prince Hassan: Benhima was seen as loyal to Hassan, while Abdelâli was critical, and reportedly disparaged Hassan's leadership in the immediate aftermath of the disaster as well as his continuing accrual of power.[138]

For the French coopérants at the Service of Urbanism, the desire for a leading French role in designing the new city was less about disaster diplomacy than about professional prestige. Their collaboration with Moroccan colleagues, like that of most coopérants, was, as described in American diplomatic assessments of the situation, "almost entirely insulated from the ups and downs which affect French relations . . . on the political or economic level."[139] Already chafing at Abdelâli's decision bring Schobinger to Rabat to oversee projects in Public Works, the hiring of Bartholomew had been a great affront. According to historian Thierry Nadau, who interviewed many of those involved, Bartholomew was "little esteemed"[140] by the French and Moroccan urban planners alike, "who would have gladly deferred to the master [Le Corbusier]" but who "protested this *urbanisme primaire.*"[141]

The French and Moroccan planners had a key factor on their side: the winds of nationalism and decolonization were with them. However many Frenchmen were involved, their plan would be produced by an agency of the newly independent Moroccan state. Ironically, the Moroccanization of the Ministry of Public Works turned out to be an asset to French influence. The Service of Urbanism was now directed by Abdesalem Faraoui.[142] There, however, as in many parts of the Moroccan government, the French and Moroccan professionals had much in common. While subtler tensions no doubt lurked beneath the surface, outside observers of Franco-Moroccan technical cooperation noted that "it was frequently next to impossible to know whether the person one was dealing with was a native or a Frenchman because of this similarity of speech and style."[143] The replacement of the French engineer-in-chief at Agadir with Mohammed Faris proved to be no loss to the French cause. Faris, who consulted regularly with the French consul in Agadir, considered the American plan too extravagant and impractical, and he was suspicious of American motives.[144] Likewise, the French-educated modernists in the Service of Urbanism were an asset for the preservation of French influence. Architect Mourad Ben Embarek, who would soon succeed Faraoui as head of the service, would present the reconstruction of Agadir as an opportunity, by proxy, for Le Corbusier to make his mark on Africa, through the impact he had made on the younger generation.[145] Yet despite the fact that it was sometimes referred to by diplomats and officials as the "Plan Français," the Public Works plan could be portrayed as an assertion of independence. Several years later, the editors of Morocco's premier architectural journal, *A + U: Revue africaine d'architecture et d'urbanisme*, would portray the reconstruction of Agadir as an example for developing nations of the utility of avoiding dependence on foreign institutions.[146] This was a bit of a stretch, considering the central role played by French nationals in the Service of Urbanism. However, both the French and Moroccan governments had, since independence, avoided publicizing the role of French coopérants in Morocco, each fearing domestic criticism of the ongoing Franco-Moroccan cooperation.[147] The public face of the Ministry of Public Works' urban planning service was Moroccan. This arrangement served the interests of both France and Morocco—to the detriment of the Americans. As in the question of responsibility for the retrieval of the dead, the formal arrangements of decolonization served the interests of the French state. Nevertheless, American influence remained an obstacle to French hopes for cultural hegemony: the "Plan Américain" seemed to be on the path to realization.

The Fall of Bartholomew

American attention to disaster diplomacy in Agadir was intensified by the Palace's decision in November 1960 to purchase Soviet MiG jet aircraft. This deal shocked the State Department and demonstrated that Yost's alarmism had been well-founded. The Soviet deal increased Moroccan leverage in extracting aid from the United States by demonstrating that, in the era of the global Cold War, the Americans were not the only alternative to dependence on France. The purchase of the MiGs also increased Abdelâli's importance to the Americans, and thus the importance of Agadir's reconstruction, since Abdelâli was seen as "not only the most pro-American among the present ministers but is also the strongest opponent in the cabinet of the Soviet arms deal."[148] Abdelâli made use of this leverage, stressing that the Agadir project was a "life and death question for US-Moroccan relations" and insisting that the Americans extend Bartholomew's contract, which had ended with the completion of the master plan in late November, to supervise and manage the organization of reconstruction.[149] The US had little choice but to approve $49,000 in additional funds but made clear that this would not imply any further commitment to fund the actual construction.[150] It seemed that this was enough and that the Americans had sealed the deal. By December 26, the Palace had officially approved the Bartholomew plan, and arrangements were made for Yost to attend an inauguration ceremony presided over by Crown Prince Hassan on January 17, 1961.[151]

Soon, however, the Agadir reconstruction project became engulfed in scandal, leading to a decisive shift in the Battle of the Plans. It began slowly. A West German newspaper, *Die Welt*, ran a story accusing the Moroccan government of diverting 2 billion francs of international earthquake relief donations to "cover the chronic deficit of the Moroccan budget."[152] The French press picked up the story. *France Observateur* connected the alleged mishandling of funds to the suffering of the displaced survivors, portrayed as freezing in tents in the middle of winter. This, in turn, linked the issue to the earlier discourse in Tangier and in *Le Figaro* about a negligent French government abandoning its colonists to the incompetence of the independent Moroccan state.[153]

The Moroccan minister of information, Alaoui, denied these charges, and stated that all foreign donations had been placed in a dedicated account, separate from Treasury funds. The government's own funds from the 1960 budget had been applied toward the 4 billion francs already spent on reconstruction and demolition; foreign aid for reconstruction totaled less than 1.5 billion francs. Complete reconstruction of the city and compensation payments to victims (necessary to spur private reconstruction) was estimated at 24 billion francs. A special tax would raise 12 billion francs, supplemented by 2 billion from the 1961 general budget.[154]

Nevertheless, the obvious gap between the estimated expense and the available revenue left unanswered questions about the project's solvency. The press scandal increased scrutiny of both the finances and pace of Agadir's reconstruction. Hassan had set an impossibly high standard with his hasty assertion amid the rubble in March 1960 when he declared that a new city would be inaugurated in a year's time. Alaoui had apparently exacerbated this problem by indicating to the foreign press that the city was already largely reconstructed, when the clearing of debris was not even completed, and the only new structures were prefabricated.[155]

The government's cost estimates also raised questions about the future of the Bartholomew Plan. Already, in December 1960, Governor Benhima had hinted to the French that the Bartholomew plan might be discarded, and that what French diplomats viewed as the *Plan Français* might be adopted after all.[156] At the end of January 1961, Chief Engineer Mohammed Faris told the French consul that the Bartholomew plan was expected to cost 200 billion francs, while the Public Works plan would cost only 25 billion.[157] In early February, High Commissioner for Reconstruction Mohamed Imani began to hold a series of meetings in Rabat; the plan for Agadir was once more up for debate.[158] This was, apparently, a hotly contested question; in mid-March, after the unexpected death of King Mohammed V and the accession of Hassan II, Faris dejectedly predicted that the pro-American factions in the government and the Palace would prevail.[159] Faris was wrong.

On June 2, Governor Benhima was suddenly appointed to the post of minister of Public Works, replacing Abdelâli. Abdelâli had been accused of embezzling from the Agadir reconstruction fund generated by the "solidarity tax," and smuggling the proceeds to the United States and Switzerland. Benhima was considered an able technocrat and a dependable supporter of the Palace, and Benhima had been openly critical of both Abdelâli and the Bartholomew plan. As Mohammed Faris saw it, this was a major reason the new king chose him for Public Works: his appointment served to indicate a clean break from both Abdelâli's corruption and his policies.[160] Of course, this scandal could have been merely a cover; Abdelâli's hostility to Hassan may well have been the real reason for his dismissal, or the high cost of the American plan, in the face of public criticism about the financing of the project, may have pushed Hassan to make a change.

In any case, Abdelâli's fall spelled the demise of the American-designed "New Agadir," although this would not be announced for several months. Benhima charged the French and Moroccan planners and architects at Public Works with the task of synthesizing elements of the two competing plans into a final plan ready for immediate implementation by the end of September.[161] The Service d'Urbanisme would also set forth the guidelines imposed on private builders;

and the Service would call upon a dream team of European and Moroccan architects to design state-owned buildings in brutalist, modernist style.[162]

Conclusion and Foreshadowing

Decolonization was not coterminous with political independence, at least not everywhere and in all respects. After independence in 1956, thousands of French continued to live, die, and be buried in Agadir. The seismic intervention of February 1960 transformed this situation, however, prompting a mass exodus of the colonists. Then, in 1961, the French military base was evacuated, and the dead rose from their graves to return to France, although not without some assistance, to join the living who had fled after the earthquake.

The disaster created opportunities for the increasingly authoritarian Moroccan state to use American assistance to lessen Moroccan dependence on France. However, both the graves of the dead and the new city built for the living would continue to be centers of contestation in Agadir. The excavation of the Ville Nouvelle and the Talborj and the end of the "Battle of the Plans" did not mean the end of controversy regarding the reconstruction of the city or the treatment of the dead. For the generation of city residents who survived the earthquake, the ruins of the Kasbah have remained a festering wound, where the resting places of the unrecovered dead are despoiled by urinating tourists, beer-drinking Moroccan youth, and most recently, cellphone towers.[163] Meanwhile, to many Moroccans longing for the cultural and architectural decolonization of Morocco, the new Agadir that rose from the ruins became a symbol of enduring French cultural hegemony.

CHAPTER 6

The Soul of a City

A S PIERRE MAS, one of the principal designers of post-earthquake Agadir, described it: "Few cities occupy a geographic position as re- markable as Agadir. After passing Cap Ghir, where the foothills of the High Atlas plunge into the Atlantic, the voyager coming from the north travels along a narrow shore of Mediterranean character for forty-some kilome- ters, before discovering the large bay of Agadir open to the south-west. The last foothill, adorned with ancient Portuguese fortifications, the Kasbah, dominates the city and the plain of the South from its height of 230 meters."[1] Despite these picturesque natural endowments, a peculiar idea about the reconstructed city of Agadir began to be expressed in the mid-1960s, one that would be frequently repeated through the 1990s and into the new millennium. Rebuilt Agadir, with its modernist architecture centered around a Mediterranean-style beach resort, became a pervasive symbol of disorientation and rootlessness. In 1967, a French writer, Péré, reported that it was already commonplace to hear people lament of Agadir that "it is a city without a soul."[2] By the 1990s, when the present author was living in Casablanca, the description of Agadir as lacking a soul had become commonplace in popular discourse in Morocco, accompanied by the assertion that Agadir was not really Moroccan. Descriptions of Agadir as a "dead city, without a soul and without a center" appeared repeatedly in the work of students graduating from the National School of Architecture in Rabat. However, this cliché has been contested by the inhabitants of Agadir, the *Gadiris*.

Do cities have souls? Recently, scholars Daniel A. Bell and Avner de-Shalit published *The Spirit of Cities: Why the Identity of a City Matters in a Global Age,* arguing that some cities, but not all, have what the authors alternately refer to as "spirit," "ethos," or "identity." Prescriptively, Bell and de-Shalit argue that this spirit, as a focus of civic pride and activity, is a desirable bulwark against the negative effects of both nationalism and globalism. Descriptively, they apply what Frederick Cooper has called a "hard" definition of identity as a collective phenomenon, "something deep, basic, abiding, and foundational."[3] The hazard of such notions of identity is that they impose theoretical unity on what is in fact a heterogeneous and plural subject: communities of many thousands of people. As Bell and de-Shalit themselves acknowledge, the life of people in a city is shaped by many factors; they mention, in particular, economics, street signs, traffic, density, segregation, hospitals, taxi drivers, and "great city planners."[4] Such diverse forces cannot create a singular entity that one can call an identity, spirit, or *ethos*—or "soul"—for an entire city. However, although cities may not have spirits or *ethea* or identities or souls, ideas about particular cities do exist, and some of these ideas circulate widely and are influential.[5]

It is clear that, in Morocco, many people, both Moroccans and foreigners, have believed that cities should have souls but that post-earthquake Agadir does not. This is due both to the particular ways in which the seismic event of 1960 reshaped the built environment in Agadir and to the cultural history of colonialism in Morocco. In neither Orléansville nor Fréjus did the destruction wrought by the catastrophes of the decolonization era become so directly associated with an imagined annihilation of cultural heritage and identity. The enduring contours of disaster extend into many aspects of human history, but disasters (and decolonizations) do not all take the same shape. It was in Agadir that disaster most dramatically transformed the cultural and architectural shape of decolonization and of local and national debates about identity and the legacy of colonialism. In part, this was because the earthquake in Agadir destroyed the old, precolonial Saadian Kasbah (misidentified by Mas as Portuguese). Following the disaster, the Kasbah district, Agadir Oufella, would be left in ruins, while urban planners rebuilt the city according to the precepts of the mid-century modernism of Le Corbusier. However, the anxiety brought about by this dramatic change in the built environment was a product of the particularities of French colonial policy and ideology in Morocco and of the post-independence monarchy's shifting approach to discourses of modernity and tradition.

In contrast to Agadir, Fréjus's disaster and the architecture and urban design of the town's reconstruction have not been central to debates about the town's cultural identity. Today, Fréjus's museums and histories mourn the disaster, while celebrating the city's heritage of Roman and French military and colonial history. But the Malpasset disaster did not seem to separate Fréjus from its past: the floodwaters bypassed the old medieval city center, thanks to the slight elevation of the hill on which it stood. Though the ancient Roman arena was not spared, it was reconstructed, and remains a major site both for tourism and civic events. Anxiety about ethnic boundaries shaped the immediate response to the disaster, and the history of the 1959 dam collapse provides important clues to the deeper roots of Fréjus's later political turmoil (as discussed in chapter 3). However, the causes of civic anxiety in Fréjus in later decades have not been associated with the flood of 1959.

Nor did the 1954 earthquake in Algeria generate the sort of discourse of cultural anomie that developed concerning Agadir. The Chélif Valley disaster did not bring about so dramatic and distinctive a transformation of urban space in Orléansville, renamed El Asnam in 1962 and Chlef in 1980. In contrast to Agadir, the rebuilt Orléansville did not become a symbol of independent Algeria: it had been the French who had built Orléansville, and the French who rebuilt it, for the city's reconstruction was largely complete by the advent of independence in 1962.[6] As was the case elsewhere in Algeria, Orléansville's pre-earthquake architecture was already influenced by the "structural classicism" of Auguste Perret and to a lesser extent, Le Corbusier's functional modernism.[7] As Aleth Picard has explained, post-earthquake urban planners transformed the city less than they would have liked: the narrowness of the city's streets contrasted with the desire, among planners and architects, for more light and space, but in much of the city the existing street grid was partially maintained, largely because numerous buildings remained usable. Chief architect Jean Bossu envisioned an architecturally cohesive "red city," distinctive among Algerian cities, but due to the very rapid pace of reconstruction and the existence of neighborhood associations of property owners who hired their own architects, guidelines for reconstruction were not consistently followed. With the exception of individual districts and buildings designed by Bossu (the Saint-Réparatus Quarter) and by Jean de Maisonseul, an admirer of Le Corbusier, the city that resulted was unremarkable from an architectural standpoint.[8] In this respect, Orléansville was quite typical among Algerian cities: the Constantine Plan's emphasis on a massive expansion of affordable housing and heavy industry had made aesthetics a low priority, and reconstruction was dominated by the directives of administrators and engineers, a pattern that would continue under the Algerian state after independence.[9] Tragically, however, in the frenzy of construction the new seismic building codes enacted in 1954 were inconsistently applied, leading to the new catastrophe of 1980.[10] After the 1980 earthquake, there were recriminations about the state's failure to implement the anti-seismic building code developed in response to the 1954 quake, but not about the city's urban plan or architectural style. Although in the 1990s, architects in Algeria, like those in Morocco, sought to connect architecture with Arab and national identity, Orléansville did not become a focal point of such concerns.[11] Like cities, streets, and other places across Algeria, Orléansville got a new name after independence, but as an architectural representation of Algerian national identity, El Asnam after 1962 had neither more nor less to offer than did Orléansville before 1954. The inhabitants of the city mourned the loss of spaces in which they had lived, but Orléansville's built environment had contained little that would allow its destruction to be portrayed as a loss of Muslim Algerian national heritage.

In Morocco, it was not just the historical accident of the seismic destruction of Agadir's Kasbah that made the 1960 earthquake central to discourses about decolonization and Moroccan identity. Morocco had a particular colonial history of the idea of an *âme*, or soul, as a desirable characteristic of Moroccans, tied to culture and implying "something deep, basic, abiding, and foundational."[12] French administrators in protectorate Morocco, terrified that colonial subjects might demand the rights of Frenchmen, as Ferhat Abbas had initially done in Algeria or as Blaise Diagne had done in Senegal, despised earlier colonial policies promoting assimilationism. Consequently, French cultural policy in Morocco sought to define what Moroccan culture was, and to control and preserve that culture, lest dangerous French notions of individualism and democracy corrupt the Moroccan population and subvert the protectorate arrangement, under which the population was ruled by "traditional" Moroccan elites who had become vassals to the French. Moroccan culture, although studied by French colonial scholars in minute, pluralistic detail, was reduced by administrators to a unitary and homogenous Moroccan "soul" or "psyche." In the French schools of protectorate Morocco, the curriculum was intended to ensure that students understood the value of preserving their Moroccanness.[13] Meanwhile, as will be discussed below, French urban planners of the early protectorate era attempted to ensure that the growing French presence in Morocco did not corrupt the Moroccan character of the kingdom's cities. More recent discourses asserting that post-earthquake Agadir was "a city without a soul" must be understood in relation to this protectorate-era fetishization of precolonial Moroccan culture. The persistence of the idea that Agadir has no soul reflects the enduring legacy of French colonial policies and discourses linking culture, urban planning, and tradition.

The post-earthquake transformation of Agadir's built environment and the legacy of cultural policy in Morocco interacted to shape the local implications of international debates about urban planning, infusing technical discussions of urban development with political and cultural meaning. Criticisms of modernist urban planning and architecture are not unique to Morocco, of course. Worldwide such criticisms have been widespread and intense. One popular

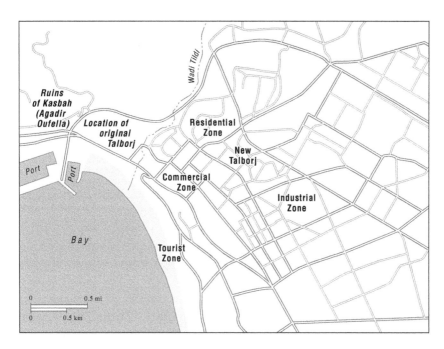

MAP 3. Reconstructed Agadir, ca. 2000. The reconstructed city grew beyond the zones envisioned by its planners. (Erin Greb Cartography.)

British writer, Theodore Dalrymple, has recently compared Le Corbusier, who was greatly admired by Agadir's urban planners, to Pol Pot, arguing, "Le Corbusier was to architecture what Pol Pot was to social reform. . . . Like Pol Pot, he wanted to start from Year Zero: before me, nothing; after me, everything."[14] With less hyperbole, the Belgian architect Jean Dethier argued in 1973 that the division of reconstructed Agadir into functional quarters "atomized" the urban environment, separating it into disconnected sections, with the result that the post-quake city was too spread out and insufficiently dense. According to Dethier, "This fragmentation [*éclatement*] of the modern city, established in all good faith in the name of hygiene, space, and circulation, annihilates in large measure the sentiment of the city, of community and animation."[15] Modernist urban planning principles emphasizing the importance of open spaces and the functional differentiation of city sections had created "a series of yawning, solemn spaces, and abstract and imperative zones."[16] In Dethier's view, these open spaces and functionally defined zones ("imperative" in the sense that they commandingly ordered city life) interfered with the activities of city residents. Such critiques were not unique to the realm of francophone urban planning. Both Dalrymple's denunciation of Le Corbusier and Dethier's more measured critique of the new Agadir resemble twentieth-century critiques of Robert Moses's urbanism in New York. Jane Jacobs, in opposition to Moses, argued that successful city life requires not dogmatic planning but a "jumping, joyous urban jumble" of mixed-use neighborhoods and spontaneous, organic growth, rooted in local history.[17] Dethier and Jacobs's visions of urban life contrasted sharply not only with Moses's work but also with the dominant principles in postwar French urban planning, which, as historian Paul Rabinow has noted, were based on "a total rejection of the organic city, which was composed, it was held, of unhealthy, inefficient, and uncontrollable accidents of history."[18] In Agadir, however, much of the "organic" city had been destroyed by another accident of history, setting the scene for a collision between the modernist desire to reshape urban life and a colonial legacy emphasizing architectural tradition as central to the preservation of Moroccan identity.

Like other developing cities, reconstructed Agadir presented many challenges to be addressed by urban planners and policymakers. Investment in tourism in later years, funded by Moroccans from outside of Agadir, focused on the speculative construction of hotels, not on the "animation" needed to appease the tourist's nagging hunger for Moroccan authenticity. As Thierry Nadau has argued, such animation was also rendered difficult by the fragmented, functionally divided layout of the city, which failed to provide a central street to draw Gadiris and tourists together for events.[19] Students at the National School of Architecture identified problems such as a lack of urban density, economic vitality, and activities for tourists, problems they hoped would be solved by a new generation of urban planners through technical means such as improvements in transportation. However, these students consistently framed such problems in terms of Agadir's alleged soullessness.[20]

Like so much else related to the disasters of the mid-twentieth century, criticisms of post-disaster Agadir were closely related to the process of decolonization. Moroccan independence in 1956 did not mean that the legacy of French colonialism had vanished or that French cultural or economic hegemony had evaporated overnight. The city's role as a vacation destination for European tourists grew, and for decades Agadir lost, to the nearby towns of Inezgane and Aït Melloul, much of the city's former role as a trade junction and depot for agricultural goods arriving from the interior.[21] The city remained economically dependent on Europe, and the most desirable spaces in Agadir became dominated by European visitors. Even commentators sympathetic to the new Agadir acknowledged that zoning according to the principle of functional division meant that a disproportionate share of the city's natural assets—access to the beach, views of the mountains—were "monopolized" by the tourist district and tended to produce a "fragmentation" that was not even unambiguously good for the tourism industry, since it tended to keep tourists "parked in their hotels," gazing at the sea rather than frequenting the town.[22]

More fundamentally, Agadir's landscape became a field of struggle over what a truly decolonized Moroccan city might be. Criticism of the new Agadir became intertwined with "an increasing search for national identity," and the reconstructed city's modernist architecture was frequently portrayed as European rather than Moroccan in character.[23] For Dethier, Agadir's urban plan constituted a form of cultural imperialism masquerading as the application of universal norms. At a conference in Agadir in 1994, scholar Mohamed Charef expanded on Dethier's denunciation of neo-imperialism in Agadir. For Charef, Agadir was "a city orphaned of its past and its memory, reconstructed by adopting the image of the Occident, in style as in organization." Charef argued that the consequences of a disregard for tradition and heritage produced not only a lack of urban vitality but psychological suffering, a direct consequence of the imposition of the vision of Agadir's urban planners, whom Charef depicted as alien: "The inhabitants find themselves with difficulty within this mechanistic conception; they feel lost, crushed, and would have certainly imagined a different city conforming to their culture, if one had asked their opinion."[24] Like Dethier, Charef connected Agadir's soullessness to its modernist use of space and to the crushing cultural violence of a neo-imperialist universalism. This was by no means a dissident perspective in the 1990s. Echoing Charef's metaphor, the director of the kingdom's state architectural service, Saïd Mouline, argued that neglect of architecture's connection to patrimony would "condemn citizens to become orphans, amnesiacs, excluded and under-developed."[25]

If we rephrase Bell and de-Shalit's terms simply as "the idea of a city," then clearly a strong idea about Agadir developed in the decades after the earthquake. If Paris, as they state, is defined as the "City of Romance," Jerusalem as the "City of Religion," and Montreal as the "City of Language(s)," then Agadir became known as the "City without a Soul." Bell and de-Shalit note that ideas about cities often develop in contrast to other cities (Jerusalem to Tel Aviv, Montreal to Toronto).[26] As historian Moshe Gershovich has pointed out, the critique of Agadir as soulless is predicated on a contrast with other Moroccan cities, such as Marrakesh and Fez, where the old city, or *medina*, has been preserved as a folkloric embodiment of Morocco's cultural heritage.[27] In comparison, one might note that Fréjus is no Paris—but no one expects it to be. Agadir, on the other hand, is implicitly faulted for not being Marrakesh, or at least Essaouira.

In Agadir, the discourse of "city without a soul" has not been universally accepted, but it has been impossible to ignore. Professor Mohamed Ben Attou, a geographer at the University Ibn Zohr in Agadir, has found it necessary to argue that today's Agadir is neither a straightforward manifestation of the vision of the city planners, "nor a city without a soul," but is developing, as all cities do, as a response to economics, demography, and the dynamic interactions between actors in the urban environment.[28] In an interview published in 2011 in the Moroccan newspaper *Libération,* M'bark Chbani asked the following of Agadir city councilman Mohammed Bajalat: "Some say that Agadir is a city without a history, without a soul: do you agree?" Bajalat, president and founding member of Forum Izorane, an organization devoted to promoting civic memory and civic pride in Agadir, responded unequivocally. Bajalat answered that the "without a soul" trope

> is revolting . . . above all, in the obstinate desire to transpose the model of the imperial cities [Rabat, Marrakesh, Fez] to Agadir. Finally, by what logic can we reduce a collective past to buildings? Certainly, the earthquake destroyed many of the buildings and their occupants, but not the memory of the city.[29]

For both Bajalat and Ben Attou, the idea of Agadir's soullessness is a misleading myth that needs to be countered. For Ben Attou, this is to be accomplished not only by means of academic rigor in the study of the actual city but through "the memory of each of its citizens" and through "a considerable effort to be deployed in order to share this collective memory."[30] Toward this end, Bajalat has taken a leading role in organizing his fellow Gadiris to preserve the memories of pre-earthquake Agadir through commemoration, while also celebrating Agadir's modernity. This approach has been endorsed by Ahmed Bouskous, an earthquake survivor who had become the rector of the Royal Institute of Amazigh [Berber] Culture. For Bouskous, the "soul of a city" was a work in progress: "To give a city a soul is the responsibility of local decision makers, and of the Gadiri population." At the same time, however, the task, for Bouskous, was to preserve the memory and heritage of Agadir through "culture, song, poetry, cinema, theater, visual arts" and especially through Amazigh culture.[31] Bouskous's approach to the question was not unlike the argument made against the "without a soul" trope by Péré in 1967: Agadir's soul, "whatever the shape of its walls," is rooted in "its location in the far south of Morocco, in its climate,

and its people, essentially *Chleuh* [Tashelhit-speaking Amazigh]."[32] Bouskous
and Péré would perhaps agree with Amazigh activists for whom the statement
that Agadir has no soul is tantamount to a denial that Berbers have culture. For
advocates of Amazigh culture, there is an alternate idea of Agadir: "Capital of
the Berber South."[33]

Tariq Kabbage (introduced in chapter 5), who became the mayor of Agadir,
has taken a different approach. "What is this soul of a city?" asks Kabbage; "We
could philosophize about this until tomorrow morning." The more pertinent
questions, for Kabbage, were whether the inhabitants of the city felt comfort-
able in relation to the place where they live, and "whether this brings them a
certain amount of pleasure, of joy." "You know," argues Kabbage, "when you
lead a life of suffering, soul or no soul, that is not the question."[34] Kabbage had
little interest in the question of whether his city had a soul; his city had people,
and it was their well-being that he cared about. Yet, as Bajalat and Bouskous
recognize, a sense of civic memory and attachment to a positive idea of a city
(Bell and de-Shalit's "civicism") helps to promote the sense of well-being desired
by Kabbage. Conversely, the widespread claim that Agadir is a "city without a
soul," if not demonstrably deleterious to urban life, has at least been a source of
anxiety for some inhabitants of the city.

Agadir before 1960

In the sixteenth and seventeenth centuries, Agadir had been a strategic fortress
and trading post of importance to the Portuguese, Saadians, and Dutch and an
outlet for the caravan trade from the Sahara and the sugar production of the
Souss Valley. The Portuguese had built an outpost, the fortress of Santa Cruz,
near the beach in an area later called Founti. After defeating the Portuguese and
destroying Santa Cruz in 1451, the Saadians constructed a larger fortress on Agadir Oufella, the mountain overlooking the bay. The Saadians also constructed
a port, and Agadir became a vital link in the sugar trade of southern Morocco.[35]
By the eighteenth century, however, the city's fortunes had declined. Agadir was
struck by a severe earthquake in 1731, with reportedly total destruction, but soon
recovered. As in the Saadian period, Agadir remained a key connection point
between the southern caravan routes, the imperial capital at Marrakesh, and
the European trade—and for this reason, was fought over in local power struggles. This changed, however, when Alaouite Sultan Mohammed ibn Abd-Allah
opened a new southern port at Mogador (Essaouira) in 1774. Finding the Souss
Valley's elites rebellious and Agadir dangerously far from Marrakesh, the sultan

closed the port of Agadir. According to some accounts, the punishment of the rebellious Souss was a major reason for the construction of Mogador in the first place.[36]

By the dawn of the twentieth century, Agadir was little more than a fishing village.[37] Then, however, Western interests began to extend southward into the Souss Valley. The revelation that the Souss region contained iron ore made Agadir a place of interest to Europeans for the first time since 1774. While the struggle for the Sultan's throne in Fez was at the center of international conflict over Morocco in 1911, competition among French and German prospectors in the south provided the pretext for the arrival of the German gunboat Panther. After German objections to a French takeover of Morocco were alleviated by the cession of a sliver of French Equatorial Africa to Germany, the French arrived in Agadir in force.[38]

French-ruled Agadir, unlike Moroccan cities such as Rabat, Casablanca, or Fez, exhibited only a superficial imprint of the French colonial philosophy of the 1910s and 1920s promoted by Morocco's first French resident-general, Hubert Lyautey, and his chief urban planner, Henri Prost. As a young officer stationed in Algeria, Lyautey had fetishized Arab culture and disdained the French impact on Algerian society. He was disgusted by the French-built cities and towns he encountered there, with their rationalist regularity and lack of any noticeably Arab or African character, beyond "shoddy goods and pastiche."[39] To this root-lessness, Lyautey contrasted the harmony of Mediterranean cities such as Rome and Naples, with architecture "well adapted to the local climate and mentality."[40] As resident-general of Morocco, Lyautey hoped that urban planning would be a remedy to the two things he hated most: French republican universalism on the one hand and cultural hybridization on the other.[41] The result was the creation of new European districts separated from the Moroccan city centers, or medinas, by greenspaces, or *cordons sanitaires*, to minimize cultural contamination. Lyautey instructed his underlings to "touch the indigenous cities as little as possible. . . . Instead, improve their surroundings where, on the vast terrain that is still free, the European city rises, following a plan that realizes the most modern conceptions of large boulevards, water and electrical supplies, squares and gardens, buses and tramways, and also foresees future extensions."[42] For Janet Abu-Lughod, Lyautist urbanism amounted to a system of "cultural and religious apartheid" based on "minimum alteration in the Moroccan quarters . . . the creation of a cordon sanitaire around these native reservations with a greenbelt of open land; and the design and construction of the most modern, efficient, elegant cities that Europe could produce."[43] Although his effort to prevent the mingling of people and cultures was unsuccessful, Lyautey's vision had a profound impact on the development of Moroccan cities such as Rabat, Marrakesh, and Fez.[44]

Far to the south, however, Agadir's growth into a medium-sized city did not begin until the late 1920s. By this time, the influence of Lyautey and Prost was waning in a new, settler-dominated Morocco. As commerce grew at the new French-built port, the mainly Tashlehit-speaking Moroccan population almost tripled, from an estimated 700 to approximately 2,000 in 1930, while the European population grew to 1,650.[45] Rampant land speculation led to the declaration of an official urban development plan in 1932.[46] The 1932 plan, in Lyautist fashion, called for a new European "Ville Nouvelle," separated spatially from the two historic Moroccan quarters: the towering heights of the Kasbah, and the fishing hamlet Founti adjacent to the beach below. The slopes of the Kasbah provided a sort of natural *cordon sanitaire*, as did two riverbeds east of the Kasbah: the Wadi Tildi, which separated the Talborj and administrative plateaus from the Ville Nouvelle, and the Wadi Tanaout, separating the Ville Nouvelle from the industrial quarter.[47]

On the Talborj plateau, however, geography and events were already producing a spatially separated commercial-residential center which attracted both Europeans and Moroccans. This district, not the Ville Nouvelle, became the heart of the city. As the Moroccan population had grown in the overcrowded Kasbah and Founti, which could not expand due to the steepness of the slope abutting the Kasbah, a new district was constructed, the Talborj. As this became the center of commerce for the city, the Moroccan inhabitants—Berber-and Arabic-speaking Muslims and Jews—were soon joined by Europeans. In cities such as Rabat and Casablanca, residential segregation eventually broke down as affluent Moroccans moved into the Ville Nouvelle, while drought and colonial land policy emptied the rural areas into new peripheral neighborhoods beyond the initial dyad of old medina and Ville Nouvelle. In Agadir, in contrast, it was the Ville Nouvelle that became peripheral, while the ethnically mixed Talborj became the center of urban life.[48]

After 1945, a new commercial boom occurred, based on the export of citrus, canned fish, and minerals. During the postwar economic recovery, construction blossomed in both the Talbordj and the Ville Nouvelle. By the early 1950s, the total population grew to around forty thousand, including close to fifteen thousand Europeans. The tourist industry also began to develop, as new hotels were constructed and the International Federation of Travel Agencies promoted Agadir as the "Moroccan Nice," "Pearl of the South," and "city of three hundred days of sunshine."[49] Boom, however, was followed by bust. Crises in agriculture and in the cannery business between 1955 and 1958 converged with political crisis, as Moroccan independence provoked an exodus of Europeans. In Agadir, the European population dropped to 4,700 by 1959. Only the small tourist industry seemed to be thriving: the city's two hundred first-class rooms and sixty second-class rooms were, reportedly, fully booked when the earthquake struck.[50] Over-construction of both buildings and roads gave observers the sense of a half-empty city: "One sees there a network of roads, often unnecessary, delimiting numerous vacant lots, interspersed with a small number of buildings."[51] To Pierre Mas, planner of the new Agadir, pre-earthquake Agadir was "inorganic, dissolute, a city with neither a center nor coherence."[52] This critique would have discursive staying power, and would be echoed in the critiques of the new, post-earthquake Agadir as well.

The postwar years had seen the rise of new approach to the use of urban planning to shape society. After Prost had departed Morocco in 1923, and Lyautey in 1925, the interests of speculators and settlers had weakened the role of statist urban planning throughout Morocco. However, it was the rise and fall of Vichy that thoroughly discredited Lyautey's culturalism. In urban planning as in colonial education, French colonial policy returned to the universalism that Lyautey had rejected. In 1944, the Office of European Habitat took on the task of housing the Moroccan population and dropped the word "European" from its name. Two years later, in 1946, Michel Ecochard was appointed to head urban planning in Morocco; Ecochard created the Service of Urbanism in 1949, which was placed within the department of Public Works. Ecochard's modernism, modeled on the principles of Le Corbusier, signaled a sharp break from Lyautey's approach.[53] Urban planning thus became divorced from the study of particular cultures. Rabinow describes Ecochard's approach as the "neglect, which bordered on contempt, of economic and political considerations" and as a "refusal to acknowledge local practices." Under Ecochard, the protectorate undertook a massive but belated effort to cope with the demographic growth of Morocco's urban populations, striving to offer the *trâme Ecochard* to the masses: a sixteen-by-eight-meter living space endowed with access to light, air, and space. "Culture" was no longer part of the equation.[54]

The Impact of the Earthquake

It has often been said that there is no such thing as a natural disaster. Despite the extent of the destruction in Agadir, the earthquake had an estimated magnitude of only 5.75 on the Richter scale, as measured on seismographs in Casablanca and in Europe. Seismologists blamed the high death toll on the location of the epicenter near the earth's surface and near population centers. Engineers blamed the prevalence of unreinforced masonry and the use of improper techniques in constructing buildings of steel-reinforced concrete.[55]

However, the growth of Agadir since the 1930s and especially during the post-war economic boom also amplified the destruction and lethality of the earth-quake, as the Kasbah population grew and, in the postwar Talborj, buildings were hastily expanded upward with additions of second and third stories made of un-reinforced concrete.[56] The lethality was not evenly distributed; the much higher survival rate of the European population was directly related to their economic domination in modern Morocco, which allowed many to live in the more expen-sive Ville Nouvelle, where a third of the buildings withstood the quake in repa-rable condition, while the Kasbah and Talborj were almost entirely destroyed. It should be noted, however, that Europeans residing in the devastated Talborj (ad-jacent to the "administrative plateau"), fared worse than Moroccan workers living in the eastern industrial quarter, which avoided much damage, due to greater distance from the epicenter, less multistory housing, and many corrugated-metal buildings.[57] Had the Lyautist model prevailed in Agadir, the discrepancy between European and Moroccan survival rates would have been greater.

After the earthquake, there was a powerful modernist consensus about the goals for reconstruction among those elites who were able to give public voice to their visions. Foreign seismologists and engineers advised that a new, bet-ter Agadir should be built of steel-reinforced concrete in the area occupied by the Ville Nouvelle and the eastern industrial district. As Daniel Williford has pointed out, this meant closing the book on traditional Moroccan architecture, on affordable, low-cost construction methods, and on the entire sections of the old city where the Kasbah, Talborj, and Founti had once housed the majority of the Moroccan population.[58] King Mohammed V endorsed this vision and sought an ambitious urban plan to create a new city that would be an "expres-sion of modern Morocco."[59] Al-Istiqlal called for the construction of "not only a new Agadir, but also, and above all, a new Morocco."[60] Rebuilding Agadir was not just about housing the survivors, mitigating risk, or restoring the port as an outlet for the agricultural produce of the Moroccan south: the city's recovery was to be a model for the future of the nation as a whole. One of the earliest enun-ciations of the idea that the Agadir disaster had created a unique opportunity (a common response to modern earthquakes) was found in a report by a West German technical assistance team, which concluded with the declaration that the unique possibility offered by the reconstruction of the new Agadir should be fully utilized. . . . Decisions concerning the reconstruction of the city, and the plans, should of course be governed by the general welfare of the city, without any consideration for certain private interests. This is the only way to build a new modern Agadir. Certain mistakes made in the past could be avoided, and the city could become an example of a modern progressive Morocco.[61]

The disregard for private interests embodied in this transnational modernist response provided an opportunity for a Moroccan monarchy interested in consolidating its power over the country. This was not at all unprecedented: ambitious urban planning had long been linked to authoritarian rule, and disasters have often provided opportunities for authoritarian modernism. The destruction and reconstruction of Lisbon in the eighteenth century had provided the opportunity for the rise of Carvalho's absolutism in Portugal. In 1830s France, cholera epidemics had spurred some intellectuals to advocate "the equivalent of a technician's coup d'état, arguing that only a planned and hierarchically coordinated effort was adequate to the crisis. Engineers could save France, but only if far-reaching changes in private property were undertaken."[62] In contrast, grand urban schemes after the London Fire of 1666 and the San Francisco earthquake of 1906 had been stymied by the assertions of property rights by the bourgeoisie.[63] Hassan would not allow this to happen in Agadir, and initiated a vast project of property expropriation and state regulation of reconstruction.

For Hassan, the architecture and urban design of Agadir was "the expression, in stone and in space, of the aspirations of the national macrocosm."[64] Just as Mohammed V in 1960 had made the monarchy the center of Moroccan humanitarian responses to the earthquake, in 1966, King Hassan II linked the modernist reconstruction of Agadir to the unity of the Moroccan nation and the nation's embodiment in the person of the king. As Hassan declared, the goal of reconstruction was

> not to simply restore the old, the replaceable, but to make a new work, alive, essentially opening on the future; to give men back reasons to live and to hope, it is necessary that these reasons merit the confidence of the dispossessed and that they are thus guaranties, sanctioned by the King and by the People as a whole, in short, that the reconstruction of Agadir be conceived as a work [that is] above all, national.
>
> And in fact Agadir was constantly for the entire country the site of a magnificent élan of solidarity, of abnegation, of union. All the nation felt

involved, challenged, all the nation, under the firm and lucid guidance, the example of our regretted father, His Majesty Mohammed V, then with We Ourselves, mobilized its means, its intelligence, its heart.[65]

The young king's embrace of modern planners' ability to build a new future without regard for the past seemed to be absolute, as he described the new Agadir as "a total city, virtually a dream city, rethought in entirety, remade by man for man, by the Moroccan for the Moroccans and Morocco." This vision of the new city was tied to a forward-looking vision for the nation as a whole: "The reconstruction of Agadir becomes as the symbol and the concrete projection of what the country wants to be, faced with any problem in the national life: the deliberate and total union of all for a better life for all and for each."[66] In this vision, the choices of individuals counted for little; the unity of the whole, under the authority of the king, counted for everything.

This approach was enthusiastically endorsed by planners such as Mourad Ben Embarek at the Moroccan Service of Urbanism. Morocco's urbanists thus joined a long line of planners, from Carvalho's chief engineer Manuel de Maia to Lyautey's Henri Prost, who found their work made easier by an autocratic state that removed the obstacle of local community resistance to a central vision.[67] This symbiosis was evident in Agadir. The link between state power and city planning was made explicit by the editors of the Moroccan architectural journal *A + U: Revue africaine d'architecture et d'urbanisme* who declared, reflecting on the reconstruction of Agadir, that "more and more, urbanism should affirm itself as a means of governing."[68] For Mourad Ben Embarek, urban design and state control went hand-in-hand in a tourist city: "user comfort" was paramount, views of the sky and the sea had to be preserved, and "commercial and speculative considerations cannot and should not affect this concept." Unregulated building would lead to overly dense construction, ruining the city's aesthetic potential and creating "regrettable chaos."[69] For Ben Embarek, one only needed to look north across the Mediterranean to Spain to see a coastline that had been "ravaged" by a lack of regulation.[70] For the monarchy, Agadir was important because of its historic role as a crucial outpost for the assertion of northern Moroccan power over the south, a role it would reprise in the 1975 Green March. After the earthquake, however, it also served as an opportune laboratory for the assertion of royal power.

And autocracy could be efficient. Even Jean Dethier, who condemned the authoritarianism of the planning process, was compelled to acknowledge that in its efficiency, Agadir was "an extraordinary success.... Regarding the financial level…

it was a tour de force. On the technical level also: 5 years after the earthquake, the new city was more than 75% constructed."[71] In lieu of a voluntary fundraising drive like those that had funded disaster relief in Orléansville and Fréjus, the state imposed a mandatory National Solidarity tax. To prevent uncontrolled reconstruction both inside and outside of the zone determined to be safe for reconstruction (mostly the area of the old Ville Nouvelle and industrial zone), the state expropriated as many as one thousand parcels of private property, covering four hundred hectares. In compensation, property owners were allowed to choose lots of equivalent size defined in the new urban plan. While the government provided grants (up to 50 percent) and loans subsidizing reconstruction costs, property owners had to submit detailed plans to the office of the High Commissariat for Reconstruction. Once an edifice was completed, the High Commissariat for Reconstruction also had to give approval before the new building could be inhabited. These measures aimed to ensure that both seismic and architectural standards were met.[72] Daniel Williford notes that the losers in the expropriation process included poorer Gadiris who lacked legal title to their homes.[73] However, the process also excluded the land speculators who had purchased land in the Ville Nouvelle during the postwar period and who had not built on their property. These absentee owners of empty lots, largely French, were ineligible for State subsidies for reconstruction, and properties considered abandoned were confiscated. The French consulate protested initially, but then relented.[74] The earthquake thus permitted another significant step in the process of decolonization, with the redistribution of French-owned land to Moroccans, under the firm control of the Moroccan state.

The centralized power of the Moroccan state in 1960 was, however, a legacy of colonial authoritarianism. Laws enacted in 1914 under French direction had precisely regulated not only the "the width of streets, the alignment of buildings, the height and construction of buildings," but also architectural style.[75] Titles to land dispensed by the state in Meknes, Fez, and Marrakesh came with the condition that construction ensue according to the urban plan. As in post-earthquake Agadir, there were to be no lots left vacant by speculators. The new protectorate in 1914 had also pioneered legislation "permitting expropriation by zones," rather than by specific lot or building, with zoning based on function.[76] Like the colonial state under Lyautey, under which all proposals for new construction in the Moroccan medinas were regulated by the Service des Antiquités, Beaux-Arts, et Monuments Historiques, and in the European districts by the Service d'Architecture et des Plans des Villes, the Moroccan state ensured that private as well as

public construction would accord with the official vision of the state's urban planners and architects.[77] Now, however, that vision was something quite different.

Designing the New Agadir

Due to efforts to "Moroccanize" the newly independent Moroccan state, the
Service of Urbanism was directed by Abdesalem Faraoui until 1961, and then
by Mourad Ben Embarek.[78] According to historian Thierry Nadau, who interviewed the principal planners and architects, however, there was no discernible
dichotomy between the service's French and Moroccan professionals in terms
of their approach to urbanism. Faroui and Ben Embarek were "little influenced
by traditional architecture, [and were] even hostile to the medinas in which
they had grown up."[79] Having received their professional training in postwar
France,[80] they had imbibed little of the Lyautist anti-assimilationism that had
been promoted in the pre-1945 schools of the protectorate and that had been
embraced by much nationalist discourse. Under the leadership of Faraoui and
Ben Embarek, the core of the team consisted of the urban planner Pierre Mas
and the landscape architect Jean Challet, who would become the primary designers of the new Agadir; together, they would lead a group of European and
Moroccan architects to design state-owned buildings in modernist style, and to
set the guidelines imposed on private builders.[81]

The new Agadir, as designed by these French and Moroccan urbanists, reflected the prevailing modernist ideas of the postwar era, ideas which diverged
from the principles of cultural preservation and segregation that had dominated
urbanism and architecture in Morocco under Lyautey and Prost, in favor of the
functionalist, universalist modernism of Ecochard and Le Corbusier.[82] Mas
and Challet aimed to preserve the city's natural assets—most notably sunlight
and the bay—but had little interest in preserving the Agadir of the past.[83] They
focused on adapting their designs to the natural environment rather than to
Moroccan culture; architecture and urbanism were viewed in reference to the
relation between universal man (an idea Lyautey had despised) and nature.[84] Architects designing individual buildings such as the new modernist city hall drew
loose inspiration from the architectural traditions of southern Morocco, but this
Moroccan-inspired modernism was a far cry from Lyautey's efforts to preserve
the traditional medina. Moreover, the urban planners, according to Mas, aimed
to "link the quarters by means of constructed elements, creating a sense of urban
unity and avoiding all social segregation."[85] This was the antithesis of Lyautism.

The new shape of the city was conditioned by tectonic as well as ideological shifts. The earthquake had destroyed the "traditional" Kasbah. It would not be rebuilt, both because seismologists had advised against rebuilding northwest of the Wadi Tildi, and because Lyautey's fetishization of Moroccan tradition was no longer in vogue among Francophone urbanists. It could be argued that, by 1960, Lyautism had been rendered irrelevant by structural and demographic changes in Moroccan cities in general and earthquake-ravaged Agadir in particular. The disaster had greatly accelerated the shrinking of the European population, a process begun by political independence and economic crisis. Tectonics had destroyed the old city; demography meant that the new city was intended for Moroccans. What place was there for the Lyautey legacy of cultural separation and modernism-for-Europeans if the old medina was gone and the Europeans were leaving? Yet demography and tectonics did not in themselves determine Agadir's fate: as Rabinow has noted, it was the culture of postwar urbanism that led the new city's designers to treat Agadir's residents as cultureless universal inhabitants of a theoretical modern world.[86]

Seismic considerations tempered the ambitions of the modernists: unlike much of housing development in Morocco since 1947, there would be no high-rises.[87] In other respects, however, the planners undertook to reshape the natural environment. A new urban unity, hitherto made impossible by geography, was to be achieved by eliminating the division created by the ravine of the Wadi Tanaout: the ravine was filled in with debris from collapsed buildings, and an aqueduct was constructed with reinforced concrete to handle the water flow. According to Mas, "This operation permitted the unification of the site of the new city, making disappear a geological accident troublesome for its development."[88] The Wadi Tildi became the new western boundary of the reconstructed city; beyond lay the bulldozed wasteland of the old Talborj and the ruins of the Kasbah. As in Bartholomew's American plan (discussed in chapter 5), a tourist district east of the port directly abutted the beach. This tourist area's hotels would largely house Europeans, but no *cordon sanitaire* would divide it from the Moroccan city. Instead, as Bartholomew had proposed, it would be immediately adjacent to the city center's commercial-administrative district, just inland to the north; a pedestrian walkway over the filled-in Tanaout ravine was to facilitate movement between the functionally distinct zones.[89]

Another aspect of the Service de l'Urbanisme's final plan was the idea of creating a "new Talborj," which, according to Mas, "posed the most delicate problems."[90] The forty-five hectare quarter was to house ten thousand to twelve thousand people, and to be the site of "traditional commerce." It would be served by two schools, a market, parks, sports fields, and a cinema. Here, in Mas's words, he and his fellow planners attempted "to recreate, within islands, by means of a network of narrow pedestrian paths, the ambiance and scale of traditional medinas."[91] Each lot had access to a road for motor vehicles as well as medina-style footpaths.[92] The idea of a modernist medina provided a solution to a practical problem. Because of the high population density of the old Talborj, each household could claim only a small indemnity from the state for their property loss, although a minimum compensation level was set at 6,000 dirhams per household to allow minimum standards to be met. Consequently, reconstruction for these families had to be extremely modest. The New Talborj was designed to bring the population of this vital commercial district back together on a scale they could afford and which would fit the designers' conceptions of urban order.[93] But there would be no *cordon sanitaire* here, either: the Talborj was immediately adjacent to the "modern" commercial and administrative sector.

This was not Lyautey's vision of the Moroccan city. In the New Talborj, there was a faint echo of the "neo-traditional" design that characterized the new Habous neighborhoods constructed in Casablanca, which had attempted to replicate the "organic image of the traditional media" but with automobile access and electrical and water infrastructure. In the Lyautist Habous, however, "all symmetry and geometricism were banned."[94] Agadir's medina-islands, in contrast, were separated from each other by a regular pattern of main roads, and bore greater resemblance to the postwar construction projects in Casablanca's Aïn Chock and Mohammedia's new medina, with their "much less literal interpretation" of the traditional medina, and the obvious "modernist influence of cubism and Bauhaus."[95] The designers of the new Talborj and of public buildings such as the new town hall may have drawn on Moroccan precedents for ideas, but this was not Lyautey's cultural preservationism; it was Corbuserian modernist planning with some local inspiration.[96]

Over time, the tourist district grew beyond its intended boundaries, driven by European demand and Moroccan investment. As Thierry Nadau has argued, the growth of the tourist sector engulfed what the planners had envisioned as the commercial center of the city, which became an area of hotels, restaurants, and shops for tourists. Moroccans shopped elsewhere, and increasing lived elsewhere, too. State control of construction prevented an increase in population density in the planned city, and the new seismic codes made officially sanctioned housing more expensive. Consequently, non-tourist commerce shifted to the southeast, pulled by the growth of residential construction beyond what had been originally conceived as the industrial zone.[97] Consequently, there was no true city center, and the New Talborj became just one neighborhood among many, never attaining the central role in city life played by its predecessor, west of the Wadi Tildi.[98]

A Return to Lyautey

Lyautism was not dead, however. In many respects, segments of Moroccan nationalism had long embraced and adapted the Lyautey legacy. The authors of the 1934 Plan de Reforms, arguably the first public articulation of protectorate-era Moroccan nationalism, had called for a renaissance of Lyautey's principles, explicitly favoring cultural dualism in education, while denouncing French policies of "two weights and two measures" in the allocation of resources.[99] Yet the nationalist embrace of Lyautey's culturalism in their denunciations of assimilation had no immediate impact on the policies or urban plans of the Moroccan and French architects and planners at the Service d'urbanisme, who "continued to follow the principles of Ecochard."[100] The resulting contrast between Lyautist-nationalist culturalism and the modernist universalism of Morocco's city planners lay at the root of emerging critiques of the new Agadir as a "city without a soul."

Beginning in the 1970s, the desire to affirm a culturally Moroccan approach to architecture and to reject Europeanization was expressed by European commentators as well as Moroccans and crossed political boundaries of the Left and Right. Dethier's 1973 critique of Agadir went beyond the notion that functionalist divisions of city districts disrupted "community and animation"[101] and defined the more fundamental problem as one of neocolonialism. Dethier argued that urban planners had imposed a Western vision of cities, and he argued for a new urbanism that would "permit the abolition of systems of mental, economic, and technical dependence on the rich countries, and favor the development of new authentic cultures in the Third World."[102] For Dethier, modernist urban planners, however well-intentioned, practiced "a new paternalism, oppressive and constraining."[103] Dethier's argument was paralleled by Abderrafih Lahbabi, writing in the Moroccan journal *Lamalif*. Citing Dethier and recognizing that decolonization was only partial, Lahbabi applied a Gramscian analysis to the problem. Lahbabi's hope was that "a new language should gradually replace the deterioration of the dominant symbolic hierarchy." This, in his view, should be the goal of architecture in Morocco. Believing that the working class needed to ally with other anti-imperialist groups, Lahbabi argued that liberation required not only social and economic emancipation, but "cultural identification."

Consequently, Lahbabi denounced the abstract humanism of the Corbusier school. Moroccan architects needed to engage in "the search for a national architectural identity" as a necessary step in the class struggle.[104]

On the opposite end of the Moroccan political spectrum, King Hassan II gave a speech in Marrakesh in 1986 addressed to architects that also called for a connection between architecture and national identity. This speech signaled an abrupt departure from the modernist ideas of architecture and city planning that Hassan had supported during the reconstruction of Agadir. The 1986 speech had two main elements: the first promoted the notion that architecture in Morocco should be tied to the maintenance of tradition and Moroccan cultural identity; the second established the monarchy as the guardian of cultural authenticity in the kingdom. Hassan, who had been a driving force behind the reconstruction of Agadir, now denounced Moroccan cities that were not recognizably Moroccan. Agadir, its ancient kasbah now nothing more than a field of ruins atop a hill, clearly no longer fit Hassan's vision of what a Moroccan city should be. Without mentioning Agadir, the king lamented that there were cities in Morocco that, if one viewed them from a helicopter, would not even be identifiable as Moroccan. Hassan contrasted such cities ("McCities" one might call them) with cities, such as Azzemour, whose historical ramparts and kasbahs identified them as unmistakably Moroccan. The king declared that new architecture in the kingdom should also "reaffirm our authenticity" and "preserve the characteristics of our country. . . . We must not renounce our mother, the land where we were born and where we live."[105] Architects, to prepare for this task, were advised to visit the kasbahs of the Moroccan south and the Atlas Mountains. To ensure that his new vision of Moroccan architecture should become a reality, Hassan proposed regulatory oversight of architectural plans for all new construction in the kingdom. Thus, Hassan extended to the entire kingdom the royal influence over architectural culture that he had exerted during the reconstruction of Agadir—but this royal influence was now directed toward very different ends.

Jennifer Roberson has argued that Hassan II's transition from his support of Corbusierian modernism to this emphasis on national cultural authenticity can be traced to his traditionalist choices in the design of his father's mausoleum in 1961, which grew into an effort to promote the "revival" of Moroccan traditional crafts skills in the 1970s. Roberson notes that by attempting to define and preserve selected aspects of Moroccan tradition, Hassan was following in the footsteps of Lyautey.[106] However, promoting tradition was also part of a broader project of justifying authoritarian rule: the Alaouite monarch's authority had to be rooted in respect for the past, as a hedge against revolutionary demands for democracy. The monarchy's change of position on architecture can be seen as part of its broader promotion of Islamic and traditionalist notions of Moroccan identity in response to political threats from the Left embodied in the 1965 student riots in Casablanca.[107] In the field of architecture, however, Hassan's new approach harmonized with the anti-colonial Left's call to challenge imperialist hegemony through an architecture of cultural identity. A new consensus was emerging that would reinforce the discourse of Agadir's soullessness.

Calls from leftist intellectuals and the Moroccan king urging architects to embrace a connection with the past and with national identity were accompanied by a trend in the architectural choices of wealthy Moroccans, who increasingly incorporated traditionalist elements, or "green tile" architecture, in new construction. The results received mixed reviews. Like Lyautey in prewar Algiers, Lahbabi and others found the results to be inauthentic "pastiche" rather than a true expression of Moroccan culture.[108] As Thierry Nadau put it in 1992, "The new buildings have nothing to do with the Moroccan. They are the palaces of a Thousand and One Nights."[109] It was not only in Agadir, apparently, where architecture failed to fulfill the dreams of those who hoped to capture the essence of the Moroccan "soul."

In the early 1990s, in the last years of his life, Hassan II sought a solution to this problem through monumental architecture. The construction of a towering new traditionalist mosque now meant that even Casablanca could pass his "helicopter test": the city's skyline became unmistakably Moroccan. Marrakesh had the Koutoubia Mosque and Fez had the Kairaouine; now Casablanca had the Hassan II Mosque. The giant new mosque in Casablanca was, however, juxtaposed in the skyline to a pair of monolithic, modernist commercial skyscrapers in the commercial district of the Maârif. Lyautey's cultural dualism lived on in the policies of the monarchy. Morocco could be both modern and traditional, but the two remained stylistically and spatially distinct; "pastiche" and hybridization were avoided.[110]

Agadir thus became an anomaly, at least among Morocco's larger and better-known cities. The planners and architects of the new Agadir, with their focus on functionalism, had rejected monumentalism.[111] There were no towering buildings to dominate the urban space: no clock tower, no royal palace, no grand mosque. Only Agadir Oufella, a vacant, vast sepulcher, stood to memorialize the past. Were it not for the inscription "God, Country, King" emblazoned on the side of the mountain to fill the need for imperial grandeur, Agadir could not pass the king's helicopter test. Consequently, the "without a soul" trope that had originated in the 1960s continued to circulate.

While this trope was distressing to some residents of the city, it did not deter Agadir's economic or demographic growth. The population of the city rebounded, rising from less than seventeen thousand after the earthquake to over sixty-one thousand in 1971.[112] By 2004 it had more than quintupled, to over 346,000. Architecture aside, this was an unquestionably Moroccan city, including just 1,925 foreigners, barely 0.5 percent. Soul or no soul, in strictly demographic terms Agadir has been more thoroughly decolonized than a number of other Moroccan cities, largely due to the effects of the earthquake.[113]

Conclusion

Unquestionably, the discourse of Agadir as a "city without a soul" would not exist if the earthquake had not destroyed the Kasbah, if tourists from Europe and vacationers from Casablanca were able to combine their beach holidays with shopping trips in a densely populated and "authentic" Moroccan fortress. In neither Fréjus nor Orléansville did disaster so greatly transform the symbolism of the architectural landscape as the 1960 earthquake did in Agadir. In Morocco, however, the lament for Agadir's soul was not just a product of the destruction of precolonial edifices; it was also the product of a colonial idea that emphasized the importance of preserving precolonial cityscapes. While Algeria's national identity was connected to the idea of a revolution, breaking from the past, the Moroccan monarchy of Hassan II, like Lyautey's colonial state, emphasized the preservation of tradition.

In Agadir, the disaster prompted an exodus of the European population and provided an opportunity for the Moroccan monarchy to assert its authority and to use American aid to lessen Moroccan dependence on France. As the victor in the "Battle of the Plans," however, France salvaged its role as Morocco's provider of technical assistance in the field of urbanism. Consequently, the destruction of Agadir permitted Morocco's French and French-educated urban planners to apply their Corbusierian ideas of universalist modernism on the scale of an entire city, untainted by the legacy of Lyautey's effort to ensure that Moroccan cities preserve an essentialist conception of Moroccan culture. For some, this new urbanism came to represent a neo-colonial continuation of French hegemony. Consequently, although the earthquake facilitated a break with traditionalist urban design in Agadir, this break fed anxiety and anomie concerning Moroccan cultural identity, and contributed to a backlash against the putatively universalist ideas of the city's planners. This backlash served the interest of the monarchy, which portrayed itself as the defender of Moroccan identity.

In Agadir, questions about the meaning of decolonization and the impact of the earthquake were intertwined and contested in the realm of architecture and urban planning over the course of decades because of Morocco's particular history and because of the specific pattern of physical damage in Agadir. However, the long impact of environmental disaster and the long struggles of decolonization also unfolded in the realms of memory, memoir, and literature.

CHAPTER 7

Rupture, Nostalgia, and Representation

T HE TASHELHIT POET IBN IGHIL memorialized the 1960 Agadir disaster in an oral poem that framed lamentations of loss in religious terms. In the poem, Ighil grapples with the fact that people and places had suddenly ceased to exist. "Where is that place of the righteous men, of the carpets and of the trays and tea?" he asks, and the poem answers: "There is nothing in it but wind."[1] Ibn Ighil's poem offered no comment on the relation between the environmental and political events of his time. As Kenneth Brown and Ahmed Lakhsassi have pointed out, the prevailing metaphor of Ighil's poem is that of the Day of Judgement, in which all is destroyed as if by a "flooding wadi," and reduced to "powder, powder."[2] In this poem, there is no nation: "Morocco" is not mentioned; nor is the city's history of French colonialism. The destroyed districts of the city are listed, and the inhabitants are identified by religion, gender, wealth, and "righteousness." The city exists in the space between God, the poet, and the dead, not "God, Country, King."[3]

Unlike Ibn Ighil, other memoirists and creative writers have memorialized the disasters of 1954 to 1960 by explicitly exploring the relation between environmental disaster and the political contexts of decolonization. These writers grappled with the realization that their loved ones had lived in colonial spaces, and that the end of colonialism had coincided with the destruction of the lives they had lived and the cities they had known. The physical places they remembered were gone, unrecognizably transformed by catastrophic movements of matter. But other transformations were also sweeping away the pre-disaster, pre-decolonization contexts of their lives.

Fiction and memoir about the disasters of 1954 to 1960 provide insight into how survivors and observers conceptualized the long aftermath of those events in later decades. Historians can make some cautious use of memoirs as sources to examine long-past events, such as the disaster response in Beni Rached, discussed in chapter 2 in light of both Belgacem Aït Ouyahia's 1999 memoir and archival documents written soon after the earthquake. But memoirs such as Aït Ouyahia's can also be used as contemporaneous primary sources, contemporaneous with the experience of surviving decade after decade with the memory of the disaster and its effects. All the sources discussed in this chapter, whether presented as fiction or memoir, differ from the archival sources relied upon in earlier chapters not in their greater or lesser veracity but rather in that they emerge from their author's reflection on an "event" that begins with the sudden onset of environmental catastrophe but that includes a long aftermath of months and years. Their object of study, in other words, is almost coterminous with the object of the present study. In contrast, archival documents such as diplomatic cables also present an author's perception, memory, and representation of an event, but the event of primary concern is often one of much smaller scope—the arrival of a shipment, or a significant conversation—in which any portrayal of the disaster as a whole is usually relegated to the background and always truncated in time. Even for reports that purport to analyze the "whole" event, like those of Marius Hautberg's account of the successes and failures of the French response to the 1954 earthquake or *Alger républicain*'s journalistic chronicle of occurrences following the disaster, the "event" they describe is one with a short duration, since they were written only weeks or months after the onset of the disaster. Unlike the archival sources produced in response to the "short" events, the memoirs and literary representations of disasters considered in this chapter offer the historian additional insight into the effects of time and memory on disasters and decolonizations, and on the perceived relations among these "long" events. The task undertaken in the present volume is to use these accounts—representations of both "short" and "long" events—to construct a critical representation of the "long" event that is transparent in its use of evidence and which reveals, from a differently informed perspective, what the individual sources cannot offer when read in isolation. The result demonstrates the ways in which, from the 1950s to the new millennium, humans experienced the events of environmental disasters through their experiences of the political, social, and cultural events of decolonization.

Nationalist Ruptures, Survivors' Nostalgia

The two earthquakes and the flood discussed in the present volume differed from the 1959 mass poisoning in Morocco and from other disasters, such as droughts, famines, and epidemics, in that they destroyed the remembered spaces of the built environment. Consequently, writers and survivors have had to grapple not only with grief for the human losses but also with the memory of lost places, a memory often shaped by nostalgic longing. "Colonial nostalgia" has most often been examined by scholars with a focus on colonizers' nostalgia for an imagined or lost past relationship between the colonizer and the imagined colonized subject, and on colonizers' nostalgia for a home they have left and may no longer have access to.[4] However, when disasters destroy cities, the disaster survivor also experiences the disruption of relationships and the loss of home. In this respect, the experience of the disaster survivor has something in common with that of the "repatriated" colonial settlers who, after decolonization, could no longer return to the physical spaces they remember. When, as in the cases considered here, the disaster occurred just before, during, or just after the process of decolonization, memories of the two were intertwined. The French of Orléansville lost the city they knew twice: once in September 1954 and once in 1962, when the violence of decolonization drove them north to France. For the French of Fréjus, the loss of Algeria was a more distant event, but because the war, like the flood, took so many lives in those years, the shock of Algeria's revolution remained connected to the memory of the flood in the work of French memoirists like Max Prado and Christian Hughes, discussed in chapter 3. For the colonized, however, the loss of Orléansville in 1954 or Agadir in 1960 exists in a different relationship to decolonization. Because of the near synchronicity of these disasters with national independence, the memorialization of the lost places and people of the pre-disaster environment implicates the representation of life under colonial rule. Nostalgia for the pre-earthquake, colonial-era city sits uneasily with the triumphalism demanded by the discourses of newly hegemonic nationalism. Yet this tension, and the desire of disaster survivors and observers to give voice to grief, inspired complex representations of the pre-disaster, pre-independence past, and of the fraught relationship between the victory of national independence and the tragedy of environmental disaster.[5]

One response to this tension between the survivor's grief and the patriot's celebration has been to celebrate the rupture brought by the disaster and to reject all nostalgia for the pre-disaster, colonial city. This was the response of Henri Kréa, discussed in the introduction to this volume, who celebrated the 1954 earthquake in Algeria as part of the rupture with the colonial past, even while acknowledging the horror of the earthquake. Kréa's treatment of the 1954 earthquake left no room for nostalgia for the city it destroyed, and one might accuse Kréa of callous indifference to the actual ties that residents of the actual Orléansville may have had to the city before independence. For Kréa, writing in 1956 and 1957, Algeria's hope lay in the future; the past held only oppression.

Orléansville as a city hardly exists in his play, which concerns the relationship between an abstract people and the imperial France/Rome.

In Kréa's play, the earthquake was a "mystical sign"[6] that awakens the people from their acquiescence. In Aït Ouyahia's personal memoir, the earthquake reveals the oppressive violence of imperialism and gives nationalist focus and meaning to the life of a young man hitherto ambivalent about his position in the colonial society. Both highlight the inequities and racial discrimination of French rule as the fundamental injustice of colonialism. The racist words of the French officer at the aid tent precipitated Aït Ouyahia's conversion from a member of the collaborationist elite to a committed representative of the "Arab" people, ready to confront the oppressor. The archival record produced in the years from 1954 to 1962 does not permit the historian to accept, as a generalized truth, Henri Kréa's and Aït Ouyahia's portrayal of the earthquake as playing a strong causal role in the Algerian people's nationalist awakening (although future research, particularly in Algerian archives, might provide additional evidence). Regardless, however, it is clear that the archival record does reveal a complex web of interactions between responses to the earthquake and responses to decolonization in those years. Works such as those by Aït Ouyahia and the other writers considered below demonstrate that this web of interconnections extends further, into the years and decades after the disaster.

Rupture in *Rocks and Lights*

Like Kréa's play, Belgacem Aït Ouyahia's chapter on the Chélif Valley earthquake, "Orléansville 54," eschews nostalgia for what the earthquake destroyed and portrays the event as triggering a positive rupture in the doctor's political itinerary from the position of a privileged and proud collaborator to that of an active, though not heroic, nationalist. However, the "Orléansville 54" chapter stands apart from the rest of the memoir. The geographical and emotional center of his memoir, *Pierres et Lumières,* lies in Kabylia, where Aït Ouyahia grew up and where he returned to practice medicine. His account of the Orléansville earthquake (introduced in chapter 2 of the present volume) thus takes the form of an excursion, a brief interlude during which he is called away from his first post in Kabylia to return to the stricken city where he had done his surgical internship. While the geography of the disaster relief expedition from Orléansville to Beni Rached is described in considerable detail, the city of Orléansville is not: beyond the walls of the hospital, the reader sees little of the colonial-era city before or after the disaster. In a nostalgic memoir full of detailed description of the places and ways of life of the narrator's personal history, the Orléansville chapter is an exception, devoid of nostalgia: the focus is on a positive rupture from the colonial past.

Aït Ouyahia's account of his Arab nationalist awakening in Orléansville is foreshadowed by his account of his work in Beni Rached in the same chapter. Aït Ouyahia's account of Beni Rached highlights the linguistic element in the young doctor's break from his French-privileged past. When Aït Ouyahia encountered the bloodied survivors of Beni Rached, he found himself, uncharacteristically, addressing them in fluent Arabic: "I was surprised to hear myself speak, not because I felt that my words would comfort these unfortunates, but because, for the first time, I pronounced six or seven sentences in a row, in Arabic, without a single word of French among them, as was my necessary habit, because of ignorance of the language."[7] Writing decades after independence, Aït Ouyahia contrasted his unexpected use of Arabic only with his more customary use of French. His native Kabyle is unmentioned in this passage, although his Kabyle roots figure prominently in his autobiography as a whole. The official nationalist ideology of the victorious Front de Libération Nationale (FLN) counted all Muslim Algerians as Arabs (as popular French colonial usage sometimes had), and Aït Ouyahia's Orléansville chapter presents a dichotomous linguistic world of Arabic and French. This is consistent with his account of his confrontation with the French officer, in which he steps forward to speak on behalf of all "Arabs." His remembered identification with Arabs was a rejection of the sense of distinction derived from his status as an *évolué*—as a French-educated medical student and the son of a French-educated schoolteacher. His Kabyle background—elsewhere a central, nostalgia-infused theme in his memoir—is left out of the story. In both accounts of his personal heroism during the earthquake, the disaster creates a bond of solidarity among Algerians, presented as Arabs. On the question of Kabyle political identity, Aït Ouyahia is silent. His nostalgia for a lost Kabylia never appears when themes of nationalist politics are present. His separation of these two themes reflects the dominant ideology under FLN rule, and the political realities of independent Algeria in 1999, where a politicized Berberism would not be tolerated. Aït Ouyahia mentions his youthful fondness for Ferhat Abbas, the most moderate of the Arab-francophone nationalist leaders to join the FLN and to survive the revolution's various purges; if the young doctor had been equally fond of the Kabyle nationalist thinkers Si Amar Boulifa and Hocine Hesnay-Lahmek, the old memoirist would not, could not, have said so.[8]

Aït Ouyahia's story of his encounter with the French officer is unique among the portions of his memoir that deal with the movement for independence in that it emphasizes the young doctor's nationalist clarity and strength of personal commitment. In other chapters, Aït Ouyahia seems interested in distinguishing his memoir from self-aggrandizing tales of nationalist heroics. The book opens with a quotation from fellow doctor Jean Bernard on the subject of writing the history of French resistance to the Nazi occupation: "Histories of the resistance seem to me to written in the form of a triptych. On the middle page, the truth; on the left-side page, the story told to the Germans after one's arrest; on the right, the description told to one's friends after the Liberation. I'm going to try to write on the page in the middle."[9] This epigraph situates Aït Ouyahia's book as a memoir of resistance, in this case resistance to French colonialism, but also fore-shadows the author's skepticism about the ideological certainties of nationalist discourse. The Bernard quotation is immediately followed by a quotation from Baudelaire and by a Kabyle proverb; together, these epigraphs frame the text's cultural landscape, shaped by the parataxis of Kabyle and French that pervades most of the memoir, though not the Orléansville chapter.

Aït Ouyahia seems to be supporting his claim to adhere to the "page in the middle" when he reveals the imperfections of his nationalist credentials through his treatment of several key events related to the nationalist struggle. In May 1945, when Muslim Algerians were protesting and rioting, and the French were committing massacres in Sétif, Aït Ouyahia was in Algiers celebrating V-E Day alongside European settlers. Later, when a friend denounced the atrocities committed by the French, the young Aït Ouyahia expressed his faith in the French *mission civilisatrice,* repeating "sempiternal clichés about schools, roads, and hospitals."[10] Eventually, Aït Ouyahia revises his position: in an anecdote set in 1955, his friend asks him to assist the nationalist cause by providing medical help to resistance fighters. Aït Ouyahia agrees, but he is distinctly nervous about keeping potentially incriminating medical supplies in his office. Soon after, when put on trial by the French for his suspected actions, he equivocates, stating, "I have never been part of a political organization" and is set free.[11] The memoirist makes no claim to be a paradigmatic nationalist hero; he explicitly seeks to move beyond such clichés.

Aït Ouyahia's accounts of his involvement with the nationalist cause outside of the Orléansville chapter do not mention the earthquake as a turning point. The Orléansville chapter also differs from those sections of his memoir in its emphasis on the strength of his newfound ideological clarity and commitment. However, Aït Ouyahia's refusal to portray himself as a hero does also appear at the end of his earthquake narrative. The militancy of his confrontation with the French officer is interrupted by the salutary intervention of a French nurse, who fabricates an emergency in the hospital in order to rescue the young doctor from an altercation that he could not win. As in his account of his trial, set many months later, Aït Ouyahia avails himself of the opportunity to escape.[12] Here, too, the author's actions are portrayed as less than heroic. Nevertheless, the earthquake chapter clearly presents a narrative in which his experience of the disaster constitutes a turning point, a nationalist epiphany.

Rupture in Khaïr-Eddine's *Agadir*

Mohammed Khaïr-Eddine's 1967 novel *Agadir* also presents seismic disaster as a point of rupture from the past, but without the embrace of nationalism conveyed by Kréa and Aït Ouyahia. Like Aït Ouyahia, Khaïr-Eddine was not a resident of the stricken city at the time of the earthquake but was part of the disaster response. Khaïr-Eddine had been born in Morocco's Tashelhit-speaking south, not far from Agadir, and as a government functionary he lived and worked in Agadir immediately following the disaster, reviewing survivors' eligibility for government assistance. His novel reveals a detailed interest in the actual impact of the earthquake on the city and the survivors. Whereas for Kréa the symbolic meaning of the Orléansville earthquake was clear, Khaïr-Eddine confronted the meaninglessness of the disaster, portraying an incomprehensible ruined landscape where, as literary scholar Ahmed Raqbi has put it, "existence has no meaning and where silence becomes king."[13] For Khaïr-Eddine, the rupture from the past is disorienting, but it is also fraught with longing, and the pull of nostalgia is powerful. Yet, for Khaïr-Eddine, nostalgia is a wound, "voracious nostalgia" from which his characters seek to be cured.[14]

Unlike Kréa, and unlike Aït Ouyahia's representation of his younger self, Khaïr-Eddine placed no faith in decolonization as a cure for the ills of the past and certainly no faith in anti-colonial political revolution. Like other writers discussed here, Khaïr-Eddine portrays the violence of the earthquake and the violence of decolonization as inseparable; Khaïr-Eddine, however, does not distinguish the violence of decolonization from the violence perpetrated by the independent Moroccan state.[15] Writing several years after political independence, an exile from an increasingly authoritarian Moroccan monarchy, Khaïr-Eddine saw the 1960 earthquake in Agadir not as the harbinger of a new order but as a horrific disruption of an oppressive totalitarianism that extended from the mythic past to the post-independence present. This disruption revealed the fundamentally unmoored situation of the human individual, but Khaïr-Eddine offers no new narrative. His novel, a bricolage of "stream-of-consciousness, splintered persona, multiple perspectives and other techniques of discontinuity"[16] prevents the privileging of a unifying interpretation of the disaster or of Moroccan history. As Larbi Touaf puts it, "As a postcolonial novel, *Agadir* leaves no place for 'fixity' or 'purity' when it comes to questions of individual or collective identity."[17] Alternating between prose and dramatic dialogue, realism and hallucinogenic fantasy, past and present, Khaïr-Eddine's novel offers no simple blueprint for understanding the catastrophe of the earthquake or the future of the individual after decolonization.

Originally from the town of Tafraout, 150 kilometers from Agadir, Khaïr-Eddine grew up in Casablanca. Discussing Khaïr-Eddine's ancestral roots is problematic as a means of understanding his perspective, however. Hédi Abdel-Jaouad observes that "Khaïr-Eddine refuses to privilege the figure of the ancestor as a potential redeemer of precolonial identity. On the contrary, he is obsessed with the present in all its complexity."[18] Yet Touaf argues that Khaïr-Eddine's rejection of grand narratives was conditioned by his Berber background, a cultural heritage which ill-fit the dominant narratives of Moroccan nationalism centered on monarchy and on Arab and Islamic identity.[19] In *Agadir*, a character called "The Corruptor" speaks to the head of the Armée de Libération Nationale: "Our tribe is historically a people . . . Lord, Berber since the placenta, before the drop of sperm."[20] Yet neither Berberism nor the military struggle against colonialism offered any solution. In the novel, the seventh-century Berber queen Kahina appears as one of a succession of tyrants including a caliph, a caïd, the dynasts of the Berber Almoravides and the Arab Saadians, as well as the modern minister of the Interior. The Romans called Kahina the "Serpent Queen of Barbary" she says, but she references Marx and calls herself a communist. Unlike Kréa's Jugartha, Kahina provides no salvation. Whether she represents heritage or a vision of the future, the narrator says he does not know her and rejects her.[21]

On the opening page, the narrator's unnamed traveling companion experiences the disaster as liberating: although he has lost his house, wife, and two daughters, he had already wanted to repudiate them. The narrator confronts the futility of his own mission to aid the people of the city, for he can recognize no city: "They lied to me. There is not the least hint of a city here. . . . Little chance that I have of returning life to the people here. They are traumatized men. I am not the Good Lord." According to the narrator, the attempt to find meaning by looking to the past, the lost city, is hopeless. The city of the past is a city of corpses: "But I sense clearly the subterranean presence of the cadaver of a city. . . .

Disturbing odors: exhalations of crushed rats, of human limbs in decomposition, the stench of disemboweled sewers." The survivors' desire to recover the traces of identity in the ruins of the past is presented as pathological, exemplified by the man who carries with him the severed finger of his dead wife. The narrator states, "The population does not want to leave the city which is, it is believed, the cradle of civilization and the matrix in which History is formed. They don't know that their history is already done. But what will and what faith they apply to get from the rubble that which is no longer usable."[22]

The novel *Agadir* demonstrates a much more intimate experience of the earthquake than does Kréa's play *Le séisme*. Like his narrator, Khaïr-Eddine was sent to Agadir after the earthquake by the Moroccan state to process survivors' applications for state aid and had direct contact with the aftermath of the disaster there. The opening section situates the narrative, such as it is, firmly in the real space of post-disaster Agadir: in a prefab trailer, eight meters long, three and a half meters wide, and three meters high, equipped with a table and chair, a typewriter and a calculator, and adorned with portraits "of the dead king and the living king"[23]—portraits which frame the novel in the aftermath of the earthquake, after the death of Mohammed V in 1961 and the ascension of Hassan II.

This realism does not pervade the entire novel. In one section, the city is dominated by talking animals, most notably a powerful parrot and a cobra. Khaïr-Eddine's feverishly depicted "ville zoologique" is defended by "brigades of monkeys" and inhabited by "alligators dozing in the infected water of rectangular pools," as well as by cigarette-smoking dogs, makeup-wearing hyenas, and "gorillas with eyeglasses."[24] Khaïr-Eddine was likely referencing accounts describing pre-Islamic Berber reverence for sacred animals, ideas later incorporated into North African Muslim beliefs that animals might testify on Judgement Day.[25] Yet Khaïr-Eddine's vision of a city of hyenas and infected pools prowled by scavengers also echoes archival accounts of the cordoned-off disaster zone, carpeted with quicklime due of fears of epidemics of cholera and typhoid, guarded by soldiers and prowled by scavenging animals. And when Khaïr-Eddine writes of corpses, he references a massive epidemiological, cultural, and political problem that occupied the Gadiri public, as well as French and Moroccan diplomats, for months after the disaster, as discussed in chapter 5 of this volume. The questions asked by Khaïr-Eddine's narrator were the great political questions of 1960 and 1961: What would happen to the bodies of the dead, buried under the ruins? Would reconstruction take place over the bodies of the victims?

Khaïr-Eddine considers two options for building a meaningful future without an oppressive attachment to the past: the construction of a new, modern city, and emigration. Both scenarios are rooted in the historical experience of the city in the months and years after the earthquake; neither offers a satisfying solution to the problems of disorientation faced by his narrator. Khaïr-Eddine mocks the Moroccan state's plans to build a new city, guided by the architectural principles of modernist rationalism. The bulldozers and soldiers of the newly independent state, as much as the earthquake itself, have produced the disorientation felt by his narrator.[26] Khaïr-Eddine juxtaposes the urban planners' vision, written in capital letters and focused on creating an orderly and comprehensible future, to the vain search of a survivor obsessed with finding his lost home (and pet gazelle). For Khaïr-Eddine, both are equally absurd, or demented: "I have not yet grasped what I am looking for, and I never find my City. ONE MUST PERHAPS BEGIN BY BUILDING HOUSES WELL-ALIGNED SIDE BY SIDE leaving a large enough space between them AND DISTANCING THEM FOLLOWING A METICULOUS GEOMETRY (but the problem that is posed in the first place is to know whether it is permitted to build on the debris of a dead city)."[27] In the political and physical realm, the solution in fact arrived at had been to excavate the bodies from the areas where rebuilding would take place—while leaving the fallen Kasbah in ruins as a mass grave of unretrieved corpses. For Khaïr-Eddine, however, this exemplified the larger issue: whether one could find one's place in the present through reference to the past. Although he rejected all attempts to tie the future to nostalgically imagined pasts, Khaïr-Eddine was equally critical of the modernist urban planners' hopes that mere architecture could solve the problems faced by the city's inhabitants simply by ensuring an environment of light and space. He mocked the utopianism of those who felt they could improve the human condition through architecture: "THE HOUSES CONSEQUENTLY WILL HAVE FORTY WINDOWS AND THE MONTHS FORTY DAYS in order to bring all into equilibrium and we will have parrots and birds." Here, Khaïr-Eddine immediately repeats the central question: "Must one build on the site of the dead city?"[28]

Khaïr-Eddine also explores the possibility of an escape from the oppressive legacy of the past through the option of emigration to France. However, Khaïr-Eddine, who in fact moved to France in 1965, presents this as a delusionary hope. The Shepard declares, "Make me a passport I want to go to France / be a simple miner / in the rectum of the black soil." His sheep (Le Troupeau), however, advise him to stay ("Don't go, France, it is terrible"), as does the Goat ("What the devil will you do in France The cold the snow the daily duties your flute will you throw it away you will be a beggar oh do not go"). The Billy-Goat, however, speaks of the fleshly pleasures available in France—"The whores come in abundance calling / who wants a whore who wants one five francs the vagina"—a perspective for which the Billy-Goat is promptly lynched by the mob.[29] At the end of the novel, the narrator's uncle tries to persuade him to leave the city, but Europe is portrayed as a negation rather than a solution: "In Europe where I lived in the slums of rain in the voices of our brothers, incomprehensible We have no brothers I am your uncle and your father but the relations only exist because we wanted them One invents them ceaselessly." Yet Khaïr-Eddine's novel finishes on a hopeful note: "One must build on the void, voilà. Keep nothing from the past . . . past . . . bad [ellipses *sic*]; if not, if a memory is possible, but a reinvented past, in the colors of a new vision, and leaving healthy (*sain*), new [. . .] I am going to a land of joy, young and gleaming, far from the cadavers. So behold me naked, simple, elsewhere." For Khaïr-Eddine, his concluding optimism (if not ironic) rests solely on the writer's ability to break down identities, ideologies, and grand narratives: "I will leave with a poem in my pocket; that suffices."[30]

Scholarly discussions of Khaïr-Eddine's *Agadir*—the most discussed of all the representations of disaster treated here—have tended to neglect the author's encounter with the historical earthquake, instead portraying his exploration of the disaster in metaphorical terms, as a symbol for the rupture of conventions. The earthquake is thus reduced to a mere instrument of the author's desire to break from literary and linguistic conventions and produce "a writing in rupture with conformism."[31] There is no doubt that Khaïr-Eddine's culturally and politically transgressive work emerged in the context of "an interdisciplinary and transnational movement" and from the sociocultural matrix of avant-garde thinkers in Morocco that in 1966 became centered around the journal *Souffles*.[32] Nevertheless, I would argue that we should approach Khaïr-Eddine's novel, like the other memoirs and representations considered here, from the point of view of its environmental as well as literary context, acknowledging the historical earthquake as a force of change. Ted Steinberg has argued that environmental history is not just the story of human thought and action inscribed on a passive environment; it is also the "less anthropocentric and less arrogant" story of nature's impact on human thought and action.[33] Touaf writes of Khaïr-Eddine's novel, "The earthquake in *Agadir* introduced a world of great instability, violence, and disruption. It aimed at shaking people's blind faith and sapping the foundations of archaic social and aesthetic order, an aim that reverberated in the literary and political discourse."[34] This, minus the intentionality, is true not only of the earthquake in *Agadir* the novel but also the earthquake in Agadir the city.

To see Khaïr-Eddine's portrayal of the earthquake as a *product* of his desire to depict a world of "instability, violence, and disruption" is to ignore the effect of the earthquake itself. His desire to enact textual violence against literary convention may have been a product of his literary milieu, but his very notion of violence itself was in no small part a product of the earthquake. Khaïr-Eddine's portrayal of a city of rotting corpses, ruined buildings, scavenging animals, and severed limbs engaged with the tectonic, social, and political forces that shaped the process of burying, excavation, reburying, and reconstruction in the months and years following February 29, 1960, processes discussed in chapter 5. The earthquake helped shape Khaïr-Eddine's novel, a major work applying new approaches to Moroccan literature and to the remembering of disaster. Literature, like memory, is "socially-framed, present-orientated, relational and driven by specific agents,"[35] but it does not exist independently of the past, that is, of the environmental, political, and social forces that shaped the history that is remembered or represented. While post-modernist and post-colonial deconstructions of narrativity were part of francophone literary culture in the early 1960s, Khaïr-Eddine's groundbreaking use of non-linearity to represent the disorientation of modernity cannot be separated from the historical experience of the disaster he both witnessed and experienced, working in the aftermath of the earthquake.[36]

In Khaïr-Eddine's novel, the earthquake eclipses any historical rupture produced by decolonization, but the two are nevertheless portrayed as overlapping events, part of the alienation of the modern individual from the past, intertwined in Khaïr-Eddine's non-narrative. Khaïr-Eddine's novel rejected nationalist and monarchist narratives that portrayed decolonization as the transformational event which was to connect the Moroccan nation both to a modern future and to a celebrated heritage. Instead, he depicted a future of uncertain meaning, and the present's violent separation from an inscrutable past.

Nostalgia in *Rocks and Lights*

Like Khaïr-Eddine's *Agadir,* Aït Ouyahia's memoir is narratively disjointed. Criticized by reviewer Dehbia Aït Mansour in the newspaper *Liberté* for the lack of a unifying structure or theme, Aït Ouyahia defended his approach in an interview in *Algérie-Littérature-Action,* stating that "the digressions came on their own; they were not calculated. When writing, I even sometimes had the impression that they had neither head nor tail. . . . I realize that one might lose one's footing sometimes. I did not want to write to be easy. And if the reader must search a bit, I can say that I chose this difficulty."[37] The tone of Aït Ouyahia's response is in keeping with the tone of his memoir, which is peppered with the aging medical professor's critiques of the ignorance of the younger generation. The chaotic structure of his narrative might also be ascribed to the fact that Aït Ouyahia in 1999 was not a professional writer but an obstetrician who had hitherto published no other work aside from medical articles, even if he did later go on to write several novels. However, Aït Ouyahia in 1999 was a highly educated member of the francophone elite, and his assertion that his fragmented narrative structure was a product of uncalculated stream-of-consciousness must be viewed with skepticism, for his book may reflect the influence of Khaïr-Eddine's work or that of other francophone and North African post-modernists, or the doctor's engagement with medical or literary scholarship on trauma. Regardless, the memoir's lack of an overall narrative structure is a boon, for it allows space for many strands of thought and experience, undeterred by the hobgoblins of consistency or the tyranny of thematic focus.

The Orléansville chapter is disconnected from other themes in the larger memoir of Aït Ouyahia's life, in which his involvement with the struggle for national independence constitutes a significant recurring theme, but that is nevertheless overshadowed by his treatment of his relationship with his father and grandfather and by his itinerary negotiating his Kabyle village background and his relationship with his French education. These themes are alluded to in the book's subtitle: "Memories and digressions of an Algerian doctor, son of a schoolteacher 'of indigenous origin.'" While the dominant narrative of Aït Ouyahia's short account of the Orléansville disaster celebrates a nationalist rupture, the book as a whole repeatedly returns to a nostalgic theme lamenting the destruction of the past and the loss of memory, while at the same time approvingly narrating the transformation of Aït Ouyahia's family from Kabyle villagers to French-educated elites.

The book begins with the destruction of Aït Ouyahia's natal village in Kabylia, and his grandmother's home, by the guns of the French Operation Jumelles in 1959, a destruction that would be made complete after independence when, long after the war, the bulldozers of the new regime arrived to pave the way for a reconstruction that disregarded the past and rendered the village unrecognizable. Like Khaïr-Eddine's novel, this narrative does not portray national independence as salvation. In James McDougall's terms, nationalist modernity imposed its own "disciplinary order" on this society, no less destructive than that of the colonizers.[38] A great ash tree of the village, a symbol of heritage for Aït Ouyahia, survived the bombs but was felled, after independence, by the bulldozers. Post-independence, the destruction of the heritage of village life, symbolized by the bombs and bulldozers, continued, as the residents' way of life was replaced by a new consumerism and a construction boom driven by expatriate remittances from France.[39] The village of nostalgia was lost, and independence did not restore it.

This nostalgia for a lost Kabylia is tempered, however, by a narrative that portrays Aït Ouyahia's family's encounter with French colonial education, from the grandfather who forbade his children from entering the French school, to the father who became a schoolteacher, to Aït Ouyahia's own education as a doctor. While Aït Ouyahia's encounters with French racism play a part in this story, and he is occasionally critical of others' aping of French ways, there is no hint of regret about his career trajectory, for that is what led to him becoming an obstetric surgeon. Aït Ouyahia describes his youthful belief in the "oeuvre civilisastrice" in which he took part as a French-educated doctor.[40] Even in the Orléansville chapter, Aït Ouyahia portrays his anti-colonialism as built upon frustration with the hypocrisy of French rule, not on the rejection of its principles. He was driven to nationalism by an accumulation of insults: first racial insults as a schoolboy ("Ils sont cons, les Kabyles"), then insubordinations by French nurses ("I did not come here to obey an Arab"), and finally those words that were "unworthy of a French officer" but that were nonetheless delivered: "All thieves, these Arabs."[41]

Mostefa Lacheraf's preface suggests that a recently revived interest among Algerian intellectuals in local histories and cultural diversity in Algeria, and particularly in Kabyle village life, was what made Aït Ouyahia's memoir publishable. This interest, as much as chronology, may account for the concentration of passages focusing on Kabylia near the beginning of the memoir. A more disciplined writer or more ruthless editor might have trimmed much of the later material from the manuscript, which would have been a loss for the historian. Aït Ouyahia's treatment of the 1954 earthquake seems disconnected both from his pervasive nostalgia for the past and from those portions of his memoir that deal with the movement for independence, and which make no mention of the earthquake as a formative experience.

His work as a doctor provides the one unifying theme throughout the memoir. Aït Ouyahia's true moment of heroism comes, not in his confrontation with the French officer outside aid tent or in his work for the FLN, but in his treatment of the injured in Beni Rached. It is as a doctor, not a political actor, that he finds his calling. Yet, as symbolized by his sudden burst of fluency in Arabic at Beni Rached, the two are never fully separated in Aït Ouyahia's writing. Elsewhere in Aït Ouyahia's larger memoir, there are instances in which, as in the Orléansville chapter, Aït Ouyahia portrayed the medical struggle against the inanimate causes of human suffering as akin to the political struggle against colonial injustice.

Writing in his own voice as a much older doctor and professor of medicine, Aït Ouyahia inveighs against his young medical students' ignorance of both the heroism of the Algerian combatants who gave their lives for independence and of the French doctor who pioneered treatment for osteomalacia, the softening of the bones, a common cause of maternal death during childbirth in colonial Algeria. Linking natural disasters, politics, and medicine, Aït Ouyahia writes, "History, when the cyclone quiets, when all the volcanos are extinguished, will someday recognize him, perhaps, as a benefactor of *des femmes indigènes*, or of *femmes musulmanes*, as they said also, indifferently, back then."[42] Aït Ouyahia portrays national independence as a crucial turning point in the medical struggle, as he states that osteomacia has disappeared in Algeria "thanks to independence . . . [*sic*] and also to socialism" which brought "bread and milk everywhere . . . [*sic*] schools and pupils, with rosy cheeks and shoes on their feet."[43]

Aït Ouyahia then launches into a story, which he recounted to his medical students, of his first breach (*siege*) delivery, in the snow, in the remote town of Icheriden. The linkage of medical work and nationalist struggle found in the Orléansville chapter appears in this anecdote as well. Aït Ouyahia, in a telling pun, labels this story "the siege of Icheriden," recalling a French colonial military victory of 1857 in which a Kabyle village was destroyed during the French conquest of the area. After describing his obstetric accomplishments at Icheriden, Aït Ouyahia explains that he resisted the temptation to lecture his students on the nineteenth-century political-military events there. Instead, he tells them the story of Hassiba Benbouali, an eighteen-year-old ALN fighter who died for the Algerian cause. This is not a non-sequitur: both the obstetric and political heroics seem to be included in Aït Ouyahia's lament that "Algeria forbids history! My country forbidden its history, forbidden its heroes!"[44] Here, the revolution itself is the object of nostalgia, accompanied by nostalgia for the medical achievements of his own youth. Aït Ouyahia looks back from the point of view of a disappointing present in which Algerian society has seemingly turned its back on both the heritage of the village and the potential of the revolution. In support of this theme, Aït Ouyahia's narrative of the earthquake serves as a vehicle to demonstrate both the political clarity and the medical heroism that he associated with the years of his early adulthood. The revolutionary struggle against injustice and the medical struggle against suffering seem to be of a kind—even if Aït Ouyahia acknowledges that his own actions were more distinguished on the latter front than on the former.[45]

Troubles in Childhood Paradise

Habib Tengour of Algeria and Jacques Bensimon of Morocco have both produced writings that combine themes of rupture and nostalgia. Their personal narratives of the earthquakes and other dramas of the era of decolonization address the ambiguous and complex relationships among events. Both are accomplished creative professionals: Tengour is a writer, while Bensimon, a Moroccan-born, Berber, Jewish writer and filmmaker, became chair of the National Film Board of Canada. Like Khaïr-Eddine, Tengour and Bensimon self-consciously confront the question of the relationship between memory and the events of the past.

Jacques Bensimon's 2012 account of life in pre-earthquake Agadir, *Agadir, un paradis dérobé* (*Agadir, a Stolen Paradise*), like Khaïr-Eddine's novel, and like the accounts of the Fréjus disaster written by Gaston Bonheur and Régina Wallet (discussed in chapter 3), exists on the boundary—always blurred—of fiction and memoir. Bensimon, unlike Bonheur and Wallet, does not claim an absolute veracity for his work; nor does Bensimon purport to simply present his memories, without questioning their accuracy. Bensimon presents his work in the form of a narrative memoir but states explicitly that he has taken license to fictionalize elements of the story, "placing real people in romanticized situations, introducing invented personages into real contexts," and he asks the survivors of Agadir to forgive him for such embellishments. Bensimon writes that his approach, inspired by his career as a filmmaker, is "to perpetrate here a 'true novel' or a 'romanticized reality.'"[46] This disclaimer, at the end of the book's introduction, deliberately invites the reader to question the truthfulness of Bensimon's account of his life in Agadir before the earthquake.

Bensimon's *Agadir* presents itself, beginning with its title, as a nostalgic text. The narrative is framed by its introduction describing Bensimon's return to Agadir in 2010 for a reunion with childhood friends on the occasion of the fiftieth anniversary of the earthquake and by a conclusion discussing the author's diagnosis with terminal colon cancer after his return to Canada. The nostalgia is clear in his opening chapters, in which Bensimon states:

> This earthquake destroyed a city that is unique in the world, the city where I lived, and I wish today to bear witness to this epoch that now survives only in my heart and in my spirit. I want to inscribe this happiness that the stones, the streets, the faces no longer embody. I want to fix this life, in order to preserve the memory of this place.[47]

Bensimon's nostalgia is also implicit in his book's closing sentence, expressing his final wish, to be buried in Agadir. In between, Bensimon provides fond descriptions of his family, his rascally friends, and the emergence of a pubescent sexual awareness. Yet this nostalgia exists in tension with other elements within the text.

Bensimon's family was not in Agadir in 1960, having emigrated during the upheaval of decolonization. Yet his opening chapters present the loss of the city where he grew up as a loss created not by decolonization or emigration but by the earthquake, which he learned of from a Radio Canada news broadcast: "With this earthquake, my life seemed to stop on February 29th, 1960 . . . a part of me is as if annihilated."[48] While for the survivors who remained in Agadir after 1960, life continued, and the city was reconstructed, for Bensimon the émigré, Agadir no longer existed. Up until that moment, he had felt that the home of his memories was still there, even in his absence. Then, the earthquake forced him to confront his irrevocable separation from the places of his past. Bensimon makes no mention of the controversies about architecture and the layout and identity of the reconstructed city, nor does he invoke the "city without a soul" discourse discussed in chapter 6 of the present volume. The city he wrote about ceased to exist in 1960; the post-disaster city is not relevant to his memoir, appearing only in his opening paragraph describing his return to Agadir in 2010, dismissed with the sentence: "I recognized nothing."[49]

There are elements in Bensimon's text that run counter to his portrayal of the earthquake as the definitive point of rupture. There are hints that, prior to the earthquake, Agadir had already begun to be "stolen" by the effects of colonialism and modernity. French education had made the boy Bensimon and his classmates "strangers in their own country"; he could barely communicate with his own grandmother.[50] The commercial success of his petty-bourgeois parents also produced alienation from the places and habits of Bensimon's early childhood, most notably when his mother's shop is moved to "a beautiful modern, avant-gardist store" owned by the Swiss company Bata.[51] Much like the critics of reconstructed Agadir, Bensimon portrayed the building's architecture as an affront to the cultural traditions of the city's "oriental" residents. For Bensimon, the alienation imposed by modernist architecture and lamented by the post-reconstruction purveyors of the "city without a soul" discourse seems to have arrived years before the earthquake, borne by the forces of global capitalism.

Despite these hints that not all was idyllic in Bensimon's pre-earthquake Agadir, the mood of nostalgic reminiscence predominates for most of the book. Near the end, however, this mood is disrupted by two scenes of startling violence. The first is the brutal rape of the narrator, aged "eight or nine" by one of his grandmother's tenants. This scene, described in bloody, graphic detail, is presented as traumatic for the boy not only because of the pain, violation, and violence of the act but also because of a feeling of shame, exacerbated by an awakened awareness of his own homosexual orientation.[52] Agadir, the reader suddenly discovers, was no "paradise" for young Bensimon, even in 1951 or 1952. The second scene of violence takes place two or three years later, in the shadow of a French-imposed curfew, among the tanks and bombings leading up to Moroccan independence. This scene also involved sexual molestation, but this time connected to the French presence, now experienced as an occupation. Young Bensimon, despite hearing explosions in the city, had broken the curfew after an argument with his father, and after staying out all night, was watching a game of pinball through the front window of a café. An eight-year-old Muslim girl stood next to him, apparently doing the same, as was a legionnaire who stood behind her. But then Bensimon noticed the girl was frozen in terror; the legionnaire had taken out his penis and was rubbing it against the girl. Bensimon exploded in rage, striking the legionnaire with his fists; when the drunken soldier stumbled and fell, he kicked him with his feet, and continued until bystanders intervened. Bensimon awoke in the hospital; it was then, abruptly, that his parents decided to emigrate.[53] Thus, in this part of Bensimon's narrative, it was sexual violence, and the abuses of colonialism, that robbed him of "his" Agadir. Nevertheless, the introduction presents the moment of the earthquake as a sudden, defining moment of loss that separates Bensimon from the subject of his nostalgic longing.

Of Fish and Bullocks

While the Agadir earthquake marked an ending for Jacques Bensimon, Algerian poet and novelist Habib Tengour has written a short memoir in which the 1954 earthquake marks the beginning of a coming of age story. Writing from the perspective of an adult academic who works and writes in both Constantine and Paris, Tengour establishes distance between his transnational adult perspective and the world of his childhood with a startling opening sentence that offers an explanation for the earthquake: "At that time, the earth was like a flat plate. It rested on the horn of a bull calf that was standing, precariously balanced on the tail of a fish. When the fish would move it made the earth tremble."[54]

For Tengour, the earthquake was not an obvious point of rupture. It struck when he was seven years old, causing minor damage and waking some of the residents of his hometown of Mostaganem, 130 kilometers from Orléansville.

Tengour's opening phrase "at that time"[55] designates a period of his Algerian childhood that continued after the earthquake, with schoolyard games and memories of his grandfather. His initial, mythological explanation of the earthquake is seamlessly followed by his grandmother's invocation of God's protection and his grandfather's search for the family Koran, and by an understated, realist description of broken crockery and an overturned water jug. His opening locates his account of this period in the realm of mythology, and he describes his childhood Algeria as a world of stories—religious stories told by his grandfather and received with slight skepticism by the precocious boy, stories told by children during recess, and stories told by the people of the town.[56]

Within this mythic context, Tengour relates local explanations of the divine causes of the disaster, as fantastic as balancing fish and bull calves. According to one story, God was punishing the city because people had brought wine into a mosque; according to another, it was because a Muslim scholar ("mufti") had gone to a brothel with a Koran in his coat pocket. There were also tales of orgies at religious sites. In yet another version of this story, it was said that prominent individuals in Orléansville "moistened their couscous with wine sauce and that, under the pretext of celebrating the Night of Error, they devoted themselves to fornicating with women and young men."[57]

It is clear that the adult Tengour, the writer, does not believe these explanations; they are stories. It is less clear whether the reader is meant to believe that these were actual explanations circulating in Mostaganem, or whether, like the story of the fish, these stories are the writer's interventions or importations from other contexts.[58] Tengour, like Khaïr-Eddine, avoids equating memory with truth.

Tengour's account of his idyllic childhood begins with the earthquake but is distant from the epicenter and destruction. This childhood idyll is later disrupted by the violence of decolonization, which produces a turning point within his narrative. It emerges, later in the account, that Tengour's father was a political prisoner of the French and was tortured, an event which apparently caused the young Tengour to suffer from stress-induced illness, interpreted as the "evil eye." Yet the nationalist struggle does not define Tengour's experience or mark the end of his childhood innocence. Rather, the intrusion of adult perspectives comes both through his awareness of his father's struggle—portrayed as distant and abstract—and his growing knowledge of individuals like the family's Jewish neighbors, the Senkmans, and French settlers, the Delages, whose decency and humanity disrupts local anti-Semitic beliefs (e.g., that "Jews wake up every morning with their mouths full of worms") and nationalist/imperialist dichotomies.[59]

Tengour portrays the Algerian Revolution as a common suffering, which brings the Muslim and Jewish families closer as fellow victims of oppression. "Today, you are like us," says Mrs. Senkman to Tengour's grandfather, relating the Jewish experience to the colonial oppression of Muslims.[60] Tengour's father is made a prisoner; the Senkmans' son Albert is conscripted for service in the French army, and the suffering of the Senkmans' son, who after his demobilization hides in his room for years, seems to parallel Tengour's stress-induced ailments related to the imprisonment of his father. The penultimate paragraph in Tengour's account of his youth ends with his grandfather's words, "Madame Delage and her husband are good human beings. You must learn to open your heart to goodness, no matter where it comes from. Your father is fighting for the country's independence. The French who torture him are not human beings."[61] This story of Algerian childhood which began with the earthquake ends with the juxtaposition of universal humanity with the inhuman violence of colonialism. The final paragraph turns to the boy's eagerness to explore Paris, where the family relocated in 1959.[62] Unlike Khaïr-Eddine, Tengour portrays this exile uncritically, as a liberation that marks the end of the story of his childhood.

Tengour's memoir, "Childhood," (in an anthology of Algerian memoirs of childhood) presents both his early years in Mostaganem and his coming of age in Paris in positive, even nostalgic terms. His nostalgia for his early childhood is uncomplicated by the knowledge that these were the years of French rule in Algeria: the pre-adolescent Tengour seems initially unaware of the French; they are not part of his world. Moreover, this whole period of his life, "at that time" is pushed into the realm of mythology by the account of the fish and the bull. Between this idyllic beginning and the happy ending, Tengour's memoir is a realistic account, focusing on his growing awareness of the meaning of colonialism and decolonization.

How does the earthquake fit into this story? Tengour's decision to begin the narrative with the 1954 earthquake stands in contrast with the narrative focus on decolonization as a coming-of-age story. Tengour's inability to omit the seismic event, which otherwise fits ill with the themes and structure of his memoir, demonstrates the prominence of the natural disaster in his memory of this time of war. For many writers, a story that begins with Algeria in 1954 is assumed to begin on the night of October 31. For Tengour, however, the earthquake of September could not be separated from the story that he tells: the earthquake marked the beginning of a series of violent events in the land of his childhood.

Sites of Nostalgia

Unlike Dr. Aït Ouyahia's nostalgic memoir, the works discussed here by Kréa, Khaïr-Eddine, Tengour, and Bensimon (three professional creative writers and one filmmaker, by trade) explicitly recognized that their representations of disaster and decolonization were creative acts. Implicitly, the pseudo-historians Bonheur, Croizard, and Wallet, who depicted the Fréjus disaster, also accepted this. Local historians Lahsen Roussafi, Yazza Jafri, Abdallah Kikr and Marie-France Dartois of Agadir, and Jacques Torres, formerly of Orléansville, have taken a different approach, going to great lengths to assemble documentation of the history of their pre-disaster cities, both textual and photographic. They have been aided by their comrades in civic organizations in Agadir or in the communities of the *pieds noirs* ("repatriated" Europeans of Algeria residing in France) from Orléansville. The fruits of this collaborative research, accompanied by (often fragmented) commentary by the authors, have been published in book form and online.[63] Maps and photographs figure prominently on these websites. Dartois's website, mfd.agadir.free.fr, systematically reconstructs the geography of pre-earthquake Agadir, district by district. While members of Agadir's Forum Izorane have aspired to the creation of a walk-through memorial park on the site of the ruined Talborj or Kasbah, this website offers a kind of virtual substitute. Both Agadir1960.com and Orleansville.free.fr also provide interactive forums for the shared construction of memory.

Academic historian Claire Eldridge, in her extensive examination of the memorialization of colonial Algeria by France's communities of pieds noirs and of *harkis* (Muslim Algerians forced to flee Algeria due to their ties with the defeated French), has argued that "many pied noir associations take the view that memory is the source of history," placing great faith in their own ability to accurately represent the past and using documentary sources only when they accord with a nostalgic view of the colonial era, rejecting outsiders' attempts to present more nuanced or contradictory views that might expose the injustices of the colonial system.[64] The work of Jacques Torres and his collaborators at Orleansville .free.fr fits this description. On the website and in a self-published book, Torres depicts an idealized pre-revolutionary Algeria built by the French but with the harmonious cooperation of Muslim Algerians. Torres's contempt for academic criticism of the colonial project is palpable:

> I am of the generation of May 1958 and I had to leave the land of my ancestors in 1962, chased by the "wind of history" declared by DeGaulle . . .

[ellipsis *sic*]. Since then, politically-correct [bien pensantes] heads spew torrents of lies about the French presence in Algeria, denying the work accomplished by our predecessors—become indigenous—in concert with the aboriginals, the "Arabs."[65]

Torres's memorialization of colonial Algeria is framed as a response to decolonization, not geoenvironmental disaster; the 1954 earthquake is treated only briefly, and serves mainly to convey the impression that Muslim Algerians were well-cared for after the disaster, and perhaps overindulged.[66] In keeping with the process of selective memorialization described by Eldridge, no trace appears of the suffering or even the complaints of the Muslim survivors in the wretched fall and winter of 1954.

Torres's erasure of colonial violence and oppression can be seen as stemming from an ideological support of colonialism deployed both in order to construct a nostalgic image of the colonial past and to defend the pieds noirs against imputed guilt. However, the effacing of colonial violence evident in Torres's memorialization of French Orléansville is also found in memorializations of pre-earthquake Agadir, and not just by French writers. Like Torres and the pieds noirs, both Moroccan and French survivors of Agadir found themselves cut off from the physical spaces they remembered because the built environment that had shaped their lives had been reduced to rubble. As in the work of Torres, the local history of Agadir compiled by Roussafi, Jafri, and Kikr is shaped by nostalgia and avoids addressing the injustices of colonialism. Roussafi, in personal communications and public lectures, speaks about the benefits that the French presence brought to Morocco, particularly in the form of the French lycée. Pre-earthquake Agadir, especially the doomed Talborj, is consistently portrayed as a site of inter-ethnic harmony; Roussafi has described life there as "perfect cohabitation."[67] Agadir1960 .com and Dartois's website, mfd.agadir.free.fr, are the interactive products of collaboration between French and Moroccan Muslim survivors of the Agadir earthquake. In these spaces, the prevailing nostalgia and grief leave little room for memories of the violence of colonialism in the city that was lost. For these survivors of the Agadir earthquake, the environmental trauma, not the trauma of decolonization, shapes an idealized memory of the pre-disaster city.

Mapping Loss

The attention to memories of spaces and places found in these internet-era memorializations of Orléansville and Agadir were prefigured in Ali Bouzar's 1985 book-length memoir. Bouzar was in Oued Fodda during the 1954 earthquake and in El Asnam (previously known as Orléansville, today, Chlef) when the 1980 earthquake struck. Bouzar had experienced the physical violence of environmental disaster even more intimately than had Aït Ouyahia or Khaïr-Eddine and much more directly than Bensimon (safe in Montreal in 1960) or Tengour (who experienced the 1954 earthquake from the safe distance of Mostaganem). While Jacques Bensimon portrayed a pre-disaster past that contained intense suffering, while nevertheless maintaining a nostalgic approach to that past, and Roussafi, Jafri, and Kikr focused on remembering an idealized, harmonious pre-disaster past, Bouzar's disaster memoir *Le Consentement du Malheur* combined his nostalgia with an emphasis on the suffering caused by disasters themselves. Bouzar's 1985 memoir is titled *Récit: Témoignage sur la Catastrophe d'El Asnam du 10 Octobre 1980* (Narrative: Testimony on the El Asnam Catastrophe of October 10, 1980), but Bouzar was also a survivor of the 1954 earthquake and of the years of the Algerian Revolution. His memoir combines "long" views of those events with his representation of his more recent memories of the 1980 earthquake. Like that of Roussafi, Jafri, and Kikr, Bouzar's narrative demonstrates that, for some, natural disaster could eclipse the violence of decolonization.

In the immediate aftermath of the earthquakes described here, a predominant early response was to call for reconstruction—the product of a pressing need for shelter for the many, and, for a privileged few, a chance to take advantage of new opportunities. When such reconstruction came too slowly, incriminations were many. For Ali Bouzar, however, the proper response to the destruction of a city is memorialization, not reconstruction: "For this city of memory, of imagination, of childhood, no bulldozer, no engine of *déblayage,* even the most powerful and effective, could destroy it, make it disappear. No new El Asnam, well-conceived, beautiful, very beautiful, flowering white city of the future could erase it."[68] In contrast to Kréa, Khaïr-Eddine, or Bensimon, Bouzar does not accept that the disaster has produced a rupture that renders the past unknowable or inscrutable or in need of creative representation. For Bouzar, the pre-disaster city still exists, indestructible in his memory, and he sets out to document it.

Unlike Aït Ouyahia, Tengour and Bensimon, whose treatments of the decolonization-era earthquakes in Algeria and Morocco are framed within the authors' memoirs of their youth, Bouzar treats his youth in the context of a memoir ostensibly focusing on an earthquake. Bouzar declares his intention to write only what he and his own family experienced themselves, organized with a "metallic path, the cold of a stopwatch," denoting the timing of events on that fateful October 10, 1980.[69] Yet Bouzar warns his readers that they must excuse his "digressions, backwards turns, evocations of certain details which existed before, which I knew before."[70] Such details are important, argues Bouzar, because they constitute an important part of his memory of the city and because they offer a bit of comfort to the grieving *asnamis*, the people of El Asnam.[71] Like Bensimon's account of Agadir and like the later memorial websites, Bouzar's memoir focuses specifically on the people and spaces of the city. After the earthquake, explains Bouzar, the city is enveloped in dust that obscures, rendering invisible buildings that have, regardless, been destroyed by the earthquake. The act of writing restores these buildings and these memories.[72] Memorialization, for Bouzar (as for Jacques Bensimon regarding Agadir) is part of a search for healing. Through his digressions, Bouzar's account folds together an account of the day of the 1980 earthquake with nostalgic memories of the Orléansville of his youth.

Bouzar's memoir is structured as a travel narrative, beginning with a tranquil morning walk through the city the morning before the 1980 earthquake and ending with his voyage, by car, out of the devastated city. Through this travel narrative, Bouzar provides the reader with a catalog of his personal landmarks in the city—cafés, stores, streets—reconstructing and memorializing the author's emotional geography, populating these spaces with memorable people both from the day of the 1980 earthquake and from his youth, "the time of French occupation."[73] Along the way the reader is presented with descriptions and anecdotes spanning decades. The places destroyed by the 1980 earthquake are sites of memories of colonialism, of the 1954 earthquake, and of decolonization. By organizing his memoir geographically, Bouzar flattens chronology.[74]

Unlike Aït Ouyahia's earthquake chapter or Kréa's polemical play from the 1950s, Bouzar's memoir shows little sign of nationalism or anti-colonialism, aside from a single statement contrasting "national liberation" with "the atrocities of colonial repression" among a catalog of misfortunes. These misfortunes are portrayed not as exceptional events but as integral parts of the human experience: "like the death of a father. The things of life."[75] Bouzar displays little resentment toward the French; writing in the 1980s, Bouzar's ire is reserved for religious ideologues. Contrasts between "the bipolarized society of the colonial period" and the "apparently egalitarian post-independence community"[76] seem effaced in the stories of the eccentric characters whom Bouzar recalls from his youth: Tifnini, the irritable merchant; Qabech, the enormous, simple-minded porter; Tonton Sennis, the black, French-naturalized, Muslim outcast; Bouzar's mischievous gang of *lycée* friends.

Decolonization is nevertheless inscribed in Bouzar's memory of the built environment of the city. The very names of the streets and buildings invoke the manner in which the memory of the events of 1980 parallels memories from the era before 1962, due to an official process of decolonizing place-names after independence: "rue Dahnane" is "ex-rue Bugeaud"; "rue des Martyrs" is "ex-rue Isly"; and Bouzar's apartment building, "al Nasri," is "ex-Le Progress." More than just names have changed; for each locale, Bouzar describes both what the place was in 1980 and what it had been before independence. For example, Bouzar recounts how, the morning before the 1980 earthquake, he leaves the café that was known "in the time of French occupation" as "La Rotonde" and walks through the An-Nasr residential development. Bouzar informs the reader that, after the 1954 earthquake, this had been nothing more than a "stony wasteland, dusty in summer and muddy in winter." Without warning, the narrative shifts to the context of the Algerian Revolution: "It was this stony wasteland that, one day, a man crossed like lightning, and passed very quickly in front of me. Then several seconds later, I heard an explosion, right on rue Isly, which I localized at the café La Rotonde. . . . In a fraction of a second, I found myself, an adolescent of 15 years." The violence of the anti-colonial struggle thus erupts into Bouzar's narrative of the 1980 disaster. The adolescent Bouzar is confronted by a "petit-blanc" (working-class) European man with a gun. Smoke and dust filled the air, and "soldiers ran in all directions." But Bouzar states that these were "already common scenes," and his youthful self carried on with his daily tasks. Similarly, Bouzar the writer brings the reader back to the post-independence era, reminding the reader that this place became the site of the An-Nasr apartment complex in 1962. Bouzar then proceeds with his narrative of October 10, 1980, leading to his arrival, after his morning stroll, back at his family foyer, shortly before the earthquake.[77] When the earthquake strikes, Bouzar and his family survive unscathed, though covered in dust and emotionally shaken; not all of their neighbors are so lucky. In his description of the moments of the 1980 disaster, Bouzar attributes his family's survival to his experience of the 1954 quake, flashing back to the events of his childhood: "By instinct, with my son held by the arms like a sheep bleating in terror, I moved toward the center of the courtyard. By instinct, and also because I had the reflex, unaware that twenty-six years had passed, previously, from September 9, 1954."[78] The seismic shock erases time; the events blur together.

Later, after the 1980 earthquake, Bouzar and his family are in their car, on their way out of the stricken city, and they stop to check on the well-being of family friends, and the narrative again swerves into the past. The car stops in a "rocky, dusty place, overlooking 'Train-Car City,'" (Cité Wagons), which, as Bouzar explains, became a site of emergency housing after the 1954 earthquake, where livestock cars were repurposed as temporary refugee quarters. Vacant after 1958, the area became a hangout for Bouzar and his young friends, and the anti-colonial struggle again erupts into the narrative: "an explosion." French soldiers posted nearby suspected an FLN bomb-making accident and responded in force, setting up roadblocks and conducting a manhunt. The explosion, however, did not originate from a revolutionary bomb-maker, but from one of Bouzar's mischievous friends, who had built a rocket (he later became, according to Bouzar, a nuclear physicist for independent Algeria). Captured and dragged to the police station, the boy scientist escaped imprisonment only because one of his classmates was the son of the police commissioner: he was protected by the privilege of attending an elite French school with elite French friends. Bouzar and his pals also seemed buoyed by the resilience of youth (or the fog of memory): they experience the anti-colonial struggle in exciting anecdotes and the 1954 earthquake as the creator of their childhood haunts.[79] But when the narrative abruptly returns to 1980, with the family driving out of the city, Bouzar's digression is immediately followed by a statement that seems to encompass the violence of the past as well as the present: "I begin to be exceedingly tired of this city in ruins, this disaster, these dead, these injured, the mad, the neurotic, these human shadows who wander through the streets of the city, these victims."[80] In his narrative, Bouzar leaves the ruined city behind, taking the road to Algiers.

Near the end of the memoir, the road passes through his natal village of Oued Fodda, twenty kilometers outside of the city, where the ten-year-old Bouzar had experienced the 1954 earthquake. Here, Bouzar recounts his experience of that trauma: the family escaped harm but their home was so severely damaged it had to be razed, and the family agonized for two days, fearing that Bouzar's father had died in Orléansville where he had been working.[81] The memoir then ends with Bouzar's successful arrival in Algiers. Yet Bouzar's desire to escape to safety in 1980 stands in contrast to the author's later nostalgia for the city of the past: through his writing, he attempts to rebuild the city that bulldozers cannot erase.

Unspeakable

Bouzar's memoir joins together his memories of the earthquakes of 1954 and 1980 and of decolonization. These experiences, in his account, are inseparable, coexisting in the places of his memory. Bouzar apologizes for the digressive and nostalgic elements of his narrative, but makes clear that his real purpose is not the orderly representation of the chronology of the 1980 earthquake but rather the therapeutic recollection of the past: "to soothe the wounded heart of the Asnami survivors (*sinistrés*)."[82] Bouzar sees the act of writing as presumptuous because the events he seeks to describe are beyond comprehension: "Unimaginable. What happened, in little more than half a minute, surpasses the limits of the imagination, defies human understanding and meaning, and returns man to his first humanity, made of fragility, of temporality."[83] Bouzar points out that that the disaster produced not only the thousands of dead and injured but also "thousands more mad, neurotic, anguished, the mentally mutilated, and hundreds of thousands more, finally, who can no longer live like others do, like before."[84] Bouzar is one of the lucky ones: "Praise be to God! The drama that I just lived struck a blow to my imagination, shook my heart, but passed over my reason and my mind. It destroyed neither my reason nor my soul."[85] Bouzar says that writing his "testimony" is an act of betrayal ("parjure") against the shame (*pudeur*) that envelops trauma in Algerian society, as well as an indiscretion in the face of the silence of the dead.[86] Nevertheless, having recovered his powers of language, Ali Bouzar sought to put down in words what others could not. Bouzar states that he offers his account of the disaster as an act of altruism, like those who donated blood in the aftermath to aid the injured, and as act of consolation, like the prayers that Muslims around the world offered for the fallen.[87]

The Algerian poet Moufdi Zakaria, in a poem introducing Roussafi, Jafri, and Kikr's locally published history of the Agadir earthquake, echoes Bouzar in describing the experience of some survivors of disasters, who find the horror and loss unspeakable, and who find the memories of the disaster, and of life before it, too painful to be expressed:

> There are many victims of the Agadir earthquake who do not want to
>   remember
> the seismic disaster, and hate all mention of it, and it causes them sadness
> and pain, and reminds them of its cruel effects, and also reminds them of
>   those who are
> lost, loved ones and friends. They are, in fact, living in the disaster
>   continuously[88]

Bouzar describes a similar phenomenon among survivors, who are silenced either by "the empire of terror and anguish, or . . . because shame, elemental, almost natural within our traditions, wants us to silence the suffering of the moment, the injury of the soul, the despair of being."[89] Bouzar recounts that there was a moment when he, too, was overwhelmed by the horror of the disaster and became speechless: "I am like an animal. . . . I imagine nothing, I no longer envision anything."[90] This experience of wordlessness and unspeakability seems to be a common response to disasters and has been identified in psychiatric and psychoanalytic contexts as a symptom of psychological trauma.

However, as Michelle Belaev argues, "the 'unspeakability' of trauma . . . can be understood less as an epistemological conundrum or neurobiological fact, but more as an outcome of cultural values and ideologies."[91] Bouzar's allusion to the silencing effect of disaster suggests that he would agree with Belaev: the shame that silences is only "almost" natural—it is the product of traditions. Such traditions need not be assumed to be of ancient origin. Paulina Grzeda has argued that responses to painful events are also conditioned by "well-established conventions," including the conventions of writing about trauma.[92] Emilie Morin has pointed out that the concept of "unspeakability" has also been "ubiquitous in accounts of the Algerian War of Independence," reflecting not only the horrors of torture, but also the silencing of voices by French censorship during what the press could only refer to as the "events" in Algeria, and the independent Algerian state's own silence about the divisions among Algerians during the war.[93] This suggests that, in the context of decolonization-era disasters in the Maghreb, invocations of unspeakability might be seen as responses conditioned by the experiences and cultures of decolonization as well as by the effects of disasters themselves.

Conclusion

For Moroccan, Algerian, and French writers, memories of the disasters that destroyed Orléansville, Fréjus, and Agadir have prompted not just traumatic silence but also a flowering of expression grappling with events and constructing memories of the places and people who were lost. The memoirs of survivors and the imaginations of creative writers reveal the enduring interconnections between the environmental and political events that began during the years of decolonization. These written works also often exhibit an appreciation of the disaster as actor; the inanimate becomes not just context but the author of human lives and deaths. These disasters, occurring during a sudden disruption of the West's political and military mastery of the planet, reveal the weakness of the putative divide between the environmental and the social. They also reveal how the events of decolonization and environmental disaster, however sudden in onset, endure over time, not only in their observable, objective "effects" but, as Zakaria points out, in the ongoing experiences of those who grappled with the meaning of these events.

The events discussed in this volume—decolonization and environmental disaster—were linked by synchronicity. In the memories of survivors, however, the environmental catastrophes of 1954, 1959, 1960, and 1980 were also linked to decolonization through categories of suffering and violence and through the geography of memory. It would be a mistake, however, to view these as mere coincidences that were later joined. To tell the story of decolonization without including these environmental catastrophes would involve assuming that political events were autarkic in shaping Franco-Maghrebi experiences in these years; to tell the story of environmental disaster without reference to political change would be to impose a division of events that is alien to the documentary record. The memoirs and literary representations discussed here suggest that those who lived the violence of decolonization and environmental disaster did not conceptualize or experience this violence in discrete categories. For these writers, narratives of decolonization could not be neatly separated from the other disasters their communities experienced, and environmental disaster could not be separated from the transformations brought about by national independence. Yet the mutual imbrication of these events does not appear suddenly, decades later, in works of literature and memory. The archival record discussed in earlier chapters reveals that decolonization and environmental disaster were tightly interrelated through webs of causation and meaning, in both the "short" and the "long" consequences and experiences of these events.

CHAPTER 8

Conclusion

Humanity and Environment

THE HISTORY OF DISASTERS is both defined and obscured by the persistent tendency to privilege the results of deliberate human action in history.[1] If Agadir or Orléansville had been leveled by a bomb, their destruction would be as widely recounted as the fate of Hiroshima and Nagasaki. But even human-induced events are often neglected if they cannot be ascribed directly to human intentions.[2] From this intention-centered perspective, our understanding of the history of Fréjus seems to hang upon the validity, or lack thereof, of the Schmidt-Eenboom hypothesis, and the history of the Moroccan oil poisoning upon the culpability of the US military. We tend to think that hundreds dead due to an attack matter in a way that hundreds dead due an unintended event do not. If the Front de Libération Nationale blew up the Malpasset Dam, if we judge the mass paralysis in Morocco to be the fault of American imperialism, then these events attract our attention in a way that "mere" accidents do not. A great chasm in historical perception separates the intended from the unintended, the human from the inanimate. I would argue, however, that this imagined chasm limits our understanding of history.

As Timothy Mitchell writes, "We have entered the twenty-first century still divided by a way of thinking inherited from the nineteenth."[3] The tendency to forget or ignore the suffering provoked by non-intentionally induced rapid-onset catastrophes is predicated on this way of thinking, which separates the human and the social from the nonhuman. For millennia, humans from a wide variety of cultural traditions tended to anthropomorphize the non-human forces that shape our lives (and voilà, the sun became Apollo, and Indra made the rain). The Western Enlightenment's rejection of such explanations was accompanied by the partitioning of the physical world from the world of human agents. The former was assigned to science and engineering; the latter to history and politics.

The notion of "environment" (etymologically, that which surrounds or encircles) is predicated on its opposition to the human agent at the center. Despite the idea of motion imbedded in the root verb, *virer,* the environment is traditionally depicted as relatively stable, changing only slowly (but for human intervention).[4] Thus twentieth-century scientists imagined the comforting stability of an eco-*system* in the dizzying activity of nature. The human actor, by contrast, moves impetuously within this "natural" environment, sometimes disrupting it. But this kinetically based distinction between agent and environment fails in the case of rapid-onset disasters, including both earthquakes and anthropogenic accidents in which human creations unleashed massive and unintended effects.

Philosopher Jane Bennett asks, "How can humans become more attentive to the public activities, affects, and effects of non-humans? What dangers do we risk if we continue to overlook the force of things?"[5] Disaster survivors were often keenly aware of the "force of things" in human history, as decolonization and disaster merged into a single perceived event: sudden motions of the inanimate physical world (rocks, water) intermingling with the actions of humans. Of course, tectonic plates, bursting dams, and paralyzing chemicals do not possess intentionality. However, as neuroscience, biochemistry, and philosophy undermine the illusion of human intentionality, this point seems less and less relevant, and what remains of the agent-environment distinction depends largely on the distinction between motion (the agent acts) and context (the environment mostly just sits there, surrounding). Gregory Clancey has recognized that rapid-onset disasters appear to possess the unpredictability, arbitrariness, and suddenness of human actors—and also are humanlike in their tendency to thwart modern imperial efforts to know-in-order-to-control.[6] This is not to say that slow-moving events like climate change, erosion, or desertification do not have agent-like characteristics: they have effects in history, and the environment never really just sits there, surrounding. The inanimate world acts constantly: it is "vital, energetic lively, quivering, vibratory, evanescent, and effluescent."[7] My point here is that this agentishness of the physical environment—what Bennett has called "thing-power"[8]—is more visible in large-scale, rapid-onset events, making "rapid-onset disasters" a useful category.

My goal in this book has not been to avoid anthropocentrism, for that would be to write an environmental history of these events that makes human history irrelevant and thus reifies the separation of the non-intentional from the historiography of the social and political. Instead, this volume has aimed to explore the role of the inanimate-in-motion in human history. Obviously, the scope of this book is delimited both spatially and temporally by human processes and anthropogenic places: the era of decolonization; the French empire since 1954; localities defined by patterns of human habitation. The focus throughout has been on the human experiences and human actions that followed the unintended movements of rock, water, and cresyl phosphates. I have sought to investigate the history of a neglected category of event, unintended disasters, and a neglected category of human suffering, the suffering of disaster victims. This book is, in other words, a humanistic project: there are no disasters for the inanimate, and rocks do not suffer. The emphasis here has been on how humans, as the social animals that we are, interpret the movements of the inanimate in the contexts of sociopolitical experiences—in this case, in the contexts of empire and decolonization. Yet this book also strives to recognize that humans are not the masters of their own fate, and that the nonhuman world shapes the human experience. Human history is, in Mitchell's words, "an alloy that must emerge from a process of manufacture whose ingredients are both human and non-human, both intentional and not, and in which the intentional or the human is always somewhat overrun by the unintended."[9]

It is easy enough to recognize, with Rousseau, that there is no "disaster" separate from the human context it affects. As Mitchell notes, however, it is more difficult to acknowledge that there is no human actor who is separate from the environment.[10] In examining the history of disasters, one might ask: which is agent, and which is environment? Do the humans act, and environmental events constrain these choices? Or does the disaster strike, and have an impact on the human environment in which it acts? Using the rich variety of available sources, the historian might tell a story in which the humans (their empires, their revolutions, their hierarchies) are the environment, and the central agent is the earthquake, flood, or poisonous compound which acts impulsively (though not omnipotently, its effects being channeled by the human environment). Or one might tell a story of human agents, occasionally jostled and challenged by unexpected events in the physical, non-human environment, but exerting power over each other and over the environment. Both of these possible narratives are based on a dichotomy between the human and the environmental which is highly problematic but difficult to avoid. Some historical sources, however, provide a glimpse of what it is like to experience the world without this dichotomy.

The "experts" who appear in this book tended to imagine themselves as free and rational agents acting upon a knowable and pliable, if occasionally recalcitrant, physical environment, an approach which was consistent with the colonial desire to master both the environment and the colonized subject. Engineers, seismologists, and urban planners were confident in their ability to respond effectively and to prepare for future environmental events by building a brave new world in a ruined landscape. They arrived on the scene in the wake of the disaster, when the motion of the inanimate had ceased or diminished. Among the experts discussed here, only the epidemiologists engaged with the inanimate as it still moved with the speed of agency, playing real-time defense against a still-advancing chemical toxin. The focus of this book, however, has been non-experts, for all of those living in disaster-stricken areas experienced the inanimate-in-motion. The poison in Morocco moved at a modest pace and was initially unknown to its victims until after the fact, but floodwaters and seismic waves moved with lightning speed through the human and human-built environment, felling buildings and bodies. This experience led to works of memory and representation—by Henri Kréa, Christian Hughes, Ali Bouzar, Mohammed Khaïr-Eddine, and others—that mark no great divide between the human and the environmental, and in particular no great divide between the environmental disaster and the other great transformation at hand, decolonization.

This book has demonstrated that the integration of environmental disasters and narratives of decolonization found in memoirs and later representations is also visible in the archival record produced in the 1950s and early 1960s. This early documentary evidence reveals that, from the moment disaster struck, the inanimate-in-motion shaped the human process of decolonization through multiple avenues of causality, as reactions to these disasters impacted reactions to decolonization. At the same time, humans interpreted the meaning of these environmental events in terms of the political and social events of decolonization. Chemical toxins impaired a superpower's pursuit of its Cold War objectives while enabling the opposition party in a newly independent state to mount new critiques of the national government. Floodwaters permitted the local reassertion of boundaries between colonizer and colonized in the provincial metropole, in contravention of new state policies. Tectonic movements revealed and exacerbated the inequities of colonialism, undermined the crumbling legitimacy of colonial rule, and inflamed the resentment of the colonized. Seismic waves also created opportunities for diplomacy and the extension of new forms of foreign influence and destroyed precolonial urban architecture and ways of life. This destruction catalyzed the expansion of domestic state power and created new lines of cultural contestation in the post-colony after independence. In both the immediate and long aftermaths of these events, survivors, witnesses, and opportunists produced representations of disasters that merged the environmental with the social and political.

Decolonization was not, therefore, a purely human story.

1. Introduction

1. Morin, "Unspeakable Tragedies"; Aït Saada, *Histoire de lieux*, 234; Ouardi, "Écriture, théâtre et engagement," 75–76; "Théâtre d'Henri Kréa," http://www .algeriades.com/henri-krea/article/theatre-d-henri-krea. The play was performed only in a reading in Paris in 1959. It was too seditious for France, and the Algerian theater after independence privileged works written in Arabic. In 1960 Kréa would become a signatory to the "Manifesto of the 121," supporting the cause of Algerian independence.

2. Kréa, *Le séisme*, 9–11; Aït Saada, *Histoire de lieux*, 234–38. As Brahim Ouardi has pointed out, Jugartha is a mythic figure who imparts a sense of timelessness to the values communicated by the play. Ouardi, "Mythe, théâtre et oralité," 212.

3. Kréa, *Le séisme*, 19.

4. Kréa, *Le séisme*, 82.

5. El Djamhouria Slimani Aït Saada has argued, "For Henri Kréa, the earthquake and the war emerge from the same 'tragedy.' . . . The ravages engendered by the natural disaster or by the war, a disaster due to human stupidity [*bêtise*], are alike, if [the latter] is not still more senseless" (Aït Saada, *Histoire de lieux*, 271). In Aït Saada's analysis, "For this poet [Henri Kréa], human savagery and the brutality of geography share the world equitably, because no God conceals himself behind these colossal manifestations" (389). However, Aït Saada's reading of Kréa is incomplete. In the context of the play as a whole, the choir's fatalistic and pessimistic conclusion might suggest ambivalence about the Algerian Revolution's prospects for success but not about its meaningfulness or righteousness. As Brahim Ouardi has noted, the play's structure does invoke the classical Greek tragedy, with a prologue, two episodes, and an "exodus," and the projection of an image of a mask of Jugartha in the exodus is an allusion to the use of the mask in Greek theater. However, the play does not portray the downfall of a tragic hero; instead, it communicates the righteousness of Algerian anticolonialism and the positive transformation of characters who embrace the cause. Ouardi, "Mythe, théâtre et oralité," 211–18.

6. Kréa, *Le séisme*, 45.

7. Krea, *Le séisme*, 43.

8. Kréa, *Le séisme*, 42.

9. Kréa, *Le séisme*, 48.

10. Kréa, *Le séisme*, 48. Yet redemption is possible: a soldier traumatized by his role loading bombs onto aircraft to bomb villages not only repents but joins the revolutionaries.

11. Kréa, *Le séisme*, 77.

12. Kréa, *Le séisme*, 81.

13. For example, E. Burke, "Transformation of the Middle Eastern Environment"; Church, *Paradise Destroyed*; Cutler, "Water Mania!"; Cutler, "Historical (f)Actors"; D. Davis, *Resurrecting the Granary of Rome*; M. Davis, *Late Victorian Holocausts*; Dias, "Famine and Disease in the History of Angola"; Taithe, "Humanitarianism and Colonialism"; Trumbull, "Environmental Turn in Middle East History." In addition to these works addressing European imperialism in Africa, South Asia, and the Middle East, the impact of environmental disasters has also been extensively and adroitly explored in works such as Clancey, *Earthquake Nation*; Ryang, "The Great Kanto Earthquake"; and Walker, *Shaky Colonialism*.

14. The recent works of Adam Guerin and George Trumbull IV are notable exceptions: Guerin, "Not a Drop for the Settlers"; and Trumbull, "Body of Work."

15. Accampo and Jackson, "Introduction to Special Issue," 168. See also Bergman, "Disaster."

16. Pfister, "Learning from Nature-Induced Disasters," 18. See also Bergman, "Disaster"; Quarantelli, "What Is a Disaster?"

17. "Nature-induced disaster" has been suggested as a term which, unlike "natural disaster," attributes only the onset of the catastrophe to nature, not its effects. Pfister, "Learning from Nature-Induced Disasters," 18.

18. Dynes, "Dialogue between Rousseau and Voltaire on the Lisbon Earthquake."

19. Bergman, "Disaster," 936. See also Cutler, "Historical (f)Actors," 7–8.

20. Steinberg, *Acts of God*, 184; see also xxii.

21. For example, Steinberg, "Down to Earth." As Gregory Clancey has phrased it, "We are used to the idea that accidents and disasters expose previously unimagined vulnerabilities. . . . Less self-evident is the way political actors (including scientists, architects, and other state-credentialed professionals) craft advantage from these same phenomena. How the unexpected natural disaster and the normative machinery of government intertwine, creating not only states of emergency but emergency-oriented states, is a topic we have only begun to explore despite a plethora of intriguing evidence." Clancey, *Earthquake Nation*, 4.

22. This intellectual hazard has been recognized by Accampo and Jackson, "Introduction to Special Issue," 170, and Bergman, "Disaster," 936.

23. T. Mitchell, *Rule of Experts*, 31. I am grateful to Mitch Aso and Brock Cutler for their enlightening discussions of Mitchell at the 2017 annual meeting of the French Colonial Historical Society.

24. Kréa, *Le séisme*, 45.

25. Trumbull, "Environmental Turn in Middle East History," 173.

26. Mann, "Locating Colonial Histories"; Saada, "More than a Turn?" See also Boittin, Firpo, and Church, "Hierarchies of Race and Gender," 61; and Goebel, "Capital of the Men without a Country."

27. Mann, "Locating Colonial Histories," 410.

28. Mann, "Locating Colonial Histories," 433.

29. Lacheraf, preface to *Pierres et lumières*.

30. McDougall, "Martyrdom and Destiny."

31. Aït Saada, *Histoire de lieux*, 20.

32. Aït Saada, *Histoire de lieux*, 29, 290. The depiction of colonized lands as harsh seems to have been a common motif in imperial culture; see Church, *Paradise Destroyed*, 63.

33. McDougall, "Savage Wars," 119, 122.

34. Millecam, a French advocate of Algerian independence, portrayed the environmental catastrophes of his youth—a devastating flood, a cyclone—as akin to the partisan and anti-Semitic violence that wracked his hometown of Mostaganem during the Second World War. Furthermore, for Millecam, "The earlier apocalypses were preparing other, even more ghastly apocalypses: those of the war of independence. . . . The first ones opened the way for the others." Millecam, "Apocalypses," 174. Novelist Mohamed Magani saw his own task as making sense of a particular disaster—the later earthquake in Orléansville in 1980—by relating it both to the universal human condition and to the other catastrophes: to "associate it with other natural calamities and man-made disasters with a view to demonstrate the fragility of the human condition; apply it as a metaphor of psychological and material demolition (or auto-demolition)," quoted in Aït Saada, *Histoire de lieux*, 281. Habib Tengour and Belgacem Aït Ouyahia are discussed further on, in chapters 2 and 3 (Aït Ouyahia) and 7 (both Tengour and Aït Ouyahia).

35. Khalfa, Alleg, and Benzine, *La grande aventure*, 181.

36. Aït Saada, *Histoire de lieux*, 287.

37. "Nouvel an amazigh: Agadir, capitale du Souss berbère fête Id-Ennayer," January 9, 2017, https://www.lereporter.ma/nouvel-an-amazigh-agadir-capitale-du-souss-berbere-fete-id-ennayer/. See also Abdallah Aourik and Tariq Kabbage, "Où étiez-vous le 29 février 1960?," *Agadir O'flla*, February 2008, 13.

38. Kréa, like the other writers discussed in this volume, also genders the suffering of disaster survivors in ways I hope will be explored in future scholarship. In Kréa's play the earthquake is juxtaposed to the labor pains suffered by a woman giving birth (the sister of the "Young Woman" in the play), as a sign of new life to come and also a hurdle in the fight for freedom. When Young Woman's sister dies in childbirth, and another woman dies in the earthquake, their deaths are equivalent: tragic only because they died before seeing the day of liberation.

39. Horowitz, "Complete Story of the Galveston Horror," 64.

40. On memory and histories of political (but not environmental) violence, see Makdisi and Silverstein, *Memory and Violence*. A landmark in the field of French

memory studies is the multivolume *Les lieux de mémoire*, edited by Pierre Nora, which, as Makdisi and Silverstein point out, moved the field beyond "the arbitrary distinction between 'history' as a universal, singular, and normative category and 'memories' as popular, disparate, and multiple" (10). The present volume makes use of excellent studies by Claire Eldridge and M. Kathryn Edwards: Eldridge, *From Empire to Exile*; and Edwards, *Contesting Indochina*.

41. LaCapra, *History and Memory after Auschwitz*, 9; Caruth, *Unclaimed Experience*, 6–7, 11; Ward, introduction to *Postcolonial Traumas*, 6–7.

42. Ward, introduction to *Postcolonial Traumas*, 1–13; Ifowodo, *History, Trauma, and Healing*. Both Ward and Ifowodo trace this argument back to Frantz Fanon.

43. Erikson, *New Species of Trouble*, 21.

44. Horowitz, "Complete Story of the Galveston Horror," 64, 75. See also Erikson, *New Species of Trouble*, 22.

45. Within the fields of memory studies and trauma studies, scholars have come to recognize that all memory—even memories of violence and injustice—should not be placed uncritically in the category of trauma. Heike Becker has discussed the implications of the "individualising and medicalizing commands of this [trauma] discourse." Dominick LaCapra has recognized that overuse of a psychoanalytic concept of trauma runs the risk of "the pathologization of historical processes"; Roger Luckhurst has explored problematic aspects of concepts of trauma ("an exemplary conceptual knot") in an increasingly post-Freudian intellectual landscape. Becker, "Beyond Trauma," 32; LaCapra, *History and Memory after Auschwitz*, 13; Luckhurst, *Trauma Question*, 14. The present volume treats a wider range of experiences and effects of disaster than would typically be included in trauma studies, such as the experiences and responses of those who are traumatized neither directly nor through the secondary transmission of trauma but who seize upon the disaster as an opportunity.

46. Kréa, *Le séisme*, 45.

2. Algeria, 1954

1. Aït Ouyahia, *Pierres et lumières*, 269.

2. Debia, *Orléansville*, 66–67. Debia's population figure of thirty thousand, based on census data, included the city and its two suburbs; forty thousand were counted in the commune as a whole. Debia, *Orléansville*, 55.

3. Debia, *Orléansville*, 76.

4. Debia, *Orléansville*, 76–77. The description of the local gourbis is from p. 87; see also Marius Hautberg, *Orléansville*, December 1954, Archives Nationales de France, Pierrefitte (hereafter ANFP), 19770120/box 15 (SNPC Opération Orléansville 1954), folder "Comptes rendu," 52. On the number of destroyed residences, see also Sous-Préfet d'Orléansville to Préfet Gromand, November 20, 1954, ANFP

F/2/box 4349, folder 1. This document specifies 7,236 "houses" and 34,792 gourbis destroyed. This conforms to the numbers given in Directeur du Service Central de Secours [Algiers], n.d. [October 1954], ANFP F/2/4349, folder "Affaires diverses." J.-P. Rothé counted 7,286 "houses" and 21,006 gourbis, but this did not include the effects of the October tremors (quoted in Hautberg, *Orléansville*, 52).

5. That is, "orogenic movement (Alpine folding)." Rothé, "Les tremblements de terre d'Orléansville." See also Hautberg, *Orléansville*, 49.

6. J.-P. Rothé reported 109 "European" dead in September (8 percent) and 1,188 Muslims (92 percent), according to Hautberg, *Orléansville*, 52. As Todd Shepard points out, it is important to recognize that in French Algeria, "Muslim" was a legal category and was not dependent on an individual's beliefs or practices. Shepard, *Invention of Decolonization*, 7. The subprefecture's November figures counted 1,240 dead for the entire arrondissement, including only 40 of European origin. Sous-Préfet d'Orléansville to Préfet Gromand, November 20, 1954, ANFP F/2/4349, folder 1. A figure of 1,412 dead is given in "Compte rendu," which was sent to the president of the republic by J. Marin, September 18, 1954, ANFP F/2/4349, folder 1. While the numbers of dead may well have been much higher than official figures indicate, it is the number rendered homeless that became a point of controversy in the 1950s. An official estimate in 1954 acknowledged 200,000 *sinistrés* (victims/survivors). Directeur du Service Central de Secours [Algiers], n.d. [October 1954], ANFP F/2/4349, folder "Affaires diverses." A dissident group, the Comité National Algérien d'Aide aux Sinistrés d'Orléansville, estimated that 235,000 were affected by the earthquakes, 80 percent of whom were effectively without shelter. Comité National Algérien d'Aide aux Sinistrés d'Orléansville to Minister of Interior, October 18, 1954, ANFP F/2/4349, folder "Affaires diverses." These figures are proportional to the estimated number of homes destroyed. James Lewis has recently estimated that 300,000 were left homeless in 1954. Lewis, "Algerian Earthquakes of May 2003."

7. Hautberg, *Orléansville*, 51; "Réunion du Comité de direction de Comité National de Secours et de Solidarité," September 22, 1954, 5, Centre d'Archives d'Outre-Mer (hereafter ANOM) FM 81F/288.

8. National Research Council, *El-Asnam, Algeria Earthquake*, 12. In 1955, as a consequence of the earthquake, new seismic codes were instituted, but they would not be rigorously applied in Algeria.

9. Hautberg, *Orléansville*, 53. See also "Simple questions," *Alger républicain*, October 2–3, 1954. As Daniel Williford has pointed out regarding the 1960 Agadir earthquake, engineering analyses tended to reach blanket conclusions about the inevitable failure of "traditional" and masonry construction while promoting the view that "modern" reinforced concrete represented the only remedy to seismic risk despite occasional failures. Williford argues convincingly that this unwillingness to investigate ways to make low-cost construction safer made risk mitigation a privilege for the well-off. Williford, "Seismic Politics," 998–1000.

10. Bekkat, preface to *Histoire de lieux*, 5. Compare Spivak, "Can the Subaltern Speak?"

11. Though political dissidents of the nationalist and leftist varieties often undertook to speak for the disempowered, these anticolonial voices were nevertheless those of elites, relatively speaking. The reader will note, in particular, the absence of voices from Muslim women and working-class Muslim men in the narratives presented here, an absence that I hope will be remedied in future work.

12. For example, Tricou, "La création d'Orléansville," 88; Pelosse, "Évolution socio-professionnelle," 1; Picard, "Orléansville," 66. For a broader view, see also McDougall, *History of Algeria*, 58–71.

13. Tricou, "La création d'Orléansville," 93; "Histoire de Chlef: El-Asnam ou Orléansville à l'époque Française," 2009, http://www.reflexiondz.net/Histoire-de-Chlef-El-Asnam-ou-Orléansville-a-l-epoque-francaise_a760.html; Bekkat, preface to *Histoire de lieux*, 5; Aït Saada, *Histoire de lieux*, 22; Effros, *Incidental Archaeologists*, 90–91.

14. Busson, "Le développement géographique," 52.

15. A reorganization of French departments in 1905 produced the numbering of the three Algerian departments within the metropolitan numeration.

16. Pelosse, "Évolution socio-professionnelle," 11–12. This chapter's treatment of the urban development of Orléansville is particularly indebted to Pelosse's work. On Orléansville's early development as a primarily European provincial town, see also Christelow, "Muslim Judge and Municipal Politics," 5.

17. Debia, *Orléansville*, 55. Julia Clancy-Smith has demonstrated that migration to Algeria from France, Spain, Malta, and Italy was part of a broader migration to North African shores and that a narrow focus on the French in Algeria risks obscuring "the largest trans-sea dispersal of peoples since the Iberian expulsions." Clancy-Smith, *Mediterraneans*, 14.

18. Debia, *Orléansville*, 55.

19. Pelosse, "Évolution socio-professionnelle," 5.

20. Pelosse, "Évolution socio-professionnelle," 2.

21. Kateb, "La gestion administrative," 410–11.

22. McDougall, "Savage Wars," 122. As McDougall argues, the *indigénat* has often been described as beginning in 1881 but was built upon earlier precedents. The Algerian-born children of non-French European settler men were made French citizens in 1889; Jewish Algerians in the *départements* were granted French citizenship by the Crémieux decree in 1870. McDougall, *History of Algeria*, 126, 107, 115.

23. According to French accounts, the Saïah had long been a leading family of the nearby Medjadja people and were of a distinguished religious lineage that, it was claimed, originated among the Idrissid royal family of Morocco. Although they had joined Abd al-Qadir's opposition to French rule, the Saïah subsequently adopted a successful new strategy of association with the French. The first member of the Saïah family to become a French-backed caïd, or rural governor, was Si Henni ben Essaïah, who was awarded the title of Grand Officer of the Legion

of Honor before his death in 1897. The Bouthiba, the other *grand famille* in Orléansville, had developed from a branch of the Saïah in Ténès but by 1954 was prominent in the Chélif Valley as well. René Yves Debia, *Monographie politique de l'arrondissement d'Orléansville*, June 1954, ANFP F/2/4350, 6–8; Pelosse, "Évolution socio-professionnelle," 57. See also McDougall, *History of Algeria*, 80–81.

24. McDougall, *History of Algeria*, 105–6; McDougall, "Impossible Republic," 780–81. Because Jewish Algerians had been naturalized by the Crémieux decree of 1870, Jewish voters of Algerian origin were included in the "European" first college.

25. McDougall, *History of Algeria*, 183–88; McDougall, "Impossible Republic," 786–89; Stora, *Algeria: 1830–2000*, 26. See also Ageron, *Modern Algeria*, 100–105; Talbott, *War without a Name*, 24; Lyons, *Civilizing Mission in the Metropole*, 27–28; Shepard, *Invention of Decolonization*, 36–41. The population figures given are for 1954. As Shepard points out, France would extend a more equal citizenship to Muslim Algerians in 1958, four years into the Algerian Revolution.

26. Debia, *Monographie politique*, 6–8.

27. Pelosse, "Évolution socio-professionnelle," 1, 8–9. A planned "quartier arabe" proposed in 1848 had never come to fruition. Picard, "Orléansville," 65–75.

28. Stora, *Algeria: 1830–2000*, 7, 13, 22; Ageron, *Modern Algeria*, 88; McDougall, *History of Algeria*, 90–99. See also Busson, "Le développement géographique," 34–35.

29. Pelosse, "Évolution socio-professionnelle," 5, 7.

30. Pelosse, "Évolution socio-professionnelle," 7–9. Pelosse states that the European population peaked at 6,289 in 1960, according to official figures. It is likely that these figures for "Europeans" included Jews of Algerian origin who had been naturalized en masse by the Crémieux decree of 1870. Pelosse points out that the European population by 1960 included a large number of "security force" personnel and their families. By 1966 only 260 foreigners remained in the city.

31. "Rapport du service départmental de l'urbanisme," 1955, quoted in Pelosse, "Évolution socio-professionnelle," 1.

32. Debia, *Orléansville*, 56.

33. Debia, *Orléansville*, 61.

34. Commissaire à la Reconstruction, "Rapport," to Ministry of Interior, February 17, 1958, ANFP F/2/4350, folder "Comité National de Secours aux Victimes," 6.

35. Debia, *Orléansville*, 62.

36. Hautberg, *Orléansville*, 75. See also Debia, *Orléansville*, 61. In Debia's account the demise of the city walls was connected with the undesirable growth of the suburbs: "The faubourgs had won: the ramparts were undergoing demolition."

37. Debia, *Monographie politique*, 5, 13–14. The other *communes de plein-exercice* in the arrondissement were Montenotte (pop. 7,000), Cavaignac (6,000), Charon (8,000), Les Attafs (4,000), and Carnot (8,000).

38. Debia, *Monographie politique*, 6.

39. "Note sur le séisme d'Orléansville," September 10, 1954, ANFP F/2/4349, folder 1.

40. Kessler, *Une expérience de protection civile en temps de paix: Le séisme d'Orléansville*, December 23, 1954, ANFP 19770120/box 15 (SNPC Opération Orléansville 1954), folder "Comptes rendu," 10.

41. Kessler, *Une expérience*, 10; Kessler, "Premiers telegrames announcant la catastrophe d'Orléansville," appendix to *Une expérience*; Lt. Col. Curie, "Opération Orléansville: Rapport," September 27, 1954, ANFP 19770120/box 15, folder "Comptes rendu," 10–11.

42. Kessler, *Une expérience*, 11.

43. Kessler, *Une expérience*, 11.

44. Aït Ouyahia, *Pierres et lumières*, 269–70. On medicine and medical training in Algeria 1900–1954, and on the poor conditions at the Orléansville hospital in the mid-1950s, see Johnson, *Battle for Algeria*, 44–47, 63.

45. Kessler, *Une expérience*, 11.

46. American helicopters were also used for reconnaissance and medical evacuations. Roger Leonard, "Débats de l'Assemblée algérienne, compte rendu in extenso 11 octobre 1954," *Journal officiel de l'Algérie* 38 (1954): 992, ANOM 81F/1851; Ministry of the Interior, SNPC to Vrolyck and Curie, September 12, 1954, ANFP F/2/4349, folder 1. See also Morris to Ministry of the Interior, September 18, 1954, ANFP F/2/4349, folder 3, "Presse, radio, mouvement de solidarité." See also Resident General of France in Morocco to Governor General of Algeria, ANFP F/2/4349, folder "Mouvements de solidarité."

47. Kessler, *Une expérience*, 11.

48. Hautberg, *Orléansville*, 7.

49. Kessler, *Une expérience*, 13–14; François Mitterand, "Aide aux victims du séisme," September 16, 1954, ANFP F/2/box 4349, folder 1.

50. Curie, "Operation Orléansville," 1–6; Comité central de secours, "Activité du service central de secours aux sinistrés," September 20, 1954, ANFP 19770120/box 15, folder "Comptes rendu journaliers"; Hautberg, *Orléansville*, 2–3. The US Air Force officer was a Captain Davis.

51. At the subprefecture there was the luxury of the running water, until the pipes were broken by the September 16 aftershock. Hautberg, *Orléansville*, 8, 13, 24. See also "Une mission de secours," *La Dépêche*, September 18, 1954, in ANFP 19770120/box 15, folder "Revue de presse."

52. Hautberg, *Orléansville*, 9–10.

53. Aït Ouyahia, *Pierres et lumières*, 270–72. Aït Ouhia's flashback to his own experience of an automobile accident is discussed in chapter 3 of the present volume.

54. Debia, *Orléansville*, 80.

55. Debia, *Orléansville*, 80–81.

56. Hautberg, *Orléansville*, 8.

57. Debia, *Orléansville*, 80–81.

58. Hautberg, *Orléansville*, 8, 13, 24. See also "Une mission de secours," *La Dépêche*, September 18, 1954, in ANFP 19770120/box 15, folder "Revue de presse."

59. Curie, "Operation Orléansville," 12. See also "Note" to Minister of Interior, September 17, 1954, ANFP 19770120/box 15, folder "Comptes rendu."

60. Curie, "Operation Orléansville," 10.

61. Sous-Préfet d'Orléansville to Préfet Gromand, November 20, 1954, ANFP F/2/4349, folder 1. See also Kessler, "Premiers telegrames annoncant la catastrophe d'Orléansville." Debia's initial telegrams counted the dead at 52 "of European origin" and 116 "français-musulmans."

62. Governor-General of Algeria to Ministry of the Interior, September 9, 1954, ANFP F/2/4349, folder 1; Ministry of the Interior, "Renseignements sur le tremblement de terre de 9 Septembre 1954," September 9, 1954, ANFP F/2/4349, folder 1. Kessler noted that, as word spread of disaster in surrounding areas, false rumors abounded, including false reports that the large dam at Oued Fodda had burst and flooded the entire valley (a smaller dam at Pontéba had indeed burst, flooding some homes); that the city of Miliana had been "mostly destroyed"; and that looting was rampant. Kessler, *Une expérience*, 12.

63. Curie, "Operation Orléansville," 10; Sous-Préfet d'Orléansville to Préfet Gromand, November 20, 1954, ANFP F/2/4349, folder 1.

64. Some word of casualties in Beni Rached had reached Orléansville previously: a typed report in the Ministry of Interior archives from September 10 contains a handwritten figure of thirty-five dead from Beni Rached, less than an eighth of the total. "Note sur le séisme d'Orléansville," September 10, 1954, ANFP F/2/4349, folder 1.

65. Curie, "Operation Orléansville," 10. In contrast, Aït Ouyahia also mentions a news story stating that "the charnel-house of Beni-Rached was discovered by chance by a young Kabyle doctor." He took mild offense at the term "par hasard" since he was deliberately searching for affected villages far from the road, but he states that they had no premeditated plan to go to Beni Rached specifically, and in that sense he accepted the term "by chance." Aït Ouyahia, *Pierres et lumières*, 279–80.

66. Kessler, *Une expérience*, 2, 12. See also Hautberg, *Orléansville*, 6.

67. Kessler, *Une expérience*, 12. Kessler would go on to become a *conseillieur technique* to the government of Cameroun and later the prefect of Mayotte.

68. Kessler, *Une expérience*, 13.

69. Fanon, *Dying Colonialism*, 121–45. As Ellen Amster has pointed out, medicine had long been a tool of colonialism, defining a state of supposed "degradation" among Algerians and providing a justification for conquest and the subjugation of in Algeria. Amster, *Medicine and the Saints*, 60.

70. Aït Ouyahia, *Pierres et lumières*, 273.

71. "Commune mixte de Chélif, Douar du Beni Rached," n.d., Service Algérien de la Presse et d'informations, ANOM FM 81F/288; "Base de données des députés français depuis 1789," accessed January 17, 2017, http://www2.assemblee-nationale .fr/sycomore/fiche/(num_dept)/6208.

72. Aït Ouyahia, *Pierres et lumières*, 274.

73. Bouzar, *Le consentement du malheur*, 139–40.

74. Aït Ouyahia, *Pierres et lumières*, 278.

75. Aït Ouyahia, *Pierres et lumières*, 279–80.

76. On the impact that historians of memory, including Pierre Nora, have on our understanding of memory, memorialization, and history as "polyvalent . . . popular, disparate and multiple," see Makdisi and Silverstein, introduction to *Memory and Violence*, 10; LaCapra, *History and Memory after Auschwitz*, 8–42.

77. The Day of Solidarity raised 436 million francs (1.2 million US dollars), and by the end of September departmental campaigns had already raised an additional 173,502,000 francs, including, in roughly equal measure, private donations (from individuals and organizations) and gifts from local governments (municipal and departmental). Commissaire à la Reconstruction, "Rapport sur la gestion administrative et financière des fonds recueillis par le Comité National," February 17, 1958, ANFP F/2/4350, folder "Comité National de Secours aux Victimes." Comité de Direction du Comité National de Secours (meeting minutes), December 17, 1954, ANFP 4350, folder "Comité National de Secours aux Victimes"; "Conférence de presse du 28 décembre de monsieur le ministre de l'intérieur," ANOM 81F/1851, folder 2; "Deuxième compte rendu d'activité des comités départementaux," September 30, 1954, ANFP 19770120/box 15, folder "Comité National de Secours." Currency conversion throughout is approximate and based on the exchange rate during the year being discussed (in this case, 350 old francs to the dollar in 1954), not on local purchasing power parity. See "Exchange Rates between the United States Dollar and Forty-One Currencies," at www.measuringworth.com.

78. "Deuxième compte rendu d'activité des comités départementaux," September 30, 1954, ANFP 19770120/box 15, folder "Comité National de Secours"; Ministre de l'intérieur, "Circulaire ministérielle no. 285," September 16, 1954, ANOM FM 81F/288.

79. François Mitterand, "Aide aux victimes du séisme," September 16, 1954, ANFP F/2/4350, folder "Comité National de Secours aux Victimes."

80. Kessler, *Une expérience*, 16.

81. Ministry of the Interior, "Communiqué à la presse," n.d., ANFP F/2/box 4349, folder 1.

82. On "developmentalism" in French colonial thought, see McDougall, "Impossible Republic."

83. Roger Moris to Ministry of the Interior, September 13, 1954, ANFP 19770120/box 15, folder "Comptes rendu."

84. Radio broadcast transcript, September 20, 1954, ANFP F/2/4349, folder 3, "Presse, radio, mouvement de solidarité."

85. Hautberg, *Orléansville*, 10.

86. Hautberg, *Orléansville*, 55.

87. Hautberg, *Orléansville*, 27. See also Comité central de secours, "Compte rendu journalier," September 21, 1954, ANFP 19770120/box 15, folder "Comptes rendu

journaliers." Curie's figures are similar: by September 24 the SNPC in Orléansville had acquired 686 tents of various types, mostly from SNPC stocks and from the Service de Santé and the Red Cross, and had distributed them all to survivors, along with additional 1,700 tents supplied by the sapeurs-pompiers and some beds. Curie, "Operation Orléansville," 5. Hautberg noted that the shortage of tents was exacerbated by the fact that recipients sometimes left the Orléansville area altogether (e.g., for Algiers or for mainland France), without returning the tents. This, however, cannot account for more than a small fraction of the shortfall. Hautberg, *Orléansville*, 17.

88. Comité central de secours, "Activité du service central de secours aux sinistrés."

89. Debia, *Orléansville*, 76–77; Sous-Préfet d'Orléansville to Préfet Gromand, November 20, 1954, ANFP F/2/4349, folder 1. Compare "Note au sujet d'Orléansville," n.d. [sometime after October 9, 1954], ANFP F/2/4349. Official estimates rose rapidly in October, as these documents show; among them, the latter "note" was an outlying estimate of the destruction from the two major tremors of September 9 and 16, claiming only twenty thousand gourbis and five thousand houses had been destroyed—but this was still an order of magnitude greater than the number of tents supplied.

90. Hautberg, *Orléansville*, 35. Hautberg was in Paris from September 27 to 30. See also Hautberg, *Orléansville*, 70; SNPC, "Note sur la distribution de matériel," n.d. [after October 9, 1954], ANFP 19770120/box 15, folder "Matériels."

91. Comité central de secours, "Compte rendu," September 30, 1954, ANFP 19770120/box 15, folder "Comptes rendu journaliers."

92. Hautberg, *Orléansville*, 26; "Conférence de presse du 28 décembre de Monsieur le Ministre de l'intérieur," ANOM 81F/1851, folder 2. See also Directeur du Service Central de Secours [Algiers], n.d. [October 1954], ANFP F/2/4349, folder "Affaires diverses"; "Compte rendu," sent to the President of the Republic by J. Marin, September 18, 1954, ANFP F/2/4349, folder 1.

93. Hautberg, *Orléansville*, 26. See also Directeur du Service Central de Secours [Algiers], n.d. [October 1954], ANFP F/2/4349, folder "Affaires diverses." On the construction of the gourbis, see Hautberg, *Orléansville*, 52.

94. Assemblée algérienne, "Compte-rendu analytique," October 14, 1954, 12–13, ANOM 81F/1851, folder 3; "Débats de l'Assemblée algérienne, Compte rendu in extenso 11 octobre 1954," *Journal officiel de l'Algérie* 38 (1954): 1014, ANOM 81F/1851.

95. Minister of the Interior to metropolitan prefects, September 9, 1954, ANFP F/2/4350, folder "Comité National de Secours aux Victimes."

96. Hautberg, *Orléansville*, 44.

97. "M. Léonard a dressé le bilan de la dévastation," *La Dépêche du Midi*, September 25, 1954, in FP 19770120/box 15, folder "Revue de presse."

98. See Shepard, *Invention of Decolonization*, 41–49; Lyons, *Civilizing Mission in the Metropole*, 2. This issue is further discussed in chapter 3 of the present volume.

99. Pelosse, "Évolution socio-professionnelle," 44.

100. Raymond LaQuière, "Débats de l'Assemblée Algérienne, Compte rendu in extenso 11 octobre 1954," *Journal officiel de l'Algérie* 38 (1954): 992, ANOM 81F/1851.

101. Quoted in Pelosse, "Évolution socio-professionnelle," 44.

102. Debia, *Orléansville*, 79. Whereas Debia used the term "brassage," Peyréga's preface in English used the term "melting pot" (8). The idea that disasters lead to "utopian" social unity seems to echo an international trope: see Horowitz, "Complete Story of the Galveston Horror," 64.

103. McDougall, "Impossible Republic."

104. For example, "Avertissement," *L'est républicain*, September 23, 1954; and "Il reste 20,000 sans-abri," *L'Aurore*, September 17, 1954; both in ANFP 19770120/box 15, folder "Revue de presse."

105. Not until 1957 would the FLN organize the Algerian Red Crescent as an alternative to the French Red Cross, an important part of the FLN'S struggle to assert sovereignty over Algeria. Johnson, *Battle for Algeria*, 99.

106. Kessler, *Une expérience*, 13–14; "Débats de l'Assemblée Algérienne, Compte rendu in extenso 11 octobre 1954," *Journal officiel de l'Algérie* 38 (1954): 1013, ANOM 81F/1851. Jennifer Johnson has stated that "chronic neglect presented an opportunity for improvement and propaganda," and this was particularly true in the Chélif after the earthquake. Johnson, *Battle for Algeria*, 197. On the MTLD and the FLN, see Stora, "La différenciation entre le FLN et le courant messaliste."

107. Ahmed Benzadi, "Aprés le séisme du Bas-Chéliff le Gouvernement général n'a pas été touché," *La République algérienne*, September 17, 1954, ANOM FM 81F/1851.

108. For example, Ahmed Akkache, "Quinze jours après la tragédie," *Liberté*, September 23, 1954, ANOM FM 81F/1851.

109. Fletcher, "Politics of Solidarity."

110. Khalfa, Alleg, and Benzine, *La grande aventure*, 154. See also Fletcher, "Politics of Solidarity," 84–90; Shepard, *Invention of Decolonization*, 78–81; and Debia, *Monographie politique*. Debia noted significant PCA activity in Orléansville and Ténès before the earthquake. As Pelosse notes, at the time of the earthquake the CGT in Algeria still had a substantial Algerian Muslim membership; this would change in February 1956 with the founding of a nationalist labor organization, the UGTA. Pelosse, "Évolution socio-professionnelle," 60.

111. Fletcher, "Politics of Solidarity," 86–90.

112. Khalfa, Alleg, and Benzine, *La grande aventure*, 157–58.

113. *Alger républicain*, October 1, 1954, 1, 6.

114. "100 Paysans font une marche sur Duperré," *Alger républicain*, October 1, 1954. See also "Un habitant du douar Chouchaoua," *Alger républicain*, November 20, 1954.

115. "La solidarité aux sinistrés d'Orléansville," *Alger républicain*, October 2–3, 1954.

116. "300 fellahs devant la Marie," *Alger républicain*, October 5, 1954.

117. "Orléansville," *Alger républicain*, October 12, 1954; "700 personnes hier," *Alger républicain*, October 15, 1954.

118. "Plus de 700 femmes," *Alger républicain*, October 29, 1954; Pelosse, "Évolution socio-professionnelle," 49.

119. André Ruiz, "Les sinistrés doivent défendre leurs droits," *Alger républicain*, October 12, 1954.

120. "Extrait du témoignage Chrétien du 8/10/1954," extract dated October 20, 1954, ANFP F/2/4349, folder "Affaires diverses." On "progressivist" French Protestant and Catholic organizations and publications in relation to Algeria in the 1950s and 1960s, see Fontaine, *Decolonizing Christianity*.

121. "Les sinistrés accusent," *La voix des sinistrés du Chéliff* 1 (December 1954): 3.

122. Paul Galea, "Cinq semaines après III," *Alger républicain*, October 17–18, 1954.

123. Hacene Benslimane, "Les Profiteurs de la misère," *La voix des sinistrés du Chéliff*, September 1955, 1. See also R. R., "Il y a un an: Orléansville," *Liberté*, September 1955, ANOM FM 81F/1851.

124. Saïah Abdelkader to René Paira, November 10, 1955, ANFP F/2/4349, folder "Affaires diverses."

125. Assemblée Algérienne, Session extraordinaire, "Compte rendu analytique," October 14, 1954, 18, ANOM 68 MiOM/1; "Compte rendu in extenso 14 octobre 1954," *Journal officiel de l'Algérie* 39 (1954): 1018, ANOM 81F/1851.

126. "Beni Rached," *Alger républicain*, October 13, 1954.

127. "Le Conseil Général d'Alger," *Alger républicain*, October 8, 1954.

128. Comité National Algérien d'aide aux Sinistrés to Minister of the Interior, October 18, 1954, ANFP F/2/4349, folder "Affairs diverses."

129. Paul Galea, "Cinq semaines après," *Alger républicain*, October 16, 1954.

130. "Nouvelle Manifestation," *Alger républicain*, October 24–25, 1954.

131. Commissaire à la reconstruction, "Rapport sur la gestion administrative et financière des fonds recueillis par le Comité National," February 17, 1958, ANFP F/2/4350, folder "Comité National de Secours aux Victimes"; Pelosse, "Évolution socio-professionnelle," 45.

132. Kessler, *Une expérience*, 13.

133. Curie, "Operation Orléansville," 10–11. As Kessler points out in his report, the Algerian counterpart to the metropolitan civil protection service had just been created by the governor-general of Algeria several months before the earthquake. Consequently, the director of disaster relief, even after his return on September 12, was "deprived of [any] means of autonomous action." Kessler, *Une expérience*, 7; see also 2.

134. Kessler, *Une expérience*, 7.

135. Kessler, *Une expérience*, 8–10. On September 19 the leadership of the response effort was reorganized: Tony Roche, secretary-general of the prefecture of the department of Algiers, became *directeur de secours* for Orléansville, and Freychet returned to Algiers. Additional administrators were borrowed from other prefectures in Algeria. Hautberg, *Orléansville*, 22.

136. Hautberg, *Orléansville*, 11, 14.

137. Roger Leonard to Minister of the Interior, October 10, 1954, ANFP F/2/4349, folder 1.
138. The metropolitan Service National de Protection Civile (SNPC) had originated in preparations for the Second World War. In 1938 a service of *Défense Passive* had been created within the defense ministry to prepare and organize local emergency and firefighting organizations in order to prepare for the possibility of the aerial bombing of French cities. In 1948 the service was placed under the authority of the ministry of the interior. After the Korean War the service was reorganized as the SNPC and given additional resources in order to prepare the French population for both wartime and peacetime calamities. Kessler, *Une expérience*, 1–4. René Debia also views the quake's aftermath through the memory of war: stocks of food had been "annihilated" and the army distributed aid "as if to a population of refugees in the middle of a war." Debia, *Orléansville*, 78.
139. Kessler, *Une expérience*, 2–4. Kessler notes that fires in Landes in 1949 and floods in the Netherlands in 1953 had produced massive solidarity campaigns that foreshadow the response to Orléansville.
140. "Avertissement," *L'est républicain*, September 23, 1954, in ANFP 19770120/box 15, folder "Revue de presse."
141. Hautberg, *Orléansville*, 58.
142. Debia, *Orléansville*, 91.
143. Debia, *Monographie*, 10, 23.
144. Hautberg, *Orléansville*, 43.
145. Hautberg, *Orléansville*, 75.
146. Hautberg, *Orléansville*, 75–78.
147. Debia, *Orléansville*, 96.
148. Quoted in Khalfa, Alleg, and Benzine, *La grande aventure*, 182.
149. Quoted in Khalfa, Alleg, and Benzine, *La grande aventure*, 182.
150. Quoted in Evans, *Algeria*, 116.
151. Evans, *Algeria*, 117; McDougall, *History of Algeria*, 192–99.
152. "Cinq militants syndicaux d'Orléansville arrêtés," *Alger républicain*, November 23, 1954; Pelosse, "Évolution socio-professionnelle," 50. This repression parallels that experienced by Christian aid workers later in the war. See Fontaine, *Decolonizing Christianity*, 81–86.
153. "La répression ne fera pas oublier le séisme," *La voix des sinistrés du Chéliff*, January 1955.
154. Pelosse, "Évolution socio-professionnelle," 50.
155. On relations between the FLN, the PCA, the PCF, and *Alger républicain*, see McDougall, *History of Algeria*, 203–4; Horne, *Savage War of Peace*, 137; Shepard, *Invention of Decolonization*, 80; and Fletcher, "Politics of Solidarity," 84–98.
156. "La repression ne règle rien," *Alger républicain*, November 3, 1954.
157. "La fédération dénonce la carence administrative," *La voix des sinistrés du Chéliff*, September 1955.

158. "La repression ne règle rien," *Alger républicain*, November 3, 1954.

159. "Il y a un démain," *Le journal d'Alger*, September 8, 1955, ANOM FM 81F/1851.

160. Debia, *Orléansville*, 5.

161. Debia, *Orléansville*, 83.

162. Debia, *Orléansville*, 88–89.

163. *Algèr républicain*, November 28, 1954; See also Pelosse, "Évolution socio-professionnelle," 49. Pelosse cites the PCA newspaper *Liberté*, December 2, 1954.

164. Nicholas Zannettacci, "L'administration a-t-elle décidé de ne pas reloger des sinis-trés?" *Algèr républicain*, October 31–November 1, 1954; "L'Assemblée algérienne discute de la situation," *Algèr républicain*, November 24, 1954. Zannetacci's figures matched those recorded in official sources—forty thousand square meters of pre-fabricated housing plus ten sheds for livestock had been ordered. See Ministère de la reconstruction et du logement, "Rapport au ministère," October 12, 1954, ANFP F/2/4350, folder "Rapport de l'inspection générale."

165. Sous-Préfet d'Orléansville to Préfet Gromand, November 20, 1954, ANFP F/2/4349, folder 1. See also "Conférence de presse du 28 decembre de Monsieur le Ministre de l'Intérieur," ANOM 81F/1851, folder 2.

166. Kessler, *Une expérience*, 17. According to Kessler, a total of 5,200 tents and 3,900 tarps were eventually distributed by Civil Protection, in addition to 1,030 tents supplied by the army, a total which, he stated "in theory," could house as many as one hundred thousand people.

167. Debia, *Orléansville*, 88; "Conférence de presse du 28 decembre de Monsieur le Ministre de l'Intérieur," ANOM 81F/1851, folder 2.

168. Comité Chrètien d'entente France-Islam, "Communiqué," October 1, 1954, ANFP F/2/4349, folder "Affaires diverses." Demands for more substantial housing in rural areas were not unfeasible in light of the financial situation: by mid-December 1954 only a tenth of the donated funds had been transferred for use in Algeria, to pay for the immediate relief effort, tents, and "Operation Gourbi." Comité de Direction du Comité National de Secours (meeting minutes), December 17, 1954, ANFP 4350, folder "Comité National de Secours aux Victimes."

169. M. H. Davis, "Restaging Mise en Valeur"; McDougall, "Impossible Republic," 777; "Le plan de Constantine," Centre de documentation historique sur l'Algérie, ac-cessed August 4, 2016, http://www.cdha.fr/le-plan-de-constantine.

170. M. H. Davis, "Restaging Mise en Valeur," 176; McDougall, "Impossible Republic," 777.

171. Debia, *Orléansville*, 88–89.

172. Debia, *Orléansville*, 5.

173. Debia, *Orléansville*, 91. Debia's sentiments echo those of the new commissioner for reconstruction, who declared in December 1954 that "improvement of the in-dividual's habitat is incomplete if it is not accompanied by an improvement in the conditions of collective life." Commissaire pour la réconstruction, "Instructon sur

l'organisation de l'amélioration de l'habitat," December 13, 1954, ANFP F/2/4350, folder "Rapport de l'Inspection générale."

174. M. H. Davis, "Restaging Mise en Valeur," 181.

175. Hautberg, *Orléansville*, 14.

176. Pelosse, "Évolution socio-professionnelle," 7. The injustice of European domination must have grown more obvious as the European proportion of the population shrank. In Orléansville in 1955 only 606 of the city's 20,000 Muslims possessed, as "evolved" Algerians, the right to vote as members of the "first college," as Europeans did; another 4,305 Muslims could cast ballots as "second college" voters. Debia, *Monographie politique*, 20.

177. Saïah Abdelkader to René Paira, November 10, 1955, ANFP F/2/4349, folder "Affaires diverses."

178. "Débats de l'Assemblée Algérienne, Compte rendu in extenso 11 octobre 1954," *Journal officiel de l'Algérie* 38 (1954): 1033–34, ANOM 81F/1851; R. Thomas, Service de la Construction, "Rapport, concernant la Commissariat de la reconstruction," November 1958, 50–52, ANOM 81F/1852; Ministry of the Interior, "Conditions dans lesquelles sont secourues les victimes," n.d., ANFP F/2/4349, folder "Presse, radio, mouvments de solidarité"; Ministry of the Interior, "Mesures financières prises à l'occasion du séisme," n.d., ANFP F/2/4349, folder "Presse, radio, mouvments de solidarité"; Pelosse, "Évolution socio-professionnelle," 48; Commissariat à la reconstruction, "Le Séisme du Valée du Chélif," April 20, 1956, ANFP F/2/4350, folder "Rapport de l'inspection générale," 83–93.

179. Pelosse, "Évolution socio-professionnelle," 49. A noted example was Saïah Abdelkader's son's share in the cement company Les cimenteries du Chélif.

180. Pelosse, "Évolution socio-professionnelle," 46–48.

181. Roger Gromand to Secrétaire général du ministère de l'interieur, December 16, 1954; Debia to Gromand, January 7, 1955, ANFP F/2/4349, folder "Affaires diverses"; "Nous vivons toujours sous des tentes," *Echo d'Alger*, [December 1954 or January 1955], in same folder.

182. Roger Gromand to Secrétaire Général du Ministère de l'Interieur, December 16, 1954; Debia to Gromand, January 7, 1955, ANFP F/2/4349, folder "Affaires diverses"; Pelosse, "Évolution socio-professionnelle," 47.

183. Commissaire à la reconstruction, "Rapport sur la gestion administrative et financière des fonds recueillis par le Comité National," February 17, 1958, ANFP F/2/4350, folder "Comité National de Secours aux Victimes." See also "Note à l'attention de M. le Secrétaire d'état à l'interieur chargé des affaires Algériennes," March 14, 1957, ANFP F/2/4350. An additional six hundred thousand francs were raised in Algeria by the Service Central de Secours. Commissariat de la reconstruction, "Le séisme de la valée du Chélif," April 20, 1956, ANFP F/2/4350, folder "Rapport de l'inspection générale du Ministère de la reconstruction et du logement."

184. Commissariat à la reconstruction, "Le séisme du valée."

185. Governor General of Algeria to Minister of the Interior, August 5, 1955, ANFP F/2/4349, folder "Affaires diverses."

186. "Une delegation de deputes communists français et algériens au Ministère de l'Intérieur," *Alger républicain*, October 28, 1954.

187. Pelosse, "Évolution socio-professionnelle," 10.

188. Pelosse, "Évolution socio-professionnelle," 33–37; R. R., "Il y a un an: Orléansville," *Liberté* (September 1955), ANOM FM 81F/1851. Pelosse bases his figures on R. R.'s hostile account, intended to show the contrast between the figure of 2,800 employed in the clearing of ruins in November 1954 and only 800 working a year later.

189. Bedjaoui, *Law and the Algerian Revolution*, 40. Bedjaoui reprints a 1956 map published by the Government-General.

190. Bedjaoui, *Law and the Algerian Revolution*, 37; Heggoy, *Insurgency and Counterinsurgency in Algeria*, 87; "L'apparition de rebelles dans la région d'Orléansville jusqu'alors préservée, accroît encore l'anxiété des Européens," *Le Monde*, March 13, 1956, www.lemonde.fr; Debia, *Monographie politique*; G. Aramu, Commissaire de police, "Rapport," June 4, 1956, ANOM ALG 914/4, folder "Rapport policier Orléansville."

191. Valette, "Le maquis Kobus," 69–88.

192. R. Thomas, "Inspecteur des finances," April 18, 1959, 2, ANOM 81F/1852.

193. Ageron, "Une dimension de la guerre d'Algérie," 565, 585. Orléansville, formerly part of the département of Algiers, became the seat of the département of Orléansville in 1957. On *regroupement*, see also McDougall, *History of Algeria*, 218; Pelosse, "Évolution socio-professionnelle," 10; Entelis, *Algeria*, 53; and Heggoy, *Insurgency and Counterinsurgency in Algeria*, 212–29. According to official figures Orléansville's population increased only by fifteen thousand between 1955 and 1960, although some of the smaller towns in the area grew more rapidly. Pelosse, "Évolution socio-professionnelle," 10.

194. Ageron, "Une dimension de la guerre d'Algérie," 567–85; Fontaine, *Decolonizing Christianity*, 155–56.

195. Commissaire à la reconstruction, "Rapport sur la gestion administrative et financière des fonds recueillis par le Comité National," February 17, 1958, ANFP F/2/4350, folder "Comité National de Secours aux Victims," 8. See also Johnson, *Battle for Algeria*, 116–25.

196. Pelosse, "Évolution socio-professionnelle," 39.

197. Pelosse, "Évolution socio-professionnelle," 26, 58–59; see also Ageron, *Modern Algeria*, 119.

198. Pelosse, "Évolution socio-professionnelle," 26.

199. Pelosse, "Évolution socio-professionnelle," 10. See also R. Thomas, "Inspecteur des finances," April 18, 1959, 2, ANOM 81F/1852.

200. Aït Ouyahia, *Pierres et lumières*, 267.

201. Only 6 percent of Algerian children attended school in 1929, and even by 1954 no more than 2 percent of rural children did. Ageron, *Modern Algeria*, 76; Stora, *Algeria: 1830–2000*, 24. See also Colonna, *Instituteurs Algériens 1883–1939*.

202. Fanon, *Dying Colonialism*, 131.

203. In one incident related by Hautberg, gendarmes were called in to maintain order on September 9 when mobs of irate "indigènes" (natives) who had received vouchers for tents from the municipal authorities demanded immediate satisfaction from SNPC aid workers, who did not speak Arabic. A few days later gendarmes were again called in to provide security when Hautberg was instructed to relocate the tents of earthquake survivors (of unspecified ethnicity) near the subprefecture to make room for administrative tents. Hautberg, *Orléansville*, 9, 22.

204. Aït Ouyahia, *Pierres et lumières*, 282.

205. Chapter 7 herein further discusses Aït Ouyahia's presentation of his relationship to the nationalist movement and his embrace of this "Arab" political identity, including an epiphany of fluency in Arabic during his intervention in Beni Rached.

3. Fréjus 1959

1. Hughes, "Souvenir d'un jour tragique," 109.

2. Donat, *La tragédie Malpasset*, 22.

3. Hughes, "Souvenir d'un jour tragique," 110.

4. Heggoy, *Insurgency and Counterinsurgency in Algeria*, 91–101; Ageron, *Modern Algeria*, 111; Stora, *Algeria: 1830–2000*, 33–56. In 1955 troop levels in Algeria increased from 56,600 to 120,000; by the end of 1956 there were 350,000 French troops in Algeria. Ageron, *Modern Algeria*, 109–12; Stora, *Algeria: 1830–2000*, 48.

5. McDougall, *History of Algeria*, 215. As McDougall recognizes, torture has long been a widely used tool of colonial terror. See also Thenault and Branche, "Le secret sur la torture pendant la guerre d'Algérie."

6. McDougall, *History of Algeria*, 217; McDougall, "Impossible Republic," 790–811.

7. Pierre Herbaut, "Le sort de milliers de jeunes français dépend de la volonté générale de conciliation en Afrique du Nord," *La voix socialiste* (Draguignan), September 17, 1955, Archives Municipales de Fréjus, 30W/60.

8. Choi, *Decolonization and the French of Algeria*, 34–35; Heggoy, *Insurgency and Counterinsurgency in Algeria*, 87.

9. House and MacMaster, *Paris 1961*, 90.

10. Ritzi and Schmidt-Eenboom, *Im Schatten des Dritten Reiches*, 185–86.

11. "Von Agenten und Attentaten," *Kölner Stadt-Anzeiger*, October 24, 2013, www.ksta.de.

12. House and MacMaster, *Paris 1961*, 90.

13. "Catastrophe de Fréjus: Benjamin Stora ne croit pas à la théorie de l'attentat," *Le Monde*, January 25, 2013, http://www.lemonde.fr/societe/article/2013/01/25/catastrophe-de-frejus-benjamin-stora-ne-croit-pas-a-la-theorie-de-l-attentat_1822884_3224.html. The television channel Arte, which aired the 2013 documentary "Le long chemin vers l'amité," publicizing Ritzi and Schmidt-Eenboom's hypothesis, has acknowledged the need for further investigation. Schmidt-Eenboom

has also acknowledged the need to confirm Christmann's statement via corroborating evidence that might exist in the records of the CIA or other nations' intelligence archives, having found no such evidence in the French or German archives. "Von Agenten und Attentaten," *Kölner Stadt-Anzeiger.*

14. Valenti and Bertini, *Barrage de Malpasset,* 25.

15. Menant, "La vielle femme et le barrage," 49.

16. Robion and Foucou, *Fréjus Ve–XXe siècle,* 176; Raymond Cartier, "Malpasset qui fit rire, qui fit peur, qui tua," *Paris Match* 258 (December 19, 1959): 59; Foucou, *Malpasset,* 5; Menant, "La vielle femme et le barrage," 47–50.

17. Valenti and Bertini, *Barrage de Malpasset,* 27–29, 46. The total volume of the reservoir was over 49 million cubic meters, with 24.5 million usable; the water level was still several meters below full when the torrential rains came in November 1959. See also Martin Ricketts, "The Malpasset Dam Disaster—Could the Var Suffer Again?," *Riviera Reporter,* n.d., http://www.rivierareporter.com/local-living/151 -the-malpasset-dam-disaster-could-the-var-suffer-again.

18. Quoted in Valenti and Bertini, *Barrage de Malpasset,* 75.

19. Archives Municipales de la Ville de Cannes, "Expositions virtuelles: Nomination des nouveaux responsables municipaux par Vichy (1K63_0005)," available online at http://expos-historiques.cannes.com/a/3516/nomination-des-nouveaux -responsables-municipaux-par-vichy-1k63-0005/.

20. Reprinted in Valenti and Bertini, *Barrage de Malpasset,* 75–79.

21. Ministère de l'Agriculture, *Final Report of the Investigating Committee,* 39.

22. René Laurin, *Report to Assemblée Nationale: Annex to Minutes of December 17,* vol. 467, 1959, ANFP F/2/box 4343, folder "Catastrophe de Fréjus"; "La catastrophe de Fréjus: Vingt-trois milliards et demi de dégats," *Figaro,* December 19–20, 1959, in ANFP F/2/box 4343, folder "Catastrophe de Fréjus," subfolder "Presse." I have converted old francs to new francs throughout, except where otherwise specified.

23. "La loi d'aide aux sinistrés de Fréjus modifiée par le sénat," *Figaro,* December 24, 1959, in ANFP F/2/box 4343, folder "Catastrophe de Fréjus," subfolder "Presse."

24. Valenti and Bertini, *Barrage de Malpasset,* 89. Eventually an additional two hundred thousand of state funds were allocated for disaster victims, for a total of one million new francs from state funds.

25. *Délibérations du conseil municipal,* January 15, 1960, Archives Municipales de Fréjus, file 7D1-31, and December 17, 1971, file 7D1-36; Cour de cassation, chambre criminel, audience publique no. 66-91852, December 7, 1967, www.legifrance.gouv .fr; Conseil d'État statuant au contentieux no. 76216, May 28, 1971, www.legifrance .gouv.fr; Conseil d'État, "Le droit à réparation d'une ville sinistrée par la rupture d'un barrage," October 22, 1971, Archives Municipales de Fréjus. After the court's eventual finding of state liability (but not fault), it was ruled that aid provided from the solidarity donations fund would be counted toward this liability, contradicting the senate's intentions from 1959. Conseil d'État statuant au contentieux no. 88356, December 3, 1975, www.legifrance.gouv.fr. In 1959 the senate had specified

that receipt of private aid would not in any way count against eligibility for state aid (although the finance minister objected that such a provision was unnecessary, complaining that "one might as well write into a law that 'the State shall not steal from the victims.' . . . The six billion collected will be left entirely at the disposition of the survivors"). "La loi d'aide aux sinistrés de Fréjus modifiée par le sénat," *Figaro*, December 24, 1959, in ANFP F/2/box 4343, folder "Catastrophe de Fréjus," subfolder "Presse."

26. By January 18, 1960, the department of Grande Kabylie alone had collected 13,125,899 old francs. "Fonds recueillis par le 'comité d'aide aux sinistrés du Var,' 18 janvier 1960," ANOM ALG ALGER 1k/1283.

27. Ministère du Développement Durable, "Rupture d'un barrage: Le 2 décembre, Malpasset [Var], France," *Analyse, recherche et information sur les accidents* 29490 (April 2009): 6, http://rme.ac-rouen.fr/29490_barrage_malpasset.pdf.

28. Valenti and Bertini, *Barrage de Malpasset*, 119. This inquiry was ordered by the Court of Appeals of Aix-en-Provence on December 12, 1959.

29. Duffaut, "Traps behind the Failure of Malpasset Arch Dam." Some remained skeptical of such conclusions. A 1946 geographical study by a Professor Corroy had declared that coal and gneiss of Reyran basin "present excellent geographical conditions from the point of view of being watertight [de son étanchéité]. In fact, the total impermeability of this reservoir is assured"; quoted in Valenti and Bertini, *Barrage de Malpasset*, 29. However, in 1972 Olivier Cousin argued that the site of the dam had been shifted two hundred meters from the location originally proposed; more recently, Valenti's and Bertini's 2003 history of the dam suggests that insufficient attention was paid to a later report by Corroy, urging further study of the site. Valenti and Bertini, *Barrage de Malpasset*, 31, 131. See also Corroy, "Geological Study of a Dam Project"; Corroy, "Dam Project with a Storage Reservoir," 55.

30. The Malpasset Dam project was thus part of an imperial "assemblage," to use Jane Bennett's term, and the 1959 disaster in Fréjus, like the mass oil poisoning in Morocco later that year (see chapter 4 of the present volume) was produced by a combination of human and nonhuman action. Bennett, *Vibrant Matter*, 21.

31. Mann, "Locating Colonial Histories," 433.

32. Mann, "Locating Colonial Histories," 414.

33. Mann, "Locating Colonial Histories," 415.

34. Bonheur, "Visitez la Pompeï provençale," 18. See also Prado, *L'imprévisible nature*, 138.

35. Edwards, *Contesting Indochina*, 98.

36. Mann, "Locating Colonial Histories," 414. See also Clancy-Smith, *Mediterraneans*, 5–13, 98.

37. A. Martin to Préfet du Var, December 3, 1954, ANFP F/2/4349, folder "Mouvements de solidarité."

38. Stora, *Algeria: 1830–2000*, 63. Stora notes that the figure of 436,000 Muslim Algerians in 1962 came from the Ministry of the Interior; the 1962 census counted 350,000.

39. A. Martin to Préfet du Var, December 3, 1954, ANFP F/2/4349, folder "Mouvements de solidarité." In the ten months following the earthquake, 341 from the Orléansville area were known to have emigrated to France. Pelosse, "Évolution socio-professionnelle," 39–40. After the earthquake in the Chélif, Fréjus came to exemplify the new trend toward family migration in the 1950s, as noted by Amelia Lyons in *Civilizing Mission in the Metropole*, 4, 10.

40. Bonheur wrote that by 1959 there were "almost 1,000 North Africans," mostly from the Orléansville region of Algeria; Croizard described an Algerian migrant estimating the North African community to be 1,100, "almost all from Orléansville." Bonheur, "Visitez la Pompeï provençale," 18; Croizard, "L'avenue de la mort," 139–40.

41. A. Martin to Préfet du Var, December 3, 1954, ANFP F/2/4349, folder "Mouvements de solidarité." See also Valeri, "La Vague," 96; Bonheur, "Visitez la Pompeï provençale," 18–19.

42. Bonheur, "Visitez la Pompeï provençale," 18–19.

43. René Paira to Directeur des Affaires d'Algérie, July 26, 1956, ANFP F/2/4350, folder "Comité National de Secours aux Victims." The exact figure was 9,039,400 francs. See also Directeur des Affaires d'Algérie to Secrétaire Général, February 14, 1956, ANFP F/2/4350, folder "Comité National de Secours aux Victims."

44. "The Malpasset Dam Disaster—Could the Var Suffer Again?," *Riviera Reporter*. The rains and the rise of the water level occurred in November.

45. Shepard, *Invention of Decolonization*, 41–49; Lyons, *Civilizing Mission in the Metropole*, 2.

46. Lyons, *Civilizing Mission in the Metropole*, 32. See also Choi, *Decolonization and the French of Algeria*, 21.

47. Mairie de Fréjus, "Décision no. 8, Français d'Afrique du Nord," December 6, 1959, ANFP F/2/4343, folder "Catastrophe de Fréjus."

48. Mairie de Fréjus, "Décision no. 4, Français d'Afrique du Nord," December 5, 1959, and "Décision no. 8," December 6, 1959, ANFP F/2/4343, folder "Catastrophe de Fréjus"; *Délibérations du Conseil Municipal*, December 7, 1959, Archives Municipales de Fréjus, file 7D1-31.

49. Mairie de Fréjus (Var), "Catastrophe de Malpasset: 1–Bilan des pertes," n.d. [ca. December 1960–December 1961], ANFP F/2/4343, folder "Catastrophe de Fréjus." Given that the December 7 telegram gave an approximate count of thirty Italians (as well as one German, three Swiss, one Moroccan, and one "dame de Pondicherry"), it appears that these "Italians" were included with the European French in the unpublished municipal report. That the thirty-nine "North African" dead were Algerian Muslims (a common usage at the time), not Moroccans or Tunisians, is also confirmed by the December 7 telegram, which counted fifty-eight "Algerians" (clearly Muslims, since Algerians were placed under the category of "foreigners") of whom only eighteen had been identified at that point, and only one Moroccan. Ministère Intérieur Service National Protection Civile Inspection Générale to Tous Préfets (métropole Algérie), December 7, 1959, ANOM FM 1389, folder

"Catastrophe de Fréjus." The lists of the identified dead and reported missing published in a January 1960 official municipal bulletin (Archives Municipales de Fréjus) includes fifty-eight names that appear to be of Muslim North African origin plus four apparently West African names.

50. Donat, *La tragédie Malpasset*, 22.

51. Mairie de Fréjus (Var), "Catastrophe de Malpasset: 1–Bilan des pertes," n.d. [ca. December 1960–December 1961], ANFP F/2/4343, folder "Catastrophe de Fréjus." Aid to non–North Africans in this report totaled 88,190,431.96; this does not include other expenses such as for transportation, for aid workers, for disposal of dead animals, for public safety, or for other disasters—for a total cost of over 172 million francs for the disaster response. The exact figure for North Africans is 360,212.12 francs. This figure for North Africans is identical in a different draft version of this report, which includes a slightly different total of 90,932,660 francs for payments to the metropolitan French in Fréjus. Neither the figures published by the city government in the December 1960 anniversary brochure nor later reports by the Prefecture du Var issued in 1966 and 1968 distinguished aid payments by ethnic origin. It is not clear whether this accounting of donations received and aid distributed included the provision of housing by nonstate actors such as the Protestant CIMADE (Comité Inter-Mouvements Auprès Des Évacués), which provided prefabricated short-term housing for Muslims, and by Secours Catholique, which also provided some housing. According to the 1960 Mairie report, 4,611,320.76 francs was spent on housing, for the general population, both short and long term (according to the 1968 report, this figure was 5,140,596.25 francs). Even if we were to assume that CIMADE'S assistance meant that Algerian Muslims received a fair share of the investment in housing, this would mean that Muslims still received only 0.43 percent of the rest of the money used for survivors (360,212 francs spent on Muslims out of 83,049,836 in nonhousing expenditures). On the role of CIMADE in Fréjus, see Ville de Fréjus, *Malpasset un an après*, 6–9; for a broader context, see Fontaine, *Decolonizing Christianity*, 58–59.

52. Ministère Interieur Service National Protection Civile Inspection Générale to Tous Préfets (métropole Algérie), December 7, 1959, ANOM FM 1389, folder "Catastrophe de Fréjus."

53. "La loi d'aide aux sinistrés de Fréjus Modifiée par le sénat," *Figaro*, December 24, 1959, in ANFP F/2/box 4343, folder "Catastrophe de Fréjus," subfolder "Presse."

54. Ville de Fréjus, *Malpasset un an après*, 12

55. Valenti and Bertini, *Barrage de Malpasset*, 75; Bonheur, "Visitez la Pompeï provençale," 18. Michael Goebel has discussed the advantages that having an embassy provided for anticolonial dissidents in Paris originating in independent former colonies or imperial mandates. Goebel, "Capital of the Menwithout a Country," 1458–59.

56. Donat, *La tragédie Malpasset*, 43. Regarding the connections between the "solidarity" fundraising campaign to raise funds to indemnify victims of the Fréjus flood

and the response to the Orléansville earthquake, see François Mitterand and Félix Faure, "Creation et composition d'un Comité National d'Action et de Solidarité des Victimes du Seisme de la Région d'Orléansville," ANFP F/2/4343, folder "Catastrophe du Fréjus."

57. Bureau des Affaires Financières et Administratifs to Préfet, Conseilleur Technique, January 29, 1961, ANFP F/2/4343, folder "Var . . . emplois des fonds."

58. Préfet du Var to Ministre d'Intérieur, Direction Général des Collectivités Locales and Ministre des Finances et Affaires Économique, Direction de la Comptabilité Publique, February 2, 1968, ANFP F/2/4343, folder "Var . . . emplois des fonds." The 1961 published anniversary report indicated 7,480,638.45 francs in leftover funds, designated as "Funds Reserved for the Readjustment of Certain Categories of Victims." Since this is less than the amount eventually used for Public Works and "diverse" expenditures, it does not seem that this fund was used for significant readjustments; the designation in 1961 seems to have been simply a placeholder. The total number of dossiers received (3,474) did not increase after 1961. Ville de Fréjus, *Malpasset, un an après*, 14.

59. Hughes, "Souvenir d'un Jour Tragique," 110. Hughes also recalled his psychosomatic sensation of "enormous thirst" in the period after the disaster, a thirst that plagued him even after the family had returned to the work of fishing and the routine of school.

60. Aït Ouyahia, *Pierres et lumières*, 282.

61. Aït Ouyahia, *Pierres et lumières*, 270–72.

62. Kai Erikson has argued that "trauma" can be defined as "the psychological process by which an acute shock becomes a chronic condition, a way of keeping dead moments alive" (Erikson, *New Species of Trouble*, 22). See also Horowitz, "Complete Story of the Galveston Horror." Erikson's and Horowitz's approaches to trauma reflects their awareness of the interconnections between memories of disasters and of the political and cultural milieu in which those disasters occur—a major focus of the present volume. This approach diverges from definitions of trauma that emphasize the radical disconnect between the traumatic event and other life experiences and memories. However, as Paulina Grzeda warns, one should be hesitant to apply a diagnosis of trauma in a medical sense based on memoirs or literary texts that make use of "anti-narrative modes, such as self-reflexivity, disruption of linear chronology, fragmentation." As Grzeda argues, it can be impossible to determine "whether such narrative patterns stem from the experience of trauma itself or whether they can be interpreted as the authors' deliberate choice to draw on well-established conventions by which trauma is recognized in literary representations." Grzeda, "Trauma and Testimony," 66.

63. Revel Renaud, "Le magicien de Paris Match," *L'Express*, October 15, 1998, http://www.lexpress.fr/informations/le-magicien-de-paris-match_630730.html.

64. Bonheur, "Visitez la Pompeï provençale," 17. As Mann explains, "In a practice known as hivernage, West African tirailleurs were removed from the front lines

and garrisoned in the south during the long winter months, where they were thought to suffer greatly from the cold." Mann, "Locating Colonial Histories," 415.

65. Bonheur's comparison of Fréjus to a colonial exposition seems to have been a local trope, dating back to 1931, according to Mann, "Locating Colonial Histories," 426.

66. Russo, "Gaby," 21. Gregory Mann has also described how the French residents of the town became accustomed to interacting with men from the colonies, and public funds were used to build a mosque for West African troops. In the interwar period a few West African soldiers retired and remained in Fréjus, and the town became a destination for some West African civilians, including a few political activists. Funds for the mosque, according to Mann, came from the city of Fréjus, the colonial government for French West Africa, and the Comité d'Assistance aux Troupes Noires. Political dissidents visiting Fréjus included in 1926 Senegal's Lamine Senghor. Mann, "Locating Colonial Histories," 427–28. See also Robion and Foucou, *Fréjus Ve–XXe siècle*, 163–65.

67. Prado, *L'imprévisible nature*, 138. Academic historian Mann also reported an emphasis on this "deeper imperial history" extending back into antiquity, in interviews with retired colonial officers in the area. Mann, "Locating Colonial Histories," 414.

68. Valenti and Bertini, *Barrage de Malpasset*, 67–68, 87. See also Pernoud, "Le Reyran se jette dans la Méditerranée," 171–72; André, "Un regard extérieur," 120.

69. Wallet, *La nuit de Fréjus*, 163.

70. Wallet, *La nuit de Fréjus*, 37.

71. Bonheur, "Visitez la Pompeï provençale," 25.

72. Wallet, *La nuit de Fréjus*, 127.

73. Wallet, *La nuit de Fréjus*, 128.

74. Croizard, "L'avenue de la mort," 139–40.

75. Croizard, "L'avenue de la mort," 140. Régina Wallet also describes this man making a similar statement and identifies him as Mohamed Azpi.

76. Wallet, *La nuit de Fréjus*, 108. Wallet repeats this on p. 138.

77. Wallet, *La nuit de Fréjus*, 138.

78. Wallet, *La nuit de Fréjus*, 194–95. There was a commission to set up to arrange for the care of and manage donations for the sixteen children orphaned in the flood; it is unclear if any of these children were of North African origin. Valenti and Bertini, *Barrage de Malpasset*, 95.

79. Wallet, *La nuit de Fréjus*, 11.

80. Eldridge, *From Empire to Exile*, 11.

81. Shepard, *Invention of Decolonization*, 82–83.

82. Lyons, *Civilizing Mission in the Metropole*, 215.

83. Wallet, *La nuit de Fréjus*, 138.

84. Eldridge, *From Empire to Exile*, 109–10. See also Choi, *Decolonization and the French of Algeria*, 130.

85. Torres, *L'Orléansvillois*, 75.

86. Robion and Foucou. *Fréjus Ve–XXe siècle*, 186. Foucou contributed only to chapter 6 of this volume.

87. Robion and Foucou, *Fréjus Ve–XXe siècle*, 186.

88. Geoffrey Bonnefoy, "Fréjus: la mosquée de la ville FN échappe à la demolition," *L'Express* and AFP, October 30, 2015, http://www.lexpress.fr/actualite/politique /fn/a-frejus-la-mosquee-de-la-discorde_1731263.html; John Lichfield, "A Far-Right Revolution as Regional France Opens Its Doors to a Reconstructed National Front," *Independent*, March 28, 2014, www.independent.co.uk; Amanda Taub, "A Small French Town Infused with Us vs. Them Politics," *New York Times*, April 20, 2017, www.nytimes.com.

89. Taub, "A Small French Town." The Rachline mairie also sought to block public funding of cultural production deemed to be unpatriotic. As Dmitri Almeida has argued, "The cultural policies of the Front National remain deeply rooted in a nativist understanding of culture and a dirigiste approach that willfully excludes postcolonial minorities." Almeida, "Cultural Retaliation," 1; see also "Le Front National assume un discours sur la culture tristement terrifiant," *Le Monde Idées*, February 17, 2017, www.lemonde.fr.

90. Edwards, *Contesting Indochina*, 88–115.

91. Mann, "Locating Colonial Histories"; Edwards, *Contesting Indochina*, 98.

92. Mann, "Locating Colonial Histories," 418.

93. "De la calomnie," *Al-Istiqlal*, March 21, 1960, 2. The government of Tunisia, under Bourguiba, also donated funds. Moalla, director general of the Banque Centrale de Tunisie, to Wilfred Baumgartner, Banque de France, December 4, 1959, ANFP F/2/4343, folder "Dons." Gregory Mann also describes a West African veterans' association that advocated "negotiated independence" while making a public show of its generosity toward the Fréjus fund, apparently attempting to use disaster donations as a means of accumulating political capital for use in their campaign for political decolonization. Mann, "Locating Colonial Histories," 429. Mann cites *L'ancien combattant soudanais*, December 1959, January 1960, and February 1960.

4. Poison, Paralysis, and the US in Morocco

1. Faraj, "Historical Background," 6. The highest contemporary tally of victims was 10,466; see Primeau, "Rehabilitation of 10,000 Victims," 1249. In 2008 *La gazette du Maroc* stated that 20,000 had been afflicted. "Scandale des huiles frélatées," *La gazette du Maroc*, February 22, 2008, http://www.lagazettedumaroc.com/articles .php?id_artl=1658.

2. Lt. Cmdr. J. D. Bailey, US Naval Attaché, November 10, 1959, United States Archives and Records Administration (hereafter NARA) RG 59, WHO Subject Files, entry UD-07D 81, box 2, folder "Morocco Oil Poisoning Case"; Faraj, "Historical Background," 9; David Nes to State, January 4, 1960, "Morocco's Toxic Oil

Disaster," NARA RG 59, WHO Subject Files, entry UD-07D 81, box 2, folder "Morocco Oil Poisoning Case"; Alfred B. Allen (AFE) to Root (AFN), March 4, 1960, NARA RG 59, entry A1 3109D, folder "M-3 Cooking Oil Incident."

3. Zinn, "Survey of Earlier Triaryl-Phosphate Intoxications."

4. Honor V. Smith and J. M. K. Spalding, "Report to World Health Organization," October 6, 1959, NARA RG 59, WHO Subject Files, entry UD-07D 81, box 2, folder "Morocco Oil Poisoning Case," 8–10. This report was later published as H. V. Smith and J. M. K. Spalding, "Outbreak of Paralysis in Morocco." See also Faraj, "Historical Background," 13.

5. Bennett, *Vibrant Matter*, 21.

6. Baum, "Jake Leg," *New Yorker* 79, no. 26 (September 15, 2003): 50–57; Zinn, "Survey of Earlier Triaryl-Phosphate Intoxications"; Tosi et al., "October 1942."

7. H. V. Smith and J. M. K. Spalding, "Report to World Health Organization," NARA, 10. It later emerged that there was also a gendered element to the distribution of suffering: women constituted 61.5 percent of the victims and women tended to be afflicted more severely. Western medical observers offered several explanations for this disparity. One is that Moroccan women consumed more oil while cooking; another is that the disparity was due to women's subordinate status within the family, as men dined first and women mopped up with bread what was left. A third explanation suggests that men often ate fewer meals at home. An exception to the general predominance of women was found in northern Morocco, where a large number of men lived away from their families and ate in canteens, where they had consumed the contaminated oil. Travers, "Results of Intoxication with Orthocresyl Phosphate"; Godfrey, "Epidemic of Triorthocresylphosphate Poisoning"; Gross, "Diagnosis and Sympomatology," 58.

8. Faraj, "Historical Background," 6.

9. H. V. Smith and J. M. K. Spalding, "Report to World Health Organization," NARA, 10.

10. Demazieres (Fransulat Meknes) to Ambafrance Rabat, September 28, 1959, Archives Diplomatiques, Ministère des Affaires Étrangères, La Courneuve (hereafter MAEC), Maroc 1956–1968, box 505, folder 505/1.

11. H. V. Smith and J. M. K. Spalding, "Report to World Health Organization," NARA, 7–8; Faraj, "Historical Background," 7–8. The Jewish community in Meknes had grown in the 1950s due to migration of members of the Moroccan Jewish minority from rural areas; it would soon begin shrinking due to emigration to Europe and Israel, as the conflicting nationalisms of Moroccan nationalism and Zionism produced a deterioration of Muslim-Jewish relations. See Laskier, *North African Jewry in the Twentieth Century*, 130.

12. H. V. Smith and J. M. K. Spalding, "Report to World Health Organization," NARA, 12. On patients' early identification of the oil as the cause of illness, see also Faraj, "Historical Background," 8.

13. David Nes to State, December 2, 1959, NARA RG 59, WHO Subject Files, entry UD-07D 81, box 2, folder "Morocco Oil Poisoning Case"; Tuyns, "Conversation

with Albert Tuyns"; Faraj, "Historical Background," 7–8. See also Krieger, "Epidemiology and the Web of Causation," 890; Buck et al., *Challenge of Epidemiology*, 85–90.

14. Charnot and Troteman, "First Toxicological Investigations in Morocco." The World Health Organization flew in Oxford neurologists Honor Smith and J. M. K. Spalding, who arrived on September 25. Smith and Spalding met with Tuyns and Baillé, and the team purchased samples of the oil from a local grocer for analysis in Rabat.

15. Faraj, "Historical Background," 9.

16. American Embassy to State, November 5, 1959, NARA RG 59, WHO Subject Files, entry UD-07D 81, box 2, folder "Morocco Oil Poisoning Case"; Agence France-Presse (AFP), October 29, 1959, MAEC, Maroc 1956–1968, box 505, folder 505/1.

17. David Nes to State, January 4, 1960, "Morocco's Toxic Oil Disaster," NARA RG 59, WHO Subject Files, entry UD-07D 81, box 2, folder "Morocco Oil Poisoning Case," 5; US Naval Attaché, November 10, 1959, NARA RG 59, WHO Subject Files, entry UD-07D 81, box 2, folder "Morocco Oil Poisoning Case"; United Press (UP), November 4, 1959, MAEC, Maroc 1956–1968, box 505, folder 505/1.

18. Consul de France à Tetuoan to Ambassadeur de France, Rabat, November 16, 1959, MAEC, Maroc 1956–1968, box 505, folder 505/1.

19. "Morocco: The Malady of Meknes," *Time*, November 30, 1959, www.time.com.

20. Faraj, "Social and Vocational Aspects," 156.

21. Denis Leroy, cited in US Naval Attaché, November 10, 1959, NARA RG 59, WHO Subject Files, entry UD-07D 81, box 2, folder "Morocco Oil Poisoning Case."

22. Villard (Geneva) to State, December 4, 1959, NARA RG 59, WHO Subject Files, entry UD-07D 81, box 2, folder "Morocco Oil Poisoning Case." See also "Intoxication de Meknes," handwritten transcription of Rabat telegram, November 5, 1959, MAEC, Maroc 1956–1968, box 505, folder 505/1; Primeau, "Rehabilitation of 10,000 Victims," 1250.

23. Gross, Robertson, and Zinn, "Organisation and Contributions," 160–61.

24. Godfrey, "Epidemic of Triorthocresylphosphate Poisoning," 690.

25. Zinn, "Treatment and Rehabilitation," 108.

26. Zinn, "Treatment and Rehabilitation," 108.

27. Amster, *Medicine and the Saints*, 2.

28. Zinn "Treatment and Rehabilitation," 108.

29. Zinn, "Treatment and Rehabilitation," 91.

30. Zinn, "Treatment and Rehabilitation," 109.

31. Amster, *Medicine and the Saints*, 2.

32. Zinn, "Treatment and Rehabilitation," 109–13.

33. Gaudefroy-Demombynes, *L'oeuvre française en matière d'enseignement au Maroc*, 98–105; Paye, "Introduction et évolution de l'enseignement moderne au Maroc," 467–77.

34. D. Gross, "Results of Treatment," 136. January 1961 figures are from a press conference by Dr. Bertrand Primeau, Canadian Health Service, described by Reuters,

"Le sort de 10,000 marocains victimes de l'huile frélatée," January 3, 1961, MAEC, Maroc 1956–1968, box 505, folder 505/1.

35. D. Gross, "Results of Treatment," 136–37. On the persistence of disability among survivors fifty-eight years later, see Touati, "Le Devenir et la qualité de vie," 48–66, 77.

36. "A Short History of the United States Naval Activities Port Lyautey Morocco," n.d. [probably pre-1958 and certainly pre-1963], Naval History and Heritage Command Archives, Shore Commands, Pre-1998, Port Lyautey, NAS 1957–1963 C.H.R. Misc., box 243.

37. The USAF bases were at Nouasser, Sidi Slimane, and Ben Guerir; two additional air force bases at Ben Slimane and Djemaa des Smahim remained unfinished. Azzou, "La présence militaire américaine au Maroc," 128–29.

38. Connelly, *Diplomatic Revolution*, 58; on Sakiet, see 160.

39. Zartman, "Moroccan-American Base Negotiations," 31–38.

40. Yost to State, December 12, 1959, NARA RG 59, WHO Subject Files, entry UD-07D 81, box 2, folder "Morocco Oil Poisoning Case."

41. Pennell, *Morocco since 1830*, 306–14.

42. Yost to State, December 12, 1959. The article did not explicitly associate the Americans with "those who take advantage of the poverty of the masses and exploit economic crises," and the fact that no Jews were affected was portrayed as evidence only of "the economic isolation in which Morocco's Jewish community lives" and of the way in which the poisoning was "engendering hatred and enmity between the communities living in Morocco." Yost noted the less inflammatory content of the article itself but quoted only the headline of the *Al Machahid* (no. 34) article; the text is translated in David Nes to State, December 17, same folder. On hostility to Moroccan Jews and to Zionism in the Moroccan press, see Laskier, *North African Jewry in the Twentieth Century*, 215–17.

43. Pennell, *Morocco since 1830*, 306–14.

44. "Motion sur les huile nocives," *Al-Istiqlal*, December 17, 1959.

45. "De Meknès à Fréjus: Les responsabilités dans l'affaire des huiles," *Al-Istiqlal*, December 17, 1959.

46. Agence Centrale de Presse, November 4, 1959, MAEC, Maroc 1956–1968, box 505, folder 505/1.

47. Don Catlett to State, November 5 and 10, 1959, NARA RG 59, WHO Subject Files, entry UD-07D 81, box 2, folder "Morocco Oil Poisoning Case"; Parodi to Affaires Étrangères, November 14, 1959, MAEC, Maroc 1956–1968, box 505, folder 505/1.

48. "Les victimes attendent un châtiment exemplaire," *Al-Istiqlal*, April 16, 1960.

49. "La Sentence avant dire droit," *Al-Istiqlal*, April 30, 1960.

50. The importance of Cold War concerns in stimulating American disaster relief in a slightly later period has been demonstrated by Drury, Olson, and Van Belle, "Politics of Humanitarian Aid."

51. Gaillard, Kelman, and Orillos, "US-Philippines Military Relations," 303.

52. Kelman, *Disaster Diplomacy*, 12.

53. Economic penetration was a secondary issue. Trade relations between Morocco and the United States increased over the next decade, due in part to the development of the Moroccan phosphate industry, but remained modest, amounting to only 7.5 percent of Moroccan imports and 4 percent of exports even in 1969. Zingg, "Cold War in North Africa," 44.

54. Osgood, *Total Cold War*, 72.

55. In the Casablanca bidonville fire, Nouasser airmen joined firefighting efforts by the Moroccan Army and the French Air Force. American relief efforts in the aftermath, which included the provision of food, tents, and the use of a generator, were accompanied by the work of the Moroccan Red Crescent, the Societé Musulman de Bienfaisance, the Casablanca municipal government, and the Ministries of Public Health and Public Works. In such a collective effort it was no wonder that the United States was not singled out by the press. On January 15, 1960, American troops again responded to another massive flood in the Gharb basin. To Navy liaison Leon Borden Blair's consternation, the publicity after the 1960 flood also lacked a focus on American heroism, and Blair blamed United States Information Service (USIS) policy, which was "to avoid 'overemphasizing' the assistance provided by the military services and to 'attempt to generate a feeling of mutual cooperation among all nations.'" The USIS had recently taken over responsibility for publicity from the armed forces themselves. According to Blair, always an advocate of the Navy, the USIS press releases lacked the vivid details that had been supplied during the floods of 1958, when Navy commanders had issued the press releases. Blair, *Western Window in the Arab World*, 268–70. Yost, in contrast, was pleased that the Moroccan press reported "an extremely laudatory story of the lunch given by the Governor of Rabat and other local Moroccan authorities" in honor of those involved in rescue efforts—but those invited included not just the US forces but also French and Moroccan troops. Yost to State, February 5, 1960, NARA RG 469, entry UD376, ICA Deputy Director Geographic Files, box 330, folder "Morocco: Disasters." See also Henry Ford, Consul General, to State, June 12, 1958, NARA RG 469, entry P-11, box 1; Comnavacts Port Lyautey to State, December 26, 1958, NARA RG 469, entry P211, box 2, folder "Morocco (Disasters)"; Nes to State, January 17, 1960, NARA RG 469, entry UD376, box 330, folder "Morocco: Disasters"; USIS, "Annual Assessment Report 1960," NARA RG 59, entry A1 5096, box 8, folder "Country Background-Rabat."

56. Don Catlett to State, November 5 and November 10, 1959, NARA RG 59, WHO Subject Files, entry UD-07D 81, box 2, folder "Morocco Oil Poisoning Case."

57. Herter to American Embassy Rabat, November 17, 1959, NARA RG 59, WHO Subject Files, entry UD-07D 81, box 2, folder "Morocco Oil Poisoning Case."

58. Yost to State, November 20, 1959, NARA RG 59, WHO Subject Files, entry UD-07D 81, box 2, folder "Morocco Oil Poisoning Case," 2.

59. David Nes to State, December 2, 1959, NARA RG 59, WHO Subject Files, entry UD-07D 81, box 2, folder "Morocco Oil Poisoning Case."

60. Villard (Geneva) to State, December 4, 1959, NARA RG 59, WHO Subject Files, entry UD-07D 81, box 2, folder "Morocco Oil Poisoning Case."

61. WHO aid was officially requested by the Moroccan Ministry of Health on September 25, 1959. Villard to State, December 8, 1959, NARA RG 59, WHO Subject Files, entry UD-07D 81, box 2, folder "Morocco Oil Poisoning Case." The formal request to the League of Red Cross Societies was made on November 17, 1959. Faraj, "Historical Background," 13.

62. "President's Good Will Trip, December 1959—Briefing Memorandum," November 27, 1959, NARA RG 59, entry 3109D, box 2, folder "22-Regional 'The President's Good Will Trip.'" See also D. A. Fitzgerald, "Briefing Memorandum," May 17, 1961, NARA RG 469, entry UD 181, International Cooperation Association, box 35, folder "Morocco FY 1961." According to Fitzgerald, American economic aid, which totaled $188.4 million from 1957 to 1961, accounted for half of all Moroccan "public investment." On military aid, see "Chronology of Events on Moroccan Arms Request," NARA RG 59, entry 3109D, box 2, folder "R-20: Tripartite Talks."

63. David Nes to State, January 4, 1960, "Morocco's Toxic Oil Disaster," NARA RG 59, WHO Subject Files, entry UD-07D 81, box 2, folder "Morocco Oil Poisoning Case," 5–6.

64. "Memorandum of Conversation," December 3, 1959, NARA RG 59, WHO Subject Files, entry UD-07D 81, box 2, folder "Morocco Oil Poisoning Case."

65. Walmsley (Tunis) to State, December 2, 1959, NARA RG 59, WHO Subject Files, entry UD-07D 81, box 2, folder "Morocco Oil Poisoning Case"; "Memorandum for the President," NARA RG 59, entry A1 3107, Office of Inter-African Affairs, box 3, folder "Morocco—Bases." See also Drury, Olson, and Van Belle, "Politics of Humanitarian Aid."

66. Yost to State, November 20, 1959, NARA RG 59, WHO Subject Files, entry UD-07D 81, box 2, folder "Morocco Oil Poisoning Case," 2. In his 1970 history of the postwar years in Morocco, Leon Borden Blair, the US Navy's political liaison officer during the crisis, wrote that "any substantial aid might very well be misinterpreted as the result of a guilty conscience. Moroccan people and those in many other parts of the world do not understand spontaneous generosity" (67). If Blair's belief, that Moroccans were inherently predisposed to see generosity as a sign of guilt, was shared by other American officials, then this would suggest that realist calculations behind the American response in 1959 may have been distorted by American racialist thinking. See also Blair, *Western Window*, 266.

67. Osgood, *Total Cold War*, 20.

68. Eisenhower quoted in Osgood, *Total Cold War*, 77.

69. Blair, *Western Window*, 303–4.

70. *Al-Istiqlal*, no. 163 (August 8, 1959), and *Al Alam* (July 29–30, 1959), quoted in Zartman, "Moroccan-American Base Negotiations," 36.

71. Blair, *Western Window*, 270.

72. Parodi to Affaires Étrangères, November 5, 1959; J. Basdevant to Ambafrance Rabat, November 6, 1959, both in MAEC, Maroc 1956–1968, box 505, folder 505/1.

73. Parodi to Affaires Étrangères, November 12, 1959, MAEC, Maroc 1956–1968, box 505, folder 505/1.

74. Premier Ministre, SDCE, "Aide F.L.N. aux victimes marocaines des huile frelatées," December 31, 1959; Premier Ministre, SDCE, "Aide F.L.N. aux victimes marocaines des huile frelatées," March 28, 1960; both in MAEC, Maroc 1956–1968, box 505, folder 505/1.

75. Primeau, "Rehabilitation of 10,000 Victims," 1250.

76. "Communique à Fransulat Meknes" to Affaires Étrangères, November 27, 1959, MAEC, Maroc 1956–1968, box 505, folder 505/1.

77. Seydoux, Affaires Étrangères, to Ambafrance Rabat, November 30, 1959; Parodi to Directeur du Service du Domaine Française au Maroc, December 12, 1959; Ministère des Finances to Ministère des Affaire Étrangères, January 27, 1960: all three in MAEC, Maroc 1956–1968, box 505, folder 505/1. See also Dispatch no. 269, "Communique of the Ministry of Public Health, January 6, 1960," NARA RG 59, WHO Subject Files, entry UD-07D 81, box 2, folder "Morocco Oil Poisoning Case"; P. M. Henry, Affaires Étrangères to Ambafrance Rabat, February 11, 1960, MAEC, Maroc 1956–1968, box 505, folder 505/1.

78. Ministère des Affaires Étrangères, Directeur, Unions Internationales, "Note pour la Direction Générale des Affaires Marocaines et Tunisiennes," December 13, 1959, MAEC, Maroc 1956–1968, box 505, folder 505/1. See also Sous-Direction du Maroc, "Note pour la Direction des Affaires Administratives et Sociales, Unions Internationales," December 29, 1959, same folder.

79. "De la calomnie," *Al-Istiqlal*, March 21, 1960.

80. Prime Minister Michel Debré, December 7, 1959, MAEC, Maroc 1956–1968, box 505, folder 505/1.

81. Osgood, *Total Cold War*, 35.

82. Connelly, *Diplomatic Revolution*, 59–60. See also Zartman, "Moroccan-American Base Negotiations," 29.

83. Sutterthwaite to Murphy, "Ambassador Yost's Suggestion," March 26, 1959, NARA RG 59, entry 3109D, box 2, folder "U.S. Policy—General."

84. US Embassy to State, November 21, 1959, NARA RG 469, entry P211, box 2, folder "Morocco (Disasters)."

85. Herter to American Embassy Rabat, November 20, 1959, NARA RG 59, WHO Subject Files, entry UD-07D 81, box 2, folder "Morocco Oil Poisoning Case."

86. Yost to State, November 23, 1959, NARA RG 59, WHO Subject Files, entry UD-07D 81, box 2, folder "Morocco Oil Poisoning Case."

87. Red Cross, "International News Release," November 26, 1959, NARA RG 59, WHO Subject Files, entry UD-07D 81, box 2, folder "Morocco Oil Poisoning Case."

88. The Leroy plan called for "20 doctors, 7 social workers, 115 nurses, 18 secretaries, and 165 service employees," as well as equipment and an ongoing budget of 800 million francs annually, "for many years." December 3 press conference, summarized in Yost to State, December 9, 1959, NARA RG 59, WHO Subject Files, entry UD-07D 81, box 2, folder "Morocco Oil Poisoning Case."

89. Herter to American Embassy Rabat, November 25, 1959, and note thereupon by "L.W.," November 27, 1959; Yost to State, November 25, 1959, NARA RG 59, WHO Subject Files, entry UD-07D 81, box 2, folder "Morocco Oil Poisoning Case."

90. William McCahon, November 25, 1959, NARA RG 469, entry P211, box 2, folder "Morocco (Disasters)"; US Embassy to State, November 21, 1959, NARA RG 469, entry P211, box 2, folder "Morocco (Disasters)"; Yost to State, November 23, 1959, NARA RG 59, WHO Subject Files, entry UD-07D 81, box 2, folder "Morocco Oil Poisoning Case."

91. Herter to American Consulate Geneva, December 1, 1959, NARA RG 59, WHO Subject Files, entry UD-07D 81, box 2, folder "Morocco Oil Poisoning Case."

92. Alfred Gruenther to State, December 11, 1959, NARA RG 59, WHO Subject Files, entry UD-07D 81, box 2, folder "Morocco Oil Poisoning Case"; Villard to State, December 12, 1959, NARA RG 469, entry P211, box 2, folder "Morocco (Disasters)."

93. Yost to State, December 12, 1959, NARA RG 59, WHO Subject Files, entry UD-07D 81, box 2, folder "Morocco Oil Poisoning Case."

94. Yost to State, December 12, 1959.

95. Osgood, *Total Cold War*, 358–60.

96. Yost to State, December 15, 1959, NARA RG 59, WHO Subject Files, entry UD-07D 81, box 2, folder "Morocco Oil Poisoning Case."

97. Yost to State, December 18, 1959, NARA RG 59, WHO Subject Files, entry UD-07D 81, box 2, folder "Morocco Oil Poisoning Case."

98. Twenty-four hundred sweatshirts and 4,320 "union suits." State to American Embassy Rabat, received December 29, 1959, in RG 469, entry P211, box 2, folder "Morocco (Disasters)."

99. Gross, Robertson, and Zinn, "Organisation and Contributions," 166–71.

100. State to American Embassy Rabat, received December 29, 1959, NARA RG 469, entry P211, box 2, folder "Morocco (Disasters)"; Henry Dunning (Secretary General of the International Red Cross), January 11, 1960, NARA RG 59, WHO Subject Files, entry UD-07D 81, box 2, folder "Moroccan Oil Poisoning Case."

101. "Responsibilities in the Oil Affair," Rabat dispatch, NARA RG 59, WHO Subject Files, entry UD-07D 81, box 2, folder "Morocco Oil Poisoning Case." This document quoted "De Meknès à Fréjus: Les responsibilités dans l'affaire des huiles," *Al-Istiqlal*, December 17, 1959.

102. Zartman, "Moroccan-American Base Negotiations," 27–40.

103. "Les relations americano-marocaines dans une phase nouvelle," *Al-Istiqlal*, December 26, 1959.

104. "Pour l'évacuation des troupes française et pour la liberation du sahara marocaine," *Al-Istiqlal*, February 27, 1960.

105. David Nes to State, January 4, 1960, "Morocco's Toxic Oil Disaster," NARA RG 59, WHO Subject Files, entry UD-07D 81, box 2, folder "Morocco Oil Poisoning Case," 7.

106. Yost to State, January 5, 1960, NARA RG 469, entry UD376, ICA Deputy Director Geographic Files, box 330, folder "Morocco: Disasters."

107. Jernegan (Baghdad) to State, January 10, 1960, NARA RG 59, WHO Subject Files, entry UD-07D 81, box 2, folder "Morocco Oil Poisoning Case." See also Premier Ministre SDECE, "Aide irakiènne aux victimes des 'huiles nocives,'" December 12, 1959, MAEC, Maroc 1956–1968, box 505, folder 505/1. The Iraqi team arrived on January 18, 1960. Nes to State, January 21, 1960, NARA RG 469, entry UD376, ICA Deputy Director Geographic Files, box 330.

108. Thompson (Moscow) to State, January 6, 1960, NARA RG 469, entry UD376, ICA Deputy Director Geographic Files, box 330, folder "Morocco: Disasters." See also R. C. Barrett, *Greater Middle East and the Cold War*, 113–15.

109. Herter (State) to American Embassy Rabat, February 8, 1960, NARA RG 469, entry UD376, ICA Deputy Director Geographic Files, box 330, folder "Morocco: Disasters." Compare Osgood, *Total Cold War*, 242.

110. Nes to State, January 20, 1960, NARA RG 59, WHO Subject Files, entry UD-07D 81, box 2, folder "Moroccan Oil Poisoning Case"; Two telegrams, Yost to State, January 26, 1960, NARA RG 469, entry UD376, ICA Deputy Director Geographic Files, box 330, folder "Morocco: Disasters." To highlight the need for vehicles, Yost pointed out that 10 to 15 percent of the patients were wholly unable to walk, and that although 3,629 of the cases were in Meknes and the surrounding towns, over 4,700 were in the Rabat Province, and hundreds of other cases were found scattered from Oujda to Casablanca, Marrakesh, and Agadir. See also Primeau, "Rehabilitation of 10,000 Victims," 1251.

111. State to Amercian Embassy Rabat, February 12, 1960, NARA RG 59, WHO Subject Files, entry UD-07D 81, box 2, folder "Moroccan Oil Poisoning Case."

112. Yost to State, February 29, 1960, NARA RG 469, entry UD376, ICA Deputy Director Geographic Files, box 330, folder "Morocco: Disasters."

113. Fordham, "Agadir"; Blair, *Western Window*, 272–76.

114. Yost to State, March 30, 1960, NARA RG 469, entry P-211, box 2, folder "Morocco: Disasters."

115. Zartman, *Morocco*, 58.

116. Blair, *Western Window*, 272.

117. Blair, *Western Window*, 297, 303.

118. *Al Alam*, March 10, 1960, quoted in Yost to Secretary of State, March 11, 1960, in RG 469, entry UD376, box 330, folder "Morocco-Disasters-Earthquake."

119. Nes to Bureau of African Affairs, April 25, 1960, NARA RG 84, entry UD 3005B, Embassy Rabat Classified General Records, box 12. From a budgetary perspective,

the State Department looked forward to a time when the bases would be evacuated, and the Moroccan government could no longer hold the bases hostage when negotiating for foreign aid. "Mutual Security Program Report on Grant Economic Assistance," February 29, 1960, in RG 84, entry UD 3005B, U.S. Embassy Rabat Classified General Records, 1956–1963, box 12, folder "500 Mutual Security Act and Program 1959–61."

120. The term "tipping point" is drawn from Pelling and Dill, "Disaster Politics." However, Pelling and Dill (like Gaillard, Orillos, and Kelman) examine the one-way influence of disasters on political phenomena; whereas here the focus is on the multiplicity of critical junctures (including both geoenvironmental and nonenvironmental inputs) shaping responses to the disaster.

121. Jones, "Otto Passman and Foreign Aid"; *Foreign Relations of the United States, 1958–1960*, v. 4, doc. 236.

122. *Hearings before the Committee on Foreign Affairs, House of Representatives*, March 1–3, 1960, Part 2, NARA Y4.F76/1:m98/11/960/pt.2.

123. For example, Sen. Albert Gore of Tennesse, in *Hearings before the Committee of Foreign Relations of the United States Senate*, March 22–April 5, 1960, NARA Y4.F76/2: M98/4/960, 13–14.

124. Dillon to American Embassy Rabat, March 3, 1960, NARA RG 469, entry UD376, ICA Deputy Director Geographic Files, box 330, folder "Morocco: Disasters: Earthquake."

125. Brent (Rabat) to ICA, March 18, 1960, NARA RG 469, entry UD376, ICA Deputy Director Geographic Files, box 330, folder "Morocco: Disasters."

126. Yost to State, March 30, 1960, NARA RG 469, entry UD376, ICA Deputy Director Geographic Files, box 330, folder "Morocco: Disasters."

127. Nes (Rabat) to State, June 23, 1960, NARA RG 59, WHO Subject Files, entry UD-07D 81, box 2, folder "Morocco Oil Poisoning Case."

128. Brent (Rabat) to ICA, March 18, 1960. Six American therapists arrived in June 1960, followed by month-long visits by several doctors. Gross, Robertson, and Zinn, "Organisation and Contributions," 171.

129. "Les victimes attendent un châtiment exemplaire," *Al-Istiqlal*, April 16, 1960. See also "La photo d'une fillette," *France Soir*, April 12, 1960; "'Les 9,000 victimes des huiles frelatées,' demande le procureur au procès," *France Soir*, April 13, 1960; "Dix mille victimes accusent: Le procès des huiles frelatées," *Combat*, April 11, 1960; "Devant la cour de justice à Rabat: Vingt-huit trafiquants d'huile," *Le Monde*, April 11, 1960; all four articles are in MAEC, Maroc 1956–1968, box 505, folder 505/1.

130. "Les victimes attendent un châtiment exemplaire," *Al-Istiqlal*, April 16, 1960. See also "Au procès des huiles de Rabat: La défense met en cause les fournisseurs américains," *Le Monde*, April 28, 1960, MAEC, Maroc 1956–1968, box 505, folder 505/1.

131. "Les victimes attendent un châtiment exemplaire," *Al-Istiqlal*, April 16, 1960.

132. "'Les 9,000 victimes des huiles frelatées,' demande le procureur au procès," *France Soir*, April 13, 1960, MAEC, Maroc 1956–1968, box 505, folder 505/1.

133. "Le Verdict: 5 condemnations à mort," *Al-Istiqlal*, April 30, 1960.
134. Arthur Allen to Root, March 4, 1960, NARA RG 59, entry A1 3109D, Records Relating to Morocco, box 2, folder "M-3 Cooking Oil Incident." The monarchy's failure to provide satisfying justice reemerged in the Moroccan press in 2008. The Ligue Marocaine des Victimes des Huiles Empoisonnées mounted a successful campaign to obtain meaningful government stipends for victims, and the resulting discussion in the press and the social media has pointed out that despite the harshness of the king's initial edict, no one was executed and the five people convicted allegedly received early release, if they were imprisoned at all. The *Ligue* itself raised the issue of American culpability, but its 2005 efforts to persuade the American government to investigate US culpability were unsuccessful. "Scandale des huiles frélatées," *La Gazette du Maroc*, February 22, 2008, http://www.lagazettedumaroc .com/articles.php?id_artl=1658.
135. B. Burke, "With the International Red Cross in Morocco," 11; Yost to State, April 29, 1960, NARA RG 469, entry UD376, ICA Deputy Director Geographic Files, box 330, folder "Morocco: Disasters"; Brent to ICA, June 21, 1960, NARA RG 59, WHO Subject Files, entry UD-07D 81, box 2, folder "Moroccan Oil Poisoning Case"; Gross, Robertson, and Zinn, "Organisation and Contribution," 173.
136. Bennett, *Vibrant Matter*, 21; Krieger, "Epidemiology and the Web of Causation."
137. See the "Friendship Day" photograph in Blair, *Western Window*, 146–47.
138. Gross, Robertson, and Zinn, "Organisation and Contributions," 165–73. Gross et al. only account for one French doctor, but an additional military doctor was also assigned to assist in the effort. P. M. Henry, Affaires Étrangères, to Ambafrance Rabat, February 11, 1960, MAEC, Maroc 1956–1968, box 505, folder 505/1.
139. Ministre des Affaire Étrangères to Ambassadeur de France, February 9, 1961, MAEC, Maroc 1956–1968, box 505, folder 505/1.

5. Death, Diplomacy, and Reconstruction

1. Benhima, "Témoignage," 3. On the death toll, see Parodi to Affaires Étrangères, March 8, 1960, Archives Diplomatiques, Ministère des Affaires Étrangères, La Courneuve (hereafter MAEC), Maroc 1956–1968, box 9/10, folder "Généralities"; Pierre de Leusse (ambassador to Morocco) to M. Couve de Murville, Affaires Étrangères, August 12, 1963, Archives Diplomatiques, Ministère des Affaires Étrangères, Nantes (hereafter MAEN), 558 PO4, box 8, folder "Inhumations"; Charef, "Agadir, une ville orpheline," 167–68.
2. Compare Kelman, *Disaster Diplomacy*, 12. Disaster diplomacy is also discussed in chapter 4 of the present work.
3. Abdallah Aourik interview of Tariq Kabbage, "Où étiez-vous le 29 février 1960?," *Agadir O'flla*, February 2008, 12.

4. Abdallah Aourik, "Où étiez-vous le 29 février 1960?" memoir, *Agadir O'flla*, February 2008, 10; Cappe, *Agadir 29 février 1960*, 49. Soon after the earthquake Governor Bouamrani and Consul Jeudy were replaced by Benhima and Jestin, respectively.

5. Yost to Secretary of State, March 7, 1960, United States Archives and Records Administration (hereafter NARA) RG 469, entry UD376, box 330, folder "Morocco: Disasters: Earthquake."

6. "Appel de S.M. le Roi," *Al-Istiqlal*, March 5, 1960.

7. Perhaps because of the royal family's prominent role in the response to the earthquake, the Istiqlali opposition was described as "moribund" in Agadir in the early 1960s, while the monarchist party, the Front pour la Défense des Institutions Constitutionnelles (FDIC) was strong. Peter Spicer, US Consulate Casablanca, "Preparations for the Communal Council Elections," August 16, 1963, NARA RG 84, entry UD 3005B, box 15, folder "350 Morocco Jan–June 1963," 4. The FDIC alliance included the Mouvement Populaire, a party with strong roots in the Berber-speaking south, which also accounts for FDIC success in Agadir. Oliver Marcy, Counselor for Political Affairs, "Moroccan Elections, Lower House; Field Analyses," May 31, 1963, NARA RG 84, entry UD 3005B, US Embassy, Rabat, Classified General Records 1956–1963, box 15, folder "350 Morocco Jan–June 1963."

8. Pennell, *Morocco since 1830*, 314.

9. Mohamed Charef cites fourteen thousand refugees in Inezgane and Aït Melloul, south of the city: Charef, "Agadir, une ville orpheline." Charles Yost reported fifteen thousand to twenty thousand in camps north and south of the city, based on communications with Oufkir. Yost to Secretary of State, March 7, 1960, NARA RG 469, entry UD376, ICA Deputy Director, box 330, folder "Morocco: Disasters: Earthquake."

10. Cappe, *Agadir 29 février 1960*, 104.

11. Abdallah Aourik interview of Ahmed Bouskous, "Où étiez-vous le 29 février 1960?," *Agadir O'flla*, no. 15, February 2008, 11. See also Benhima, "La Renaissance d'Agadir."

12. Mas, "Plan directeur et plans d'aménagement," 10. Daniel Williford notes that some survivors resisted these state-led initiatives, "preferring instead to remain in tents." Williford, "Seismic Politics," 1010.

13. Cappe, *Agadir 29 février 1960*, 104. According to Cappe, a military chaplain (Abbé Bernel) was on hand to perform last rites, and military honors were also given.

14. Cappe, *Agadir 29 février 1960*, 115.

15. Directeur du Service de Anciens Combattants to French Ambassador, December 5, 1960, MAEN 558 PO4, box 8, folder "Inhumations."

16. Le Roy to Affaires Étrangères, March 11, 1960, MAEC, Maroc 1956–1968, box 9/10, folder "Reconstruction"; Yost to State, March 7, 1960, NARA RG 469, entry UD376, ICA Deputy Director, box 330, folder "Morocco: Disasters: Earthquake"; Le Roy to Affaires Étrangères, March 12, 1960, no. 305/308, MAEC, Maroc 1956–1968, box 9/10, folder "Diplomatie Franco-Marocaine."

17. Homer Bigart, "Tubman Assails South Africans," *New York Times*, March 24, 1960.

18. Le Toullec, *Agadir 1960*, 90.

19. Le Roy to Affaires Étrangères, March 12, 1960, no. 305/308, MAEC, Maroc 1956–1968, box 9/10, folder "Diplomatie Franco-Marocaine."

20. Ambafrance to Affaires Étrangères, March 7, 1960, MAEC, Maroc 1956–1968, box 9/10, folder "Diplomatie Franco-Marocaine."

21. Bouffanais to Affaires Étrangères, March 4, 1960, MAEC, Maroc 1956–1968, box 9/10, folder "Généralités 1960." See also, same folder, Jean Basdevant, Minister of Foreign Affairs, to French Embassy, Washington, March 9, 1960. The American diplomat is referred to as "mon collègue américain"; this is most likely Yost.

22. Guiringaud, in Accra, to Affaires Étrangères, March 4, 1960, MAEC, Maroc 1956–1968, box 9/10, folder "Diplomatie Franco-Marocaine."

23. "Morocco Assails French on Quake," *New York Times*, March 13, 1960.

24. Le Roy to Affaires Étrangères, March 12, 1960, no. 1314/1316, MAEC, Maroc 1956–1968, box 9/10, folder "Diplomatie Franco-Marocaine."

25. "Des problèmes qui demeure," *Al-Istiqlal*, March 12, 1960. The atomic testing occurred in what is now Algerian territory. After Algerian independence, Moroccan nationalist belief in a historical "greater Morocco" would become a major stumbling block in relations between the two North African states.

26. Bouffanais to Parodi, March 3, 1960, MAEC, Maroc 1956–1968, box 9/10, folder "Diplomatie Franco-Marocaine." According to the French Embassy, French-language radio broadcasts did report the king's statement of thanks during his second trip to Agadir from March 6 to 7. The press and the Arabic and Berber-language broadcasts remained silent, however, with the result "that listeners could remain ignorant of the existence of an aeronaval base in Agadir." Ambafrance to Affaires Étrangères, March 7, 1960, MAEC, Maroc 1956–1968, box 9/10, folder "Diplomatie Franco-Marocaine." By March 9, however, French authorities expressed satisfaction that, due to their diplomacy, the Moroccan press coverage had become more objective. Le Roy, March 9, 1960, same folder.

27. Bouffanais to Affaires Étrangères, March 8, 1960, MAEC, Maroc 1956–1968, box 9/10, folder "Diplomatie Franco-Marocaine."

28. Bouffanais to Parodi, March 3, 1960, MAEC, Maroc 1956–1968, box 9/10, folder "Diplomatie Franco-Marocaine."

29. Bouffanais to Parodi, March 3, 1960. There were reports that the Moroccan Jewish community of Agadir, which had suffered twelve hundred to fifteen hundred deaths, was being excluded from the distribution of government relief. Parodi advised Affaires Étrangères to suggest that Israeli Prime Minister Ben Gurion solicit aid from the Jewish population of New York, where he was soon to visit. Parodi to Affaires Étrangères, March 8, 1960, MAEC, Maroc 1956–1968, box 9/10, folder "Généralities."

30. Bouffanais to Parodi, March 3, 1960.

31. For example, Hardy, *Une conquête morale*, 189.

32. Bernard Lafay, "Question Écrite No. 708," March 12, 1960, MAEC, Maroc 1956–1968, box 9/10, folder "Diplomatie Franco-Marocaine."

33. Le Roy to Affaires Étrangères, March 12, 1960, no. 305/308, MAEC, Maroc 1956–1968, box 9/10, folder "Diplomatie Franco-Marocaine."

34. "De la calomnie," *Al-Istiqlal*, March 21, 1960.

35. "Scandale à Agadir," *Europe-Magazine*, March 30, 1960, extract in MAEN, Agadir Consulat, 15 PO1, box 378, folder "Divers notes."

36. André Figueres, "Le crime d'Agadir," n.d. [ca. March 30, 1960], extract in MAEN, Agadir Consulat, 15 PO1, box 378, folder "Divers notes."

37. Le Roy to Affaires Étrangères, March 12, 1960, no. 305/308, MAEC, Maroc 1956–1968, box 9/10, folder "Diplomatie Franco-Marocaine." Attempts to remedy the situation followed. See Parodi to Affaires Étrangères, March 16, 1960, MAEC, Maroc 1956–1968, box 9/10, folder "Diplomatie Franco-Marocaine"; "Passage à Paris du Colonel Hafid," March 15, 1960, MAEC, Maroc 1956–1968, box 9/10, folder "Diplomatie Franco-Marocaine"; A. Benkirane to Radiodiffusion-Télévision Française, March 22, 1960; R. Thibault to Benkirane, March 28, 1960, MAEC, Maroc 1956–1968, box 9/10, folder "Généralités 1960."

38. Jestin to Ambafrance Rabat, April 30, 1960, MAEN, Rabat Ambassade 1956–1989, box 827.

39. General Bethouart, "Le scandale d'Agadir," *Figaro*, April 5, 1960, in MAEC, Maroc 1956–1968, box 9/10, folder "Généralités 1960."

40. General Bethouart, "Le scandale d'Agadir."

41. Choi, *Decolonization and the French of Algeria*, 53–154.

42. General Bethouart, "Le scandale d'Agadir."

43. See Shepard, *Invention of Decolonization*, 6–7.

44. "La France, va-t-elle abandonner les sinistrés français d'Agadir," *Combat*, April 16–17, 1960; "Des français sinistrés d'Agadir," *Le Monde*, April 14, 1960, both in MAEC, Maroc 1956–1968, box 9/10, folder "Généralités 1960."

45. Response to Senator Lafay's Question Écrite No. 780, posed April 15, 1960, MAEC, Maroc 1956–1968, box 9/10, folder "Généralités 1960."

46. Royaume du Maroc, Haut-Commissariat de la Reconstruction, "Agadir: Information, Urbanisme, Aide de l'État-1962," 1962, Centre de Documentation, Ministère d'Intérieur Direction de l'Urbanisme et de l'Habitat, Rabat; Ambassadeur de France au Maroc to Ministre des Affaires Étrangères, January 2, 1962, MAEN 558 PO4, box 8, folder "Agadir divers "; Telegram 1013-15 to Affaires Étrangères, March 16, 1962, in MAEC, Maroc 1956–1968, box 9/10, folder "Généralités 1960."

47. Response to Senator Lafay's Question Écrite No. 780.

48. Giscard d'Estang, Minister of Finances, to Affaires Étrangers, May 17, 1960, MAEC, Maroc 1956–1968, box 9/10, folder "Généralités 1960."

49. Response to Senator Lafay's Question Écrite No. 780; Direction Générale des Affaires Marocaines et Tunisiennes, draft letter to Deputé Lacaze, June 29, 1960,

MAEC, Maroc 1956–1968, box 9/10, folder "Généralités 1960"; Agadir Consulat, "Note sur la situation à Agadir," June 15, 1960, MAEN 15 PO1, box 378, folder "Divers notes."

50. Ambassadeur de France au Maroc to Ministre des Affaires Étrangères, January 2, 1962, MAEN 558 PO4, box 8, folder "Agadir divers."

51. Telegram 1013-15 to Affaires Étrangères, March 16, 1962, in MAEC, Maroc 1956–1968, box 9/10, folder "Généralités 1960."

52. Ambassadeur de France au Maroc to Ministre des Affaires Étrangères, January 2, 1962, MAEN 558 PO4, box 8, folder "Agadir divers."

53. Pierre de Leusse, Ambassadeur de France au Maroc, to Couve de Murville, Ministre des Affaires Étrangères, May 22, 1963, MAEN 558 PO4, box 8, folder "Agadir divers."

54. Comité d'Action Marocaine, *Plan de réformes marocaines*, 23, 41.

55. Archives Nationales du Royaume du Maroc, box D606.

56. Sécrétaire Général du Protectorat, May 25, 1951, MAEN 558 PO4, box 8, folder "Inhumations."

57. An exception was the tomb of Resident-General Lyautey, the French protectorate's first resident general. The departure of the French military from Agadir coincided with the transfer to France of Lyautey's ashes. After Lyautey's death in France in 1934, his ashes, according to his wishes, had been enshrined in a Moroccan-style mausoleum or *qubba* in Rabat. Moroccan nationalists, however, had objected to the transfer of the grand imperialist's remains to Morocco, and after independence the Lyautey mausoleum, like the French military bases, was another remnant of colonialism to be removed. For many French in Morocco the departure of Lyautey's remains in April 1961 was a symbolic event, demonstrating the full extent of decolonization. For the French of Agadir, however, the repatriation of Lyautey's ashes was overshadowed by the exhumations and repatriations of the earthquake victims and by the ceremonies surrounding the final closure of the base. Stacy Holden, "Contested Memories in Colonial Morocco: The Construction of Hubert Lyautey's Mausoleum in Rabat, 1935," Middle East Studies Association Annual Meeting 2012; James Mokhiber, "(Re) Burying and Memorializing Marshal Hubert Lyautey in Morocco and France, 1934–61," American Historical Association Annual Meeting 2014; Jestin, Fransulat Agadir, to Ambafrance Rabat, April 22, 1961, MAEN, Rabat Ambassade, 558 PO1, box 828; Cappe, *Agadir 29 février 1960*, 190–94.

58. Minkin, *Imperial Bodies*. Other scholarship on the relationship between burial, imperialism, and Euro-Maghrebi migration has focused on Muslim migrants to Europe, or on the cultural symbolism of the burial of great men in the colonies or the metropole, e.g., Nunez, "La gestion publique des espaces confessionnels"; Ansari, "Burying the Dead"; Burkhalter, "Négociations autour du cimetière musulman"; Holden, "Contested Memories"; Kevin Martin, "Commemorating the 'Martyr of Arabism' in Post-Colonial Time and Space: The Symbolic Power of Syrian Staff Colonel Adnan al-Malki's Remains," American Historical Association Annual

Meeting 2014; Katherine Hickerson, "Body Politics and Postmortem Power: The After-Lives of the Sudanese Mahdi and General Charles Gordon," American Historical Association Annual Meeting 2014. The religious significance of the burial of holy women and men in Muslim North Africa is treated in Clancy-Smith, *Rebel and Saint*.

59. Pierre de Leusse (ambassador to Morocco) to M. Couve de Murville, Ministry of Foreign Affairs, August 12, 1963, MAEN 558 PO4, box 8, folder "Inhumations."

60. Directeur du Service de Anciens Combattants to French Ambassador, December 5, 1960, MAEN 558 PO4, box 8, folder "Inhumations." Three hundred seventy-one burials were recorded.

61. Willi Cappe recounts the story of a father ("Juan," possibly Spanish) who at the Ihchach cemetery persuaded the Moroccan civil police to delay the burial of his son until he could personally arrange for the delivery of a proper coffin. Cappe, *Agadir 29 février 1960*, 84–85. Munier states that the burials of the European victims at Ihchach had been overseen by a French municipal engineer who recalled that fingerprints and photographs had been taken—yet Munier was unable to obtain these records from the Moroccan authorities. V. A. Munier, Casablanca, August 6, 1962, "Mémoire au sujet de la recherche et l'exhumation des corps de Monsieur et Madame Bordeaux," MAEN PO4, Service Juridique, folder "Inhumations."

62. Pierre de Leusse (ambassador to Morocco) to M. Couve de Murville, Ministry of Foreign Affairs, August 12, 1963, MAEN 558 PO4, box 8, folder "Inhumations"; Martin (Agadir Consulat) to French Embassy, March 10, 1960, MAEN, Rabat Ambassade 1956–1989, box 827. One of these was the body of the French consul's son, Philippe Jeudy, brought to the cemetery in Rabat. Jeudy was replaced soon thereafter. French Embassy to Agadir Consulate, March 5, 1960, MAEN 558 PO1, box 853. On the process of filling out death certificates and recording identifying information, see Cappe, *Agadir 29 février 1960*, 70–72.

63. Sous-Direction du Maroc, Direction des Affaires Marocaines et Tunisiennes, "Note pour la Direction des Affaires Administratives et Sociales," May 14, 1960, MAEC, Maroc 1956–1958, box 9/10, folder "Exhumation." The document indicates that both *directions* were in agreement on Moroccan sovereignty in this matter.

64. Martin (interim French consul) to French Embassy Rabat, March 12, 1960, MAEN, Rabat Ambassade 1956–1989, box 827.

65. Parodi, French Ambassador in Rabat, to Affaires Étrangères, May 3, 1960, MAEC, Maroc 1956–1958, box 9/10, folder "Exhumation."

66. Agadir Consulate to Rabat Embassy et al., May 4, 1960, MAEN, Rabat Ambassade 1956–1989, box 827.

67. Parodi, French Ambassador in Rabat, to Affaires Étrangères, May 3, 1960, MAEC, Maroc 1956–1958, box 9/10, folder "Exhumation."

68. Parodi, French Ambassador in Rabat, to Affaires Étrangères, May 3, 1960, MAEC, Maroc 1956–1958, box 9/10, folder "Exhumation"; Jestin to Rabat Embassy, May 4, 1960, MAEN, Rabat Ambassade 1956–1989, box 827.

69. Jestin to Rabat Embassy, May 7, 1960, MAEN, Rabat Ambassade 1956–1989, box 827.

70. Le Roy (Ministre Conseiller, French Embassy), May 8, 1960, MAEN 15 PO1, box 378, folder "Divers Notes." Le Roy described Klein's meeting with Red Cross representatives.

71. French Agadir Consulat to French Embassy Rabat, n.d. [May 1960], MAEN, Rabat Ambassade 1956–1989, box 827.

72. French Embassy Rabat to Conseilleur Juridique, May 14, 1960, MAEN, Rabat Ambassade 1956–1989, box 827.

73. French Embassy to Conseilleur Juridique, May 14, 1960.

74. Jestin to French Embassy Rabat, June 3, 1960, MAEN, Rabat Ambassade 1956–1989, box 827.

75. "Agadir. Où les travaux de déblaiement se poursuivent 1,400 corps ont été sortis depuis le 5 juillet," Le petit Marocain, October 31, 1960, MAEC, Maroc 1956–1968, box 9/10, folder "Reconstruction."

76. By August 1960, the Moroccan government had created a more formal process. Whereas before, one consular representative and a Moroccan gendarme had supervised the excavation of the Saada and a few other buildings, now the health and justice departments (services) and the municipal government had to have representatives present. Jestin to French Embassy Rabat, August 8, 1960, MAEN, Rabat Ambassade 1956–1989, box 827. Once disinterred, corpses categorized as "semi-identifiable" would be placed in a depository for several days, to allow for witnesses to identify the body. Jestin to French Embassy Rabat, August 9, 1960, MAEN, Rabat Ambassade 1956–1989, box 827.

77. "Agadir. Où les travaux de déblaiement se poursuivent 1,400 corps ont été sortis depuis le 5 juillet," Le petit Marocain, October 3, 1960, MAEC, Maroc 1956–1968, box 9/10, folder "Reconstruction." Ninety-eight percent of the bodies in the Ville Nouvelle were eventually identified, compared with only half of the bodies in the Talborj, which, once excavated, became a site of mass graves. "Agadir. Où les travaux de déblaiement se poursuivent 1,400 corps ont été sortis depuis le 5 juillet," Le petit Marocain, October 31, 1960, MAEC, Maroc 1956–1968, box 9/10, folder "Reconstruction."

78. Jestin to French Embassy Rabat, August 22, 1960, MAEN, Rabat Ambassade 1956–1989, box 827.

79. Agadir Consulat, "Note sur la situation à Agadir," June 15, 1960, MAEN 15 PO1, box 378, folder "Divers notes."

80. Agadir Consulat, "Note sur la situation à Agadir." See also Direction Générale des Affaires Marocaines et Tunisiennes, draft letter to Deputé Lacaze, June 29, 1960, MAEC, Maroc 1956–1968, box 9/10, folder "Généralités 1960."

81. Sous-Direction du Maroc to Direction des Affaires Administrative et Sociales, April 23, 1960, MAEC, Maroc 1956–1958, box 9/10, folder "Exhumation."

82. Agadir Consulat, "Note sur la situation à Agadir." Of the French population, a total of 404 bodies were eventually identified. Another 131 were missing and presumed dead. Pierre de Leusse (ambassador to Morocco) to M. Couve de Murville, Ministry of Foreign Affairs, August 12, 1963, MAEN 558 PO4, box 8, folder "Inhumations." See also Péré, "Agadir, ville nouvelle," 48.

83. An equally significant migration of the Moroccan Jewish community—to Casablanca, Canada, France, and Israel—also took place. Like the migration of Europeans, this exodus had begun in the years preceding the earthquake but was accelerated by the catastrophe. Baziz, "L'exode des rescapés juifs."

84. Zartman, "Moroccan-American Base Negotiations," 38.

85. Pierre de Leusse (ambassador to Morocco) to M. Couve de Murville, Ministry of Foreign Affairs, August 12, 1963, MAEN 558 PO4, box 8, folder "Inhumations."

86. Cappe, *Agadir 29 février 1960*, 189; Seydoux to Affaires Étrangères, July 23, 1961, MAEC, Maroc 1956–1968, box 9/10, folder "Exhumation." Journalist Willi Cappe recounted the heart-wrenching story of Michel Pourrut's search for his daughter Chantal, who, in the end, he believed to be a body recorded as "fillette inconnue" ("unknown girl") in the Ihchach cemetery's "mass grave one." Cappe, *Agadir 29 février 1960*, 84–85.

87. V. A. Munier, Casablanca, August 6, 1962, "Mémoire au sujet de la recherche et l'exhumation des corps de Monsieur et Madame Bordeaux," MAEN PO4, Service Juridique, folder "Inhumations."

88. Munier, "Mémoire au sujet de la recherche et l'exhumation des corps."

89. Pierre de Leusse (ambassador to Morocco) to M. Couve de Murville, Ministry of Foreign Affairs, August 12, 1963, MAEN 558 PO4, box 8, folder "Inhumations."

90. René Cader to French Ambassador, July 8, 1963, MAEN 558 104, Service Juridique, box 8, folder "Inhumations."

91. René Cader to French Ambassador, July 8, 1963, MAEN 558 104, Service Juridique, box 8, folder "Inhumations."

92. "Des problèmes qui demeure," *Al-Istiqlal*, March 12, 1960. See also "Il faut reconstruire Agadir," *Al-Istiqlal*, March 5, 1960.

93. André Figueres, "Le crime d'Agadir" [ca. March 30, 1960], extract in MAEN, Agadir Consulat, 15 PO1, box 378, folder "Divers notes."

94. Riddleberger to ICA Rabat, March 15, 1960, NARA RG 469, entry UD376, ICA Deputy Director, box 330, folder "Morocco: Disasters: Earthquake"; Basdevant, March 15, 1960, MAEC, Maroc 1956–1968, box 9/10, folder "Reconstruction."

95. Breecher to Gordon (State), April 6, 1960, NARA RG 469, entry UD376, ICA Deputy Director, box 332, folder "Morocco ABCD."

96. Department of State, "Memorandum of Conversation," March 14, 1960, NARA RG 59, entry A1 3109D, folder "M-2 Agadir Reconstruction." See also US Embassy Rabat to State, March 25, 1960, NARA RG 469, entry UD376, ICA Deputy Director, box 330, folder "Morocco: Disasters: Earthquake."

97. Houghton (Paris) to State, March 28, 1960, RG 59, entry 81 UD-07D, WHO Subject Files, box 10, "Emergency Aid to Agadir."

98. Hoisington, *Casablanca Connection*, 194–218.

99. Hoisington, *Casablanca Connection*, 221.

100. Azzou, "La présence militaire américaine au Maroc," 126.

101. Hoisington, *Casablanca Connection*, 241.

102. Dartois, *Agadir et le Sud marocain*, 553. However, economic ties to the United States should not be overstated, especially in terms of their impact on US policy, which emphasized geostrategic goals. According to one US State Department report, "Economically, Morocco has no great importance to the U.S. nor is it likely to have in the foreseeable future. Trade with the U.S. is small." International Cooperation Agency, "Country Economic Program," September 2, 1959, NARA RG 84, UD 3005B, box 12. Even by 1969, when phosphate exports had increased, trade with the US made up under 4 percent of Moroccan exports and 7.5 percent of Moroccan imports. Zingg, "Cold War in North Africa," 44n14.

103. Agence de Coopération Technique Internationale, "a/s aide française pour la réconstruction d'Agadir," March 7, 1960, MAEC, Maroc 1956–1968, box 9/10, folder "Reconstruction." The figure of thirteen thousand "technicians" is from D. A. Fitzgerald, "Memorandum for the Deputy Director," May 2, 1960, NARA RG 469, entry UD376, ICA Deputy Director, box 332, folder "Political Affairs."

104. Agence de Coopération Technique Internationale, "a/s aide française pour la réconstruction d'Agadir."

105. Yost to State, March 4, 1960, NARA RG 469, entry UD376, ICA Deputy Director, box 330, folder "Morocco: Disasters: Earthquake."

106. Brent (ICA Rabat), March 10, 1960, NARA RG 469, entry UD376, ICA Deputy Director, box 330, folder "Morocco: Disasters: Earthquake." The request for the seismologist and architect was later rescinded, due to the contributions of German and Japanese experts and the presence of French architects in Morocco. Marcus Gordon (State) to Joseph Brent (ICA Rabat), May 2, 1960, NARA RG 469, entry UD376, ICA Deputy Director, box 332, folder "Programs."

107. Le Roy to Affaires Étrangères, March 12, 1960, no. 305/308, MAEC, Maroc 1956–1968, box 9/10, folder "Diplomatie Franco-Marocaine."

108. State to US Embassy Rabat, March 3, 1960, NARA RG 469, entry UD376, ICA Deputy Director, box 330, folder "Morocco: Disasters: Earthquake"; Yost to State, May 25, 1960, NARA RG 469, entry UD376, ICA Deputy Director, box 332, folder "Political Affairs."

109. Brent (ICA Rabat), March 8, 1960, NARA RG 469, entry UD376, ICA Deputy Director, box 330, folder "Morocco: Disasters: Earthquake."

110. Pierre Sudreaux, Minister of Construction, March 16, 1960, MAEC, Maroc 1956–1968, box 9/10, folder "Reconstruction." The suggested planners were Royer, DuBrelle, and Rotivalle.

111. Parodi to Affaires Étrangères, March 21, 1960, MAEC, Maroc 1956–1968, box 9/10, folder "Reconstruction."

112. Parodi to Affaires Étrangères, March 28, 1960, MAEC, Maroc 1956–1968, box 9/10, folder "Reconstruction."

113. AFNOR, Casablanca, March 25, MAEC, Maroc 1956–1968, box 9/10, folder "Reconstruction."

114. Nadau, "La reconstruction d'Agadir," 150.

115. Parodi to Affaires Étrangères, March 28, 1960, MAEC, Maroc 1956–1968, box 9/10, folder "Reconstruction."

116. Yost to State, March 24, 1960, NARA RG 469, entry UD376, ICA Deputy Director, box 330, folder "Morocco: Disasters: Earthquake."

117. Yost to State, March 24, 1960.

118. Yost to State, March 25, 1960, NARA RG 469, entry UD376, ICA Deputy Director, box 330, folder "Morocco: Disasters: Earthquake."

119. Thompson (Moscow) to State, March 26, 1960, NARA RG 469, entry UD376, ICA Deputy Director, box 330, folder "Morocco: Disasters: Earthquake."

120. Riddleberger to Rabat, March 28, 1960, NARA RG 469, entry UD376, ICA Deputy Director, box 330, folder "Morocco: Disasters: Earthquake."

121. Rabat to ICA, March 31, 1960, NARA RG 469, entry UD376, ICA Deputy Director, box 330, folder "Morocco: Disasters: Earthquake."

122. Riddleberger to Rabat, April 1, 1960, NARA RG 469, entry UD376, ICA Deputy Director, box 330, folder "Morocco: Disasters: Earthquake."

123. US Embassy/ICA Rabat to State, April 19, 1960, NARA RG 469, entry UD376, ICA Deputy Director, box 330, folder "Morocco: Disasters: Earthquake."

124. Yost to State, April 15, 1960, NARA RG 469, entry UD376, ICA Deputy Director, box 330, folder "Morocco: Disasters: Earthquake." See also Saccio, "Agadir Reconstruction," May 2, 1960, same folder.

125. State to US Embassy Rabat, March 3, 1960, NARA RG 469, entry UD376, ICA Deputy Director, box 330, folder "Morocco: Disasters: Earthquake"; Yost to State, May 25, 1960, NARA RG 469, entry UD376, ICA Deputy Director, box 332, folder "Political Affairs."

126. Brent, Rabat, to ICA, April 1, 1960, NARA RG 469, entry UD376, ICA Deputy Director, box 330, folder "Morocco: Disasters: Earthquake."

127. Bartholomew, "Program for the City Plan for Agadir," April 15, 1960, NARA RG 469, entry UD376, ICA Deputy Director, box 330, folder "Morocco: Disasters: Earthquake." Bartholomew recommended that an "inventory" be developed detailing all the structures that would be needed for a city of thirty-five thousand, with plans for further metropolitan expansion. He also advised zoning for land use "according to type and logical use," with planning for adequate roads, parks and green spaces, and new administrative and commercial centers.

128. International Cooperation Administration, Rabat Mission, "Monthly Summary—May 1960," June 9, 1960, NARA RG 469, entry UD376, ICA Deputy Director, box 331.

129. "Le Discours de S.M. le Roi," *La Vigie*, June 30, 1960, MAEC, Maroc 1956–1968, box 9/10, folder "Reconstruction."

130. Yost to State, April 1, 1960, NARA RG 469, entry UD376, ICA Deputy Director, box 330, folder "Morocco-Agreements-Bilateral."
131. Marcus Gordon (State) to Joseph Brent (ICA Rabat), May 2, 1960, NARA RG 469, entry UD376, ICA Deputy Director, box 332, folder "Programs."
132. Jestin to Ambafrance Rabat, August 26, 1960, MAEN, Rabat Ambassade 1956–1989, box 827.
133. Jestin to Ambafrance Rabat, June 28, 1960, MAEN, Rabat Ambassade 1956–1989, box 827.
134. ICA Rabat to Riddelberger, ICA, July 29, 1960, NARA RG 469, entry UD376, ICA Deputy Director, box 330, folder "Morocco: Disasters: Earthquake."
135. Yost to State, July 29, 1960, NARA RG 469, entry UD376, ICA Deputy Director, box 330, folder "Morocco: Disasters: Earthquake."
136. Brent (ICA Rabat) to ICA, July 29, 1960, NARA RG 469, entry UD376, ICA Deputy Director, box 330, folder "Morocco: Disasters: Earthquake."
137. ICA, "Monthly Summary—August," September 2, 1960, NARA RG 469, entry UD376, ICA Deputy Director, box 331, folder "Morocco: Reports."
138. Hamer (ICA Rabat), "Plans for Reconstruction of Agadir," August 30, 1961, NARA RG 469, entry UD376, ICA Deputy Director, folder "D-G"; Jestin to Ambafrance Rabat, January 11, 1961, MAEN, Rabat Ambassade, 558 PO1, box 828. Jestin got his information concerning Abdelâli from Mohammed Faris.
139. "Memorandum of Conversation," Thomas Larsen (US Foreign Service) with André Millot, Affaires Étrangères, February 14, 1961, NARA RG 59, entry 3109D, box 1.
140. Nadau, "La reconstruction d'Agadir," 150.
141. Nadau, "La reconstruction d'Agadir," 150. See also Marcus Gordon (State) to Joseph Brent (ICA Rabat), May 2, 1960, NARA RG 469, entry UD376, ICA Deputy Director, box 332, folder "Programs."
142. Nadau, "La reconstruction d'Agadir," 150.
143. "Memorandum of Conversation," Thomas Larsen (US Foreign Service) with André Millot, Affaires Étrangères, February 14, 1961, NARA RG 59, entry 3109D, box 1.
144. Jestin to Ambafrance Rabat, March 17, 1961, MAEN, Rabat Ambassade, 558 PO1, box 828.
145. Ben Embarek, "Chroniques africaines," i.
146. "Agadir 1960–1965," 4.
147. "Memorandum of Conversation," Thomas Larsen (US Foreign Service) with André Millot, Affaires Étrangères, February 14, 1961, NARA RG 59, entry 3109D, box 1.
148. Yost to State, November 29, 1960, NARA RG 469, entry UD376, ICA Deputy Director, box 330, folder "Morocco: Disasters: Earthquake."
149. Brent to Rabat, November 29, 1960.
150. Riddleberger (ICA) to Rabat, December 5, 1960, and December 9, 1960, NARA RG 469, entry UD376, ICA Deputy Director, box 330, folder "Morocco: Disasters."
151. Brent to ICA, January 17, 1960, NARA RG 469, entry UD376, ICA Deputy Director, box 355, folder "Morocco: D-G."

152. Quoted and translated in Seydoux (Ambafrance Rabat) to Agadir Consulate, January 25, 1961, MAEN, Rabat Ambassade, 558 PO1, box 853.

153. Seydoux (Ambafrance Rabat) to Agadir Consulate, January 25, 1961, MAEN, Rabat Ambassade, 558 PO1, box 853.

154. "La reconstruction d'Agadir coûtera 240 millions de DH," *La vigie marocaine*, January 20, 1961, MAEC, Maroc 1956–1968, box 9/10, folder "Reconstruction." Morocco's minister of national economy and finance, Diouri, confirmed these figures; "Reconstruction d'Agadir," February 9, 1961, MAEC, Maroc 1956–1968, box 9/10, folder "Généralités 1960." Hassan repeated the figures of 24 and 12 billion several weeks later. "Agadir Survit," *Le Monde*, March 2, 1961, MAEC, Maroc 1956–1968, box 9/10, folder "Reconstruction."

155. Jestin to Ambafrance Rabat, February 13, 1961, MAEN, Rabat Ambassade, 558 PO1, box 828.

156. Roger Seydoux, December 14, 1960, MAEC, Maroc 1956–1968, box 9/10, folder "Reconstruction."

157. Jestin to Ambafrance Rabat, January 31, 1961, MAEN, Rabat Ambassade, 558 PO1, box 828.

158. Jestin to Ambafrance Rabat, February 12, 1961, MAEN, Rabat Ambassade, 558 PO1, box 828. See also "Agadir Survit," *Le Monde*, March 2, 1961, MAEC, Maroc 1956–1968, box 9/10, folder "Reconstruction."

159. Jestin to Ambafrance Rabat, March 17, 1961, MAEN, Rabat Ambassade, 558 PO1, box 828.

160. Jestin to Ambafrance Rabat, June 8, 1961, MAEN, Rabat Ambassade, 558 PO1, box 828. See also SDECE, "Stagnation des projets de reconstruction d'Agadir," MAEC, Maroc 1958–1968, box 9/10, folder "Généralités 1960."

161. Hamer (ICA Rabat), "Plans for Reconstruction of Agadir," August 30, 1961, NARA RG 469, entry UD376, ICA Deputy Director, folder "D-G."

162. Dethier, "60 ans d'urbanisme au Maroc," 49.

163. For example, Raspoutine, March 17, 2011, "Agadir Ofella (la Kasbah) est un cimetière pour le moment," thread in "Souvenirs d'Agadir," https://www.agadir1960.com.

6. The Soul of a City

1. Mas, "Plan directeur et plans d'aménagement," 6.

2. Péré, "Agadir, ville nouvelle," 89. For a more recent echo of this trope, see Benjelloun, *Happy Marriage*, 49.

3. Cooper, *Colonialism in Question*, 65.

4. Bell and de-Shalit, *Spirit of Cities*, 2, 13. The hazard of reductionism is particularly evident in the chapter on Paris, which neglects the city's diversity and the legacy of colonialism and migration, reducing Paris to "the city of romance."

5. Cooper has suggested that the ambiguously multivalent term "identity" would be better replaced by more precise terms. The same could be said of "spirit" and "ethos." Two candidates suggested by Cooper as suitable for some contexts are "self-understanding" or "self-representation." These might well fit what Bell and de-Shalit describe, and prescribe, as "civicism," a city-based alternative to nationalism. However, since such understandings are often advocated by outside agents, as Bell and de-Shalit recognize, this might be simplified to "representations" or, as I have put it here, "ideas." Cooper, *Colonialism in Question*, 74; Bell and de-Shalit, *Spirit of Cities*, 13.

6. Valentin Pelosse has argued that the Orléansville earthquake had an important effect on post-independence politics in Algeria because reconstruction had produced a large supply of good-quality, newly constructed, or newly renovated, housing. Although the nationalization of industry and agricultural land drew more attention, the new authorities' ability to control the redistribution of residential property was particularly significant in Orléansville, where 850–900 residential housing units were declared "vacant properties." These housing units became tools of patronage and repression: when Wilaya IV was defeated, units inhabited by its supporters were declared "illegally occupied" by the new government in Algiers, and the state initiated evictions. Under Ben Bella's leadership, the state in Algiers now controlled this housing stock and pursued a policy that aimed to keep rents low. This suggests that, as in Agadir, seismic disaster in the Chélif Valley may have facilitated the trend toward post-independence authoritarianism by making it easier for the central state to control property and dispense patronage. This effect was likely marginal in Algeria, in comparison to the effects of the armed struggle that concentrated political legitimacy in the military, but James McDougall has contested the view that the struggle for independence was the single determining factor shaping independent Algeria, emphasizing the multiplicity of causal factors and continuities. More research is needed to evaluate the implications of Pelosse's argument. Pelosse, "Évolution socio-professionnelle," 82–89. Regarding the use of "abandoned" land, industries, and housing, see also Ageron, *Modern Algeria*, 133; McDougall, *History of Algeria*, 249. On the role of the war of independence, compare Stora, *Algeria: 1830–2000*, 113; and McDougall, "Impossible Republic," 124.

7. Malverti, "Méditerranée, soleil, et modernité," 39–41.

8. Picard, "Orléansville," 70–75.

9. Malverti, "Méditerranée, soleil, et modernité," 46, 51, 55.

10. National Research Council, *El-Asnam, Algeria Earthquake*, section 4, 29.

11. Malverti, "Méditerranée, soleil, et modernité," 55.

12. Cooper, *Colonialism in Question*, 65.

13. Hardy, *L'âme marocaine d'après la littérature française*; Hardy and Brunot, *L'enfant marocain*.

14. Dalrymple, "Architect as Totalitarian."

15. Dethier, "60 ans d'urbanisme au Maroc," 35, 48.

16. Dethier, "60 ans d'urbanisme au Maroc," 48.

17. Douglas Martin, quoted in Seaman, "Jumping Joyous Urban Jumble," 139.

18. Rabinow, *French Modern*, 5.

19. Nadau, "La reconstruction d'Agadir," 154–56.

20. Sbenter, "Eléments d'articulation urbaine," 58; Achehaifi, "L'urbanisme moderne à l'épreuve," 161. See also Saoudi, "Le passé, le présent, et peut-être le futur," 66.

21. Achehaifi, "L'urbanisme moderne à l'épreuve," 131.

22. Nadau, "La reconstruction d'Agadir," 154.

23. Nadau, "La reconstruction d'Agadir," 165.

24. Charef, "Agadir, une ville orpheline," 173. Even Péré, not generally critical of the new city, believed that the high vacancy rates in the newly constructed urban center were due to the fact that the European-style architecture was alien to the Moroccan population. Péré, "Agadir, ville nouvelle," 87.

25. Mouline, "Éditorial," in *Urbanités en recomposition*, 10.

26. Bell and de-Shalit, *Spirit of Cities*, 13.

27. Gershovich, "Long Shadow of Lyautey."

28. Ben Attou, "Agadir gestion urbaine," 79.

29. Interview of Mohamed Bajalat in *Libération*, March 12, 2011, www.libe.ma and https://terriermichel.wordpress.com/2011/03/16/mohamed-bajalat-president-du-forumizorane-pour-nous-agadir-ofella-est-un-cimetiere-nous-ne-pouvons-donc-plus-accepter-quil-soit-profane/.

30. Ben Attou, introduction to *Le grand Agadir*, 11.

31. Abdallah Aourik, interview of Ahmed Bouskous, "Où étiez-vous le 29 février 1960?," *Agadir O'flla*, February 2008, 11.

32. Péré, "Agadir, ville nouvelle," 89.

33. "Nouvel an amazigh: Agadir, capitale du Souss berbère fête Id-Ennayer," *Le Reporter*, January 9, 2017, https://www.lereporter.ma/nouvel-an-amazigh-agadir-capitale-du-souss-berbere-fete-id-ennayer/. See Abdallah Aourik interview of Tariq Kabbage, "Où étiez-vous le 29 février 1960?," *Agadir O'flla*, February 2008, 13.

34. Aourik interview of Kabbage, *Agadir O'flla*, February 2008, 12.

35. Roussafi, Jafri, and Kikr, *Mémoires d'Agadir*, 11; R. Smith, *Ahmad Al-Mansur*, 1–42.

36. Abu-Lughod, *Rabat*, 79–80; Péré, "Agadir, ville nouvelle," 44; Roussafi, Jafri, and Kikr, *Mémoires d'Agadir*, 12–17. According to Abu-Lughod, while the development of the port at Mogador also contributed to the decline of Rabat in the late 1700s, the 1755 Lisbon-Meknes earthquake may also have played a significant role: a tidal wave seems to have exacerbated tidal and/or sandbar inconveniences of the Bou Regreg Harbor. Abu-Lughod notes that there is some uncertainty about both the environmental and political events.

37. Péré, "Agadir, ville nouvelle," 43–44; Dartois, *Agadir et le Sud marocain*, 476.

38. Roussafi, Jafri, and Kikr, *Mémoires d'Agadir*, 20–28; Brown, Lakhsassi, and Ighil, "Every Man's Disaster," 126. As Brown, Lakhsassi, and Ighil put it, "In 1911, Agadir entered international history."

39. Rabinow, *French Modern*, 112.

40. Rabinow, *French Modern*, 116.

41. As Rabinow has argued, Lyautey's approach to urban planning was influenced by "a slow but sustained move toward separating cultures [that] had already begun in other parts of North Africa," particularly Tunisia. Rabinow, *French Modern*, 298. Lyautey was also influenced by the work of thinkers on the French left, whose ideas of technocratic elitism and social paternalism crossed political divides to form a new hegemony. Urbanism, argues Rabinow, was no longer about arranging space but creating social relations. See also 261–62.

42. Quoted in Wright, "Tradition in the Service of Modernity," 302.

43. Abu-Lughod, *Rabat*, 142, 145.

44. Both Dethier and Rabinow have offered important critiques of Abu-Lughod's description of the *cordon sanitaire* as an effective barrier to intra-urban migration and cultural mixing. Dethier, "60 ans d'urbanisme au Maroc," 11; Rabinow, *French Modern*, 300–301.

45. Péré, "Agadir, ville nouvelle," 45; Dartois, *Agadir et le Sud marocain*, 527.

46. Mas, "Plan directeur et plans d'aménagement," 7.

47. Péré, "Agadir, ville nouvelle," 45.

48. Péré, "Agadir, ville nouvelle," 45; Dartois, *Agadir et le Sud marocain*, 539–41; Roussafi, Jafri, and Kikr, *Mémoires d'Agadir*, 73–80. On Rabat and Casablanca see: Abu-Lughod, *Rabat*, 201; Dethier, "60 ans d'urbanisme au Maroc," 11; Rabinow, *French Modern*, 300–301.

49. Péré, "Agadir, ville nouvelle," 48.

50. Péré, "Agadir, ville nouvelle," 48–49, 57.

51. Péré, "Agadir, ville nouvelle," 50.

52. Mas, "Plan directeur et plans d'aménagement," 7.

53. Dethier, "60 ans d'urbanisme au Maroc," 27–34; Rabinow, *French Modern*, 3. In Rabinow's words, "Ecochard's blithe attitude toward historical and ethnographic realities was not untypical of high modernism in architecture. History was imaginary; nineteenth-century European imperialism had ceased to exist.... Following high modernist principles, Ecochard held that human needs were universal."

54. Rabinow, *French Modern*, 4.

55. Williford, "Seismic Politics," 990–1002; Despeyroux, "Agadir Earthquake of February 29th, 1960," 522; Committee of Structural Steel Producers, *Agadir, Morocco Earthquake*, 38. Williford has shown that these conclusions were conditioned by a circular set of assumptions: the exact epicenter could not be located solely based on the output of distant seismographs and therefore depended on engineering analyses of damage—which depended, in turn, on engineers' beliefs about the vulnerability of different types of construction.

56. Péré, "Agadir, ville nouvelle," 52.

57. Péré, "Agadir, ville nouvelle," 48–55.

58. Williford, "Seismic Politics," 990–1002.

59. Rabat to ICA, March 31, 1960, United States Archives and Records Administration (hereafter NARA) RG 469, entry UD376, ICA Deputy Director, box 330, folder "Morocco: Disasters: Earthquake."

60. "Des problèmes qui demeure," *Al-Istiqlal*, March 12, 1960.

61. Lehman, *Reconstruction of Agadir.* Due to the earthquake Agadir became immediately connected to an international network of information, individuals, and organizations interested in the study of earthquakes and the mitigation of earthquake hazards. While the French and Moroccan press often portrayed the catastrophe in Agadir as akin to those of Fréjus and Orléansville, in the global press Agadir was suddenly discussed in terms of a category of cities that included San Francisco, Santiago, Messina, and Tokyo. The team of West German engineers and scientists who accompanied the German consul to Casablanca on an inspection of the site were soon followed by engineers from the American Iron and Steel Institute. The Japanese Ministry of Reconstruction sent two seismologists in April 1960 for a one-month mission; Japanese interest was also expressed by cosponsoring a UN resolution exhorting member countries to assist the people of Agadir, and one of the French coopérants in Agadir, engineer Robert Ambroggi, was invited to participate in a conference in Toyko in July 1960. Jean Daridan to Affaires Étrangères, April 7, 1960, MAEC, Maroc 1956–1968, box 9/10, folder "Généralités 1960"; Agenda Item 21, April 6, 1960, United Nations Economic and Social Council, 29th session, NARA RG 59, UD-07D, box 10, folder "Emergency Aid to Agadir"; Hamer to ICA, May 25, 1960, NARA RG 469, UD376, box 330, folder "Morocco—Disasters: Earthquake."

62. Rabinow, *French Modern*, 39.

63. Shrady, *Last Day*, 156–60; Rozario, "What Comes Down Must Go Up."

64. King Hassan II, "Message de S. M. Hassan II," 2.

65. King Hassan II, "Message de S. M. Hassan II," 2.

66. King Hassan II, "Message de S. M. Hassan II," 2.

67. Rabinow, *French Modern*, 76, 235. See also Dethier, "60 ans d'urbanisme au Maroc," 37. For a comparative example, see Schencking, "Great Kanto Earthquake."

68. "Agadir 1960–1965," 4. See also Nadau, "La reconstruction d'Agadir," 148.

69. Ben Embarek, "Tourisme et urbanisme," 64.

70. Ben Embarek, "Tourisme et urbanisme," 65.

71. Dethier, "60 ans d'urbanisme au Maroc," 49.

72. Royaume du Maroc, Haut-Commissariat de la Reconstruction, *Agadir: Information, urbanisme, aide de l'état-1962*, pamphlet, Centre de Documentation, Ministère d'Intérieur Direction de l'Urbanisme et de l'Habitat, Rabat; Hicks, "Rebuilt Agadir," 295. See also Lasky, "La renaissance d'Agadir"; Williford, "Seismic Politics," 1004.

73. Williford, "Seismic Politics," 1006. Williford also notes that attempts by residents to play a role in shaping the future of their city through an organized "Victim's Committee" in 1961 had little impact. Williford, "Seismic Politics," 1010.

74. Nadau, "La reconstruction d'Agadir," 151.

75. Rabinow, *French Modern*, 290–91.

76. Dethier, "60 ans d'urbanisme au Maroc," 13; Rabinow, *French Modern*, 290–93; Abu-Lughod, *Rabat*, 168–69.

77. Wright, "Tradition in the Service of Modernity," 300.

78. Nadau, "La reconstruction d'Agadir," 150.

79. Nadau, "La reconstruction d'Agadir," 160. A concurring assessment is found in "Memorandum of Conversation," Thomas Larsen (US Foreign Service) with André Millot, Affaires Étrangères, February 14, 1961, NARA RG 59, entry 3109D, box 1.

80. Dethier, "60 ans d'urbanisme au Maroc," 37.

81. Nadau, "La reconstruction d'Agadir," 49, 150. The team of architects included Faraoui, Castelnau, Riou, Verdugo, Zévaco, Demazières, and Azaury.

82. Roberson, "Changing Face of Morocco," 65–67; Dethier, "60 ans d'urbanisme au Maroc," 49.

83. Nadau, "La reconstruction d'Agadir," 154.

84. Rabinow, *French Modern*, 3.

85. Mas, "Plan directeur et plans d'aménagement," 11.

86. Rabinow, *French Modern*, 3.

87. Dethier, "60 ans d'urbanisme au Maroc," 39.

88. Mas, "Plan directeur et plans d'aménagement," 10. As Mas noted, aspects of the natural setting considered more desirable were made integral to the new plan, such as the elevation of the area that became the "new Talborj," hills in the residential areas, and dunes and valleys in the beachfront zone designated for hotels.

89. Mas, "Plan directeur et plans d'aménagement," 10.

90. Mas, "Plan directeur et plans d'aménagement," 15. Compare to Harland Bartholomew and Associates, "Agadir Master Plan" (Rabat: 1960): 26–44.

91. Mas, "Plan directeur et plans d'aménagement," 15.

92. Mas, "Plan directeur et plans d'aménagement," 15.

93. Mas, "Plan directeur et plans d'aménagement," 15; Nadau, "La reconstruction d'Agadir," 152.

94. Dethier, "60 ans d'urbanisme au Maroc," 18.

95. Dethier, "60 ans d'urbanisme au Maroc," 28.

96. I thank Mohamed Bajalat for pointing out the echoes of the southern Moroccan *agadir* (*ksar* or fortress) in the new Hôtel de Ville.

97. Nadau, "La reconstruction d'Agadir," 159; Williford, "Seismic Politics," 110.

98. Nadau, "La reconstruction d'Agadir," 154–56.

99. Regarding education, see Segalla, *Moroccan Soul*, 222–35; regarding urbanism, see Rabinow, *French Modern*, 287. For a case study in the synergy between the colonial

theories of the French right and the goals of anticolonial nationalists outside of Morocco, see Jennings, "Conservative Confluences."

100. Dethier, "60 ans d'urbanisme au Maroc," 37.
101. Dethier, "60 ans d'urbanisme au Maroc," 48.
102. Dethier, "60 ans d'urbanisme au Maroc," 6.
103. Dethier, "60 ans d'urbanisme au Maroc," 35.
104. Lahbabi, "Changement social et aliénation," 54. Lahbabi cited Gramsci's *Literatura y vida nacional* as well as Althusser's *Pour Marx*.
105. King Hassan II, "Texte du discours royal pronouncé," 4, 8. See also Barber, "Jihad vs. McWorld."
106. Roberson, "Changing Face of Morocco," 97.
107. See Wainscott, "Opposition Failure or Regime Success?"
108. Lahbabi, "Changement social et aliénation," 52.
109. Nadau, "La reconstruction d'Agadir," 165.
110. After Hassan's death in 1999, King Mohammed VI's ambitious urban renewal projects in Rabat-Salé and Tangier suggest a shift away from the Lyautey-Hassan model promoting traditionalism and toward the model of the Persian Gulf's "global cities," particularly Doha and Dubai. Lee, "Urban Politics of the Bouregreg Project."
111. Nadau, "La reconstruction d'Agadir," 160.
112. For the region of greater Agadir, the population rose from 39,000 to 114,000. Kidou, "Les changements démographiques d'une nouvelle grande ville."
113. Royaume du Maroc, Haut Commissariat au Plan, "Recensement général de la population et de l'habitat de 2004: Population légale du Maroc," 2004. In contrast, Tangier had 2,323 foreigners out of 173,477 (1.3 percent). Mohammedia, a beach town located between Casablanca and Rabat, housed 1,240 foreigners out of 188,619 (0.7 percent).

7. Rupture, Nostalgia, and Representation

1. Ighil, "Tale of Agadir," 128.
2. Brown, Lakhsassi, and Ighil, "Every Man's Disaster," 129, 131.
3. There is a clear contrast in social backgrounds between Ibn Ighil, a rural, Tashelhit-speaking oral poet, and the urban, middle-class, French-literate writers considered in this chapter. However, further study would be needed before one could make generalizations about a correlation between social class or origins and the degree to which environmental events are interpreted through a political lens.
4. For example, Marsh and Frith, *France's Lost Empires*.
5. William Bissell has explored issues of nostalgia among the formerly colonized in "Engaging Colonial Nostalgia."
6. Kréa, *Le séisme*, 81.
7. Aït Ouyahia, *Pierres et lumières*, 279.

8. James McDougall has warned against reading hegemonic strands of nationalist narratives as teleological and monolithic rather than multiple and continually contested. McDougall, *History and the Culture of Nationalism*, 12–14. On competing visions of the role of "Arabs" and "Berbers" in the Algerian nation, see 74–86 and 184–216. On French colonial discourse and policy regarding Berbers and Arabs, see Lorcin, *Imperial Identities*; Gellner and Micaud, *Arabs and Berbers*.

9. Jean Bernard, *C'est de l'homme qu'il s'agit*, quoted in Aït Ouyahia, *Pierres et lumières*, 9.

10. Aït Ouyahia, *Pierres et lumières*, 74.

11. Aït Ouyahia, *Pierres et lumières*, 65.

12. Aït Ouyahia, *Pierres et lumières*, 285.

13. Raqbi, "Agadir dans les écrits de Tahar Benjelloun," 130.

14. Khaïr-Eddine, *Agadir*, 119.

15. For example, Khaïr-Eddine, *Agadir*, 87.

16. Abdel-Jouad, "Mohammed Khaïr-Eddine," 145.

17. Touaf, "Legacy of Dissent," 53.

18. Abdel-Jouad, "Mohammed Khaïr-Eddine," 148.

19. Touaf, "Legacy of Dissent," 53.

20. Khaïr-Eddine, *Agadir*, 25.

21. Khaïr-Eddine, *Agadir*, 48–58.

22. Khaïr-Eddine, *Agadir*, 11–16.

23. Khaïr-Eddine, *Agadir*, 14.

24. Khaïr-Eddine, *Agadir*, 35.

25. Saïgh-Bousta, "Une vie, un rêve," 21. See also McDougall, *History and the Culture of Nationalism*, 195.

26. Khaïr-Eddine, *Agadir*, 15.

27. Khaïr-Eddine, *Agadir*, 106. See also 98.

28. Khaïr-Eddine, *Agadir*, 110.

29. Khaïr-Eddine, *Agadir*, 20–30.

30. Khaïr-Eddine, *Agadir*, 119–23.

31. Saïgh-Bousta, "Une vie, un rêve," 20. See also McNeece, "Le jour de la très grande violence," 148.

32. El Younssi, "*Souffles-Anfas* and the Moroccan Avant-Garde," 35.

33. Steinberg, "Down to Earth," 819.

34. Touaf, "Legacy of Dissent," 53.

35. Eldridge, *From Empire to Exile*, 11.

36. In this respect I would argue that Khaïr-Eddine's 1967 novel differs from Nina Bouraoui's 1999 *Le jour du séisme* (Paris: Éditions Stock), which reflects a new generation's concerns. As Slimani Aït Saada has argued, literary representations of Algerian earthquakes written since 1990 have tended to associate seismic disruption with the civil unrest and atrocious civil war that afflicted Algeria in that decade, as well as with "social malaise, an ill-being [*mal-être*] of the youth in search

of identity markers in the face of unrestrained globalization and the upheaval of values in a fragmented society" (Aït Saada, *Histoire De Lieux*, 283). With the passage of time, connections between the 1980 earthquake, the war of independence, and earthquake of 1954 have become less prominent—in sharp contrast to their marked presence in Ali Bouzar's 1985 memoir. Yet for Bouraoui, removed from the epicenters of this violence, even the more recent trauma of the 1980 earthquake can produce a powerful evocation of more universal struggles. The interpretation of Bouraoui's depiction of the 1980 earthquake as a metaphorical disruption is valid: Bouraoui uses the earthquake to express themes of violent transformation related to adolescence, gender, and sexual orientation. As Karima Yahia Ouahmed has argued, Bouraoui's decision to locate the earthquake in Algiers, where Bouraoui lived as a child, rather than in Orléansville/El Asnam, distances the narrative from the events of the actual earthquake (Ouahmed, "De la double origine à l'être-deux," 221–29). See also Agar-Mendousse, "Fracturing the Self."

37. Chaulet-Achour, "Itinéraires de mémoire." Chaulet-Achour quotes Dehbia Aït Mansour's review in *Liberté*, April 4, 2000.
38. McDougall, *History and the Culture of Nationalism*, 8.
39. Aït Ouyahia, *Pierres et lumières*, 13–24.
40. Aït Ouyahia, *Pierres et lumières*, 64.
41. Aït Ouyahia, *Pierres et lumières*, 282–83.
42. Aït Ouyahia, *Pierres et lumières*, 289.
43. The values of the Algerian Revolution are here portrayed nostalgically and contrasted to more recent attacks on socialism by "our grand market economists and the goals of our 'structural readjustors.'" Aït Ouyahia, *Pierres et lumières*, 291.
44. Aït Ouyahia, *Pierres et lumières*, 297.
45. In contrast to the Vietnamese doctors, discussed by Michitake Aso and Annick Guénel, whose practice of medicine came to serve the anti-imperial war effort even as the war effort facilitated medical research, Aït Ouyahia seems to have viewed the struggle for independence as instrumental to or even subordinate to the medical struggle against suffering. Aso and Guénel, "Itinerary of a North Vietnamese Surgeon"; Aso, "Learning to Heal the People."
46. Bensimon, *Agadir, un paradis*, 17.
47. Bensimon, *Agadir, un paradis*, 21.
48. Bensimon, *Agadir, un paradis*, 21.
49. Bensimon, *Agadir, un paradis*, 13.
50. Bensimon, *Agadir, un paradis*, 36, 30.
51. Bensimon, *Agadir, un paradis*, 59.
52. Bensimon, *Agadir, un paradis*, 161–66.
53. Bensimon, *Agadir, un paradis*, 198–201.
54. Tengour, "Childhood," 201.
55. Tengour, "Childhood," 201.
56. Aït Saada, *Histoire de Lieux*, 276.
57. Tengour, "Childhood," 202.

58. It is also possible that these were responses to the 1980 earthquake anachronistically projected back to 1954; Ali Bouzar's account of the 1980 earthquake mentions (and dismisses) beliefs that the disaster was God's punishment for the mixing of wine and couscous, and although Bouzar refers to this as a "famous" slur, it is not clear from Bouzar's account whether it became so before or after 1980. Bouzar, *Le consentement du malheur*, 150.

59. Tengour, "Childhood," 207.

60. Tengour, "Childhood," 208.

61. Tengour, "Childhood," 211.

62. The Tengour family's move to France was reportedly for "political reasons." "Habib Tengour," Poetry International Rotterdam, accessed June 17, 2015, http://www.poetryinternationalweb.net/pi/site/poet/item/24142/Habib-Tengour.

63. Roussafi, Jafri, and Kikr, *Mémoires d'Agadir*; Roussafi, Jafri, and Kikr, *Dhakirat Agadir fi al qarn al ashrin*; Torres, *L'Orléansvillois*; Dartois, *Agadir et le Sud marocain*. The websites are "Agadir" at mfd.agadir.free.fr; "Site d'Orléansville et sa région" at Orleansville.free.fr; and "Agadir 1960" at agadir1960.com. It should be noted that Dartois's book takes the form of an academic work of historical writing, whereas her website falls into the genre of the other books and websites discussed in this section, as a reconstruction of the past through the arrangement of documentary evidence.

64. Eldridge, *From Empire to Exile*, 260.

65. Torres, *L'Orléansvillois*, 7.

66. Torres, *L'Orléansvillois*, 175.

67. Lahsen Roussafi, lecture at Dar Si Hmad, Agadir, May 14, 2015.

68. Bouzar, *Le consentement du malheur*, 17.

69. Bouzar, *Le consentement du malheur*, 11.

70. Bouzar, *Le consentement du malheur*, 9.

71. Bouzar, *Le consentement du malheur*, 17.

72. Bouzar, *Le consentement du malheur*, 49.

73. Bouzar, *Le consentement du malheur*, 14.

74. Aït Ouyahia uses the framework of a road trip to similar effect, in chapter titled "Itineraire," integrating anecdotes and reflections on local history spanning two generations as the characters travel across Algeria. Aït Ouyahia, *Pierres et lumières*, 171–263.

75. Bouzar, *Le consentement du malheur*, 162.

76. Bouzar, *Le consentement du malheur*, 20.

77. Bouzar, *Le consentement du malheur*, 27–29.

78. Bouzar, *Le consentement du malheur*, 43.

79. Bouzar, *Le consentement du malheur*, 129–30. Bouzar is inconsistent with regard to stating his age in relation to the years in which his childhood memories occurred. He says he was ten in 1954 (139) but says he and his friends were age ten to twelve when they began playing in cité Wagons, "around 1958" (128).

80. Bouzar, *Le consentement du malheur*, 130.

81. Bouzar, *Le consentement du malheur*, 139–40.
82. Bouzar, *Le consentement du malheur*, 17.
83. Bouzar, *Le consentement du malheur*, 8
84. Bouzar, *Le consentement du malheur*, 8
85. Bouzar, *Le consentement du malheur*, 161.
86. Bouzar, *Le consentement du malheur*, 8
87. Bouzar, *Le consentement du malheur*, 12.
88. Moufdi Zakaria, "Agadir Alshahida," in Roussafi, Jafri, and Kikr, *Dhakirat Agadir fi al qarn al ashrin*, 3:4 (Segalla translation).
89. Bouzar, *Le consentement du malheur*, 8.
90. Bouzar, *Le consentement du malheur*, 120.
91. Balaev, "Trends in Literary Trauma Theory," 157.
92. Grzeda, "Trauma and Testimony," 66. See also Ward, *Postcolonial Traumas*, 5–7.
93. Morin, "Unspeakable Tragedies," 21–22.

8. Conclusion

1. T. Mitchell, *Rule of Experts*, 29–36. Mitchell, in his chapter "Can the Mosquito Speak?," extends Spivak's question to the nonhuman agent. Spivak, "Can the Subaltern Speak?" See also Cutler, "Can the North Atlantic Oscillation Speak?," and Aso, "Wrong Place, Wrong Time."
2. Kai Erikson has called into question the usefulness of intentionality as a concept for understanding the atomic destruction of these cities. Erikson, *New Species of Trouble*, 189.
3. T. Mitchell, *Rule of Experts*, 1.
4. Rowe, "What on Earth Is Environment?"
5. Bennett, *Vibrant Matter*, 111.
6. Clancey, *Earthquake Nation*, 5.
7. Bennett, *Vibrant Matter*, 112.
8. Bennett, *Vibrant Matter*, 2. For discussions of agency, "actants," deodands, and "agentic" powers, see 9, 35. I have here introduced the term "agentishness" to better convey the necessarily imprecise nature of such concepts.
9. T. Mitchell, *Rule of Experts*, 42–43.
10. T. Mitchell, *Rule of Experts*, 299. See also Bennett's discussion of Bruno Latour. Bennett, *Vibrant Matter*, 115.

BIBLIOGRAPHY

Archives and Repositories

ANFP. Archives Nationales de France, Pierrefitte.
ANOM. Centre d'Archives d'Outre-Mer, Aix-en-Provence.
Archives Départementales du Var, Draguignan.
Archives Municipales de Fréjus.
Archives Nationales du Royaume du Maroc, Rabat.
Bibliothèque de la Faculté des Lettres et Sciences Humaine, Université Ibn Zohr, Agadir.
Bibliothèque de l'Institut Royal de la Culture Amazigh, Rabat.
Bibliothèque Nationale du Royaume du Maroc, Rabat.
Centre de Documentation, École Nationale de l'Architecture, Rabat.
Centre de Documentation, Ministère d'Intérieur Direction de l'Urbanisme et de l'Habitat, Rabat.
Centre Multimedia de la Direction de l'Urbanisme, Rabat.
Fondation du Roi Abdul-Aziz Al Saoud pour les Études Islamiques et les Sciences Humaines, Casablanca.
MAEC. Archives Diplomatiques, Ministère des Affaires Étrangères, La Courneuve.
MAEN. Archives Diplomatiques, Ministère des Affaires Étrangères, Nantes.
NARA. United States Archives and Records Administration, College Park and Washington DC.
Naval History and Heritage Command Archives. Washington Navy Yard, Washington DC.
Service de la Gestion des Archives et de la Documentation du Ministère de l'Habitat, de l'Urbanisme et de l'Aménagement de l'Espace, Rabat.

Published Works

Abdel-Jouad, Hédi. "Mohammed Khaïr-Eddine: The Poet as Iconoclast." *Research in African Literatures* 23, no. 2 (1992): 145–50.
Abu-Lughod, Janet. *Rabat: Urban Apartheid in Morocco*. Princeton NJ: Princeton University Press, 1980.
Accampo, Elinor, and Jeffrey H. Jackson. "Introduction to Special Issue on 'Disaster in French History.'" *French Historical Studies* 36 (2013): 166–74.

Achehaifi, My Ahmed. "L'urbanisme moderne à l'épreuve: Cas d'Agadir." Fin d'études diplôme thesis, École Nationale d'Architecture, Rabat, 1994.

"Agadir 1960–1965." *A + U: Revue africaine d'architecture et d'urbanisme* 4 (1966): 4.

Agar-Mendousse, Trudy. "Fracturing the Self: Violence and Identity in Franco-Algerian Writing." In *Violent Depictions: Representing Violence across Cultures*, edited by Susanna Scarparo and Sarah McDonald, 18–31. Newcastle: Cambridge Scholars, 2006.

Ageron, Charles-Robert. *Les Algériens musulmans et la France.* Paris: Presses Universitaires de France, 1968.

———. *Modern Algeria: A History from 1830 to the Present.* Translated by Michael Brett. 9th ed. London: Hurst, 1990.

———. "Une dimension de la guerre d'Algérie: Les 'regroupements' de populations." *Histoire du Maghreb* (2005): 561–86.

Aït Ouyahia, Belgacem. *Pierres et lumières: Souvenirs et digressions d'un médecin algérien, fils d'instituteur d'origine indigène.* Algiers: Casbah, 1999.

Aït Saada, El Djamhouria Slimani. *Histoire de lieux: El Asnam, Miliana, Ténès.* Algiers: Hibr, 2013.

Alaoui, Moulay Abdelhadi. *Le Maroc du traité de Fes à la Libération, 1912–1956.* Rabat: Éditions la Porte, 1994.

Albertini, A., D. Gross, and William M. Zinn, eds. *Triaryl-Phosphate Poisoning in Morocco 1959: Experiences and Findings.* New York: Intercontinental Medical, 1968.

Almeida, Dmitri. "Cultural Retaliation: The Cultural Policies of the 'New' Front National." *International Journal of Cultural Policy* (2017). https://doi.org/10.1080/10286632.2017.1288228.

Amster, Ellen. *Medicine and the Saints: Science, Islam, and the Colonial Encounter in Morocco, 1877–1956.* Austin: University of Texas Press, 2013.

André. "Un regard extérieur." In *Souvenirs intimes: Malpasset,* edited by Fabienne Russo and Michel Suzzarini. Brignoles: Association du Cinquantième de la Catastrophe de Malpasset/Éditions Vivre Tout Simplement, 2009.

Ansari, Humayun. "'Burying the Dead': Making Muslim Space in Britain." *Historical Research* 80 (2007): 545–66.

Aso, Michitake. "Learning to Heal the People: Socialist Medicine and Education in Vietnam, 1945–54." In *Translating the Body: Medical Education in Southeast Asia,* edited by Hans Pols, C. Michele Thompson, and John Harley Warner, 146–72. Singapore: National University of Singapore Press/University of Chicago Press, 2017.

———. "The Wrong Place, the Wrong Time: Pests, Weeds, and Other Unwanted Colonial Actors." Paper presented at the French Colonial Historical Society annual meeting, 2017.

Aso, Michitake, and Annick Guénel. "The Itinerary of a North Vietnamese Surgeon: Medical Science and Politics during the Cold War." *Science, Technology and Society* 18 (2013): 291–306.

Azzou, El-Mustapha. "La présence militaire américaine au Maroc, 1945–1963." *Guerres mondiales et conflits contemporains* 210 (2003/2): 125–32.

Balaev, Michelle. "Trends in Literary Trauma Theory." *Mosaic: A Journal for the Inter-disciplinary Study of Literature* 4 (2008): 149–66.

Balandier, Georges. "La situation coloniale: Ancien concept, nouvelle réalité." *French Politics, Culture, and Society* 20, no. 2 (2002): 4–10.

———. "La situation coloniale, approche théorique." *Cahiers internationaux de sociologie* 11 (1951): 44–79.

Barber, Benjamin. "Jihad vs. McWorld." In *Globalization and the Challenges of a New Century: A Reader*, edited by Patrick O'Meara, Howard Mehlinger, and Matthew Krain, 23–33. Bloomington: Indiana University Press, 2000.

Bargach, Jamila. "Rabat: From Capital to Global Metropolis." In *The Evolving Arab City: Tradition, Modernity and Urban Development*, edited by Yasser Elsheshtawy, 99–117. London: Routledge, 2008.

Barrett, H. R., H. Fox, and L. Stanier. "Agadir: Thirty Years since the Earthquake." *Geography Review* 4, no. 3 (1991): 35–39.

Barrett, Roby C. *The Greater Middle East and the Cold War: US Foreign Policy Under Eisenhower and Kennedy.* New York: I. B. Tauris, 2007.

Baum, Dan. "Jake Leg." *New Yorker* 79, September 15, 2003.

Baziz, Orna. "L'exode des rescapés juifs d'Agadir après le séisme de 1960." In *La bienvenue et l'adieu*, vol. 3, *Migrants juifs et musulmans au maghreb (XVe–XXe siècle)*, edited by Frédéric Abécassis, Karima Dirèche, and Rita Aouad, 57–65. Casablanca: Centre Jacques-Berque, 2012. http://books.openedition.org/cjb/161.

Becker, Heike. "Beyond Trauma: New Perspectives on the Politics of Memory in East and Southern Africa." *African Studies* 70 (2011): 321–35.

Bedjaoui, Mohammed. *Law and the Algerian Revolution.* Brussels: International Association of Democratic Lawyers, 1961.

Bekkat, Amina Azza. Preface to *Histoire de lieux: El Asnam, Miliana, Ténès*, edited by El Djamhouria Slimani Aït Saada, 5–7. Algiers: Hibr, 2013.

Belaev, Michelle. "Trends in Literary Trauma Theory." *Mosaic: A Journal for the Inter-disciplinary Study of Literature* 4 (2008): 149–66.

Bell, Daniel A., and Avner de-Shalit. *The Spirit of Cities: Why the Identity of a City Matters in a Global Age.* Princeton NJ: Princeton University Press, 2013.

Ben Attou, Mohamed. "Agadir gestion urbaine, stratégies d'acteurs et rôle de la société civile: Urbanisme opérationnel ou urbanisme de fait?" *Geomaghreb* 1 (2003): 77–93.

———. Introduction to *Le grand Agadir: Memoire et défis du futur*, edited by Mohamed Ben Attou and Hassan Benhalima, 11–13. Agadir: Université Ibn Zohr, 2004.

Ben Embarek, Mourad. "Chroniques africaines." *A + U: Revue africaine d'architecture et d'urbanisme* 4 (1966): i.

———. "Tourisme et urbanisme." *A + U: Revue africaine d'architecture et d'urbanisme* 4 (1966): 65.

Benhima, Mohamed Taïba. "La renaissance d'Agadir." *Bulletin économique et social du Maroc* 89 (1961): 5–21.

———. "Témoignage." *A + U: Revue africaine d'architecture et d'urbanisme* 4 (1966): 3.

Benjelloun, Tahar. *The Happy Marriage.* Translated by André Naffis-Sahely. Brooklyn NY: Melville House, 2016 (2012).

Bennett, Jane. *Vibrant Matter: A Political Ecology of Things.* Durham NC: Duke University Press, 2009.

Bensimon, Jacques. *Agadir, un paradis dérobé.* Paris: Harmattan, 2012.

Bergman, Jonathan. "Disaster: A Useful Category of Historical Analysis." *History Compass* 10 (2008): 934–46.

Bernard, Stéphane. *The Franco-Moroccan Conflict, 1943–1956.* Translated by Marianna Oliver, Alexander Baden Harrison Jr., and Bernard Phillips. New Haven CT: Yale University Press, 1968.

Berque, Jacques. *Le maghreb entre deux guerres.* 3rd ed. Paris: Éditions du Seuil, 1962.

Betts, Raymond. *France and Decolonisation, 1900–1960.* New York: St. Martin's, 1991.

Bissell, William. "Engaging Colonial Nostalgia." *Cultural Anthropology* 20 (2005): 215–48.

Blair, Leon Borden. *Western Window in the Arab World.* Austin: University of Texas Press, 1970.

Blake, G. H., and R. I. Lawless. *The Changing Middle Eastern City.* New York: Harper and Row, 1980.

Boittin, Jennifer, Christina Firpo, and Emily Church. "Hierarchies of Race and Gender in the French Colonial Empire." *Historical Reflections* 37 (2011): 60–90.

Bonheur, Gaston, ed. *À Fréjus ce soir là.* Paris: Julliard, 1960.

———. "Visitez la Pompeï provençale." In *À Fréjus ce soir là,* edited by Gaston Bonheur, 9–33. Paris: Julliard, 1960.

Bouraoui, Nina. "Ecrire, c'est retrouver ses fantômes." Interview with Dominique Simonnet. *L'Express,* May 31, 2004. http://www.lexpress.fr/culture/livre/ecrire-c-est-retrouver-ses-fantomes_819681.html.

———. *Le jour du séisme.* Paris: Stock, 1999.

Bouzar, Ali. *Le consentement du malheur: Récit: Témoignage sur la catastrophe d'El Asnam du 10 octobre 1980.* Algiers: Entreprise Nationale du Livre, 1985.

Branch, M. C. "Physical Aspects of City Planning." *Annals of the Association of American Geographers* 441, no. 4 (1951): 269–84.

Brown, C. L. *From Madina to Metropolis: Heritage and Change in the Middle Eastern City.* New Jersey: Darwin, 1973.

Brown, Kenneth. *People of Salé: Tradition and Change in a Moroccan City, 1830–1930.* Manchester: Manchester University Press, 1976.

Brown, Kenneth, Ahmed Lakhsassi, and Ibn Ighil. "Every Man's Disaster, the Earthquake of Agadir: A Berber (Tashelhit) Poem." *Maghreb Review* 5, nos. 5–6 (1980): 125–33.

Buck, Carol, Alvaro Llopis, Enrique Najera, and Milton Terris, eds. *The Challenge of Epidemiology.* Washington DC: Pan American Health Organization, 1989.

Burke, Beatrice. "With the International Red Cross in Morocco." *Australian Journal of Physiotherapy* 7, no. 1 (1961): 8–13.

Burke, Edmund, III. "The Image of the Moroccan State in French Ethnological Literature: A New Look at the Origins of Lyautey's Berber Policy." In *Arabs and Berbers*, edited by Ernest Gellner and Charles Micaud, 175–99. Lexington M A: D. C. Heath, 1972.

———. "The Transformation of the Middle Eastern Environment, 1500 BCE–2000 CE." In *The Environment in World History*, 81–117. Berkeley: University of California Press, 2009.

Burkhalter, Sarah. "Négociations autour du cimetière musulman en Suisse: Un exemple de recomposition religieuse en situation d'immigration." *Archives de sciences sociales des religions* 113 (2001): 133–48.

Busson, Henri. "Le développement géographique de la colonisation agricole en Algérie." *Annales de géographie* 7, no. 31 (1898): 34–54.

Cappe, Willi. *Agadir 29 février 1960: Histoire et leçons d'un catastrophe*. Marseille: Presses de G. Cholet, 1967. www.agadir1960.com.

Caruth, Cathy. *Unclaimed Experience: Trauma, Narrative, and History*. Baltimore: Johns Hopkins University Press, 1996.

Charef, Mohamed. "Agadir, une ville orpheline de son passé: Mesure le présent, stimuler le futur." In *La ville d'Agadir: Reconstruction et politique urbaine*, 167–80. Agadir: Royaume du Maroc, Université Ibn Zohr, 1997.

Charnot, A., and S. Troteman. "First Toxicological Investigations in Morocco." In *Morocco 1959: Experiences and Findings*, edited by A. Albertini, D. Gross, and William M. Zinn, 16–27. New York: Intercontinental Medical, 1968.

Chaulet-Achour, Christiane. "Itinéraires de mémoire." *Algérie littérature/action* 39 (March–April 2000). http://www.Revues-Plurielles.org/_uploads/pdf/4_39_2.pdf.

Cherkaoui, T. E., F. Medina, and D. Hatzfeld. "The Agadir Earthquake of February 29, 1960." *Mongrafia: Instituto geográfico nacional* 8 (1991): 133–48.

Choi, Sung-Eun. *Decolonization and the French of Algeria: Bringing the Settler Colony Home*. New York: Palgrave Macmillan, 2016.

Choubert, Georges, and Anne Faure-Muret. *Le séisme d'Agadir, ses effets et son interprétation géologique: Extrait du séisme d'Agadir*. Casablanca: Service de la carte géologique, Rabat, 1962.

Christelow, Allan. "The Muslim Judge and Municipal Politics in Colonial Algeria and Senegal." *Comparative Studies in Society and History* 24 (1982): 3–24.

Church, Christopher. *Paradise Destroyed: Catastrophe and Citizenship in the French Caribbean*. Lincoln: University of Nebraska Press, 2017.

Clancey, Gregory. *Earthquake Nation: The Cultural Politics of Japanese Seismicity, 1868–1930*. Berkeley: University of California Press, 2006.

Clancy-Smith, Julia. "Algeria as Mère-Patrie: Algerian Expatriates in Tunisia." In *Identity, Memory, and Nostalgia*, edited by Patricia Lorcin. Syracuse N Y: Syracuse University Press, 2005.

———. *Mediterraneans: North Africa and Europe in an Age of Migration, c. 1800–1900*. Berkeley: University of California Press, 2011.

———. *Rebel and Saint: Muslim Notables, Populist Protest, Colonial Encounters (Algeria and Tunisia, 1800–1904)*. Berkeley: University of California Press, 1994.

Coen, Deborah R. *The Earthquake Observers: Disaster Science from Lisbon to Richter*. Chicago: University of Chicago Press, 2013.

Colonna, Fanny. *Instituteurs algériens 1883–1939*. Travaux et recherches de science politique 36. Paris: Presses de la Fondation Nationale des Sciences Politiques, 1975.

Comité d'Action Marocaine. *Plan de réformes marocaines*. Paris: Imprimerie Labor, 1934.

Committee of Structural Steel Producers. *The Agadir, Morocco Earthquake February 29, 1960*. New York: American Iron and Steel Institute, 1962.

Connelly, Matthew. *A Diplomatic Revolution: Algeria's Fight for Independence and the Origins of the Post-Cold War Era*. New York: Oxford University Press, 2002.

Cooper, Frederick. *Colonialism in Question: Theory, Knowledge, History*. Berkeley: University of California Press, 2005.

Cooper, Frederick, and Ann Laura Stoler, eds. *Tensions of Empire: Colonial Cultures in a Bourgeois World*. Berkeley: University of California Press, 1997.

Corroy. "Dam Project with a Storage Reservoir on the Reyran River (Var)." In *Final Report of the Investigating Committee of the Malpasset Dam, Paris 1960*. Translated by D. Ben-Yakov, 55–56. Jerusalem: Israel Program for Scientific Translations, n.d. (original May 11, 1949).

———. "Geological Study of a Dam Project with Reservoir on the Reyran River North of Fréjus (Var)." In *Final Report of the Investigating Committee of the Malpasset Dam, Paris 1960*. Translated by D. Ben-Yakov, 50–54. Jerusalem: Israel Program for Scientific Translations, n.d. (original November 15, 1946).

Croizard, Maurice. "L'avenue de la mort." In *À Fréjus ce soir là*, edited by Gaston Bonheur, 131–68. Paris: Julliard, 1960.

Culot, Maurice, and Jean-Marie Thiveaud, eds. *Architectures françaises d'outre-mer*. Liège: Mardaga, 1992.

Cutler, Brock. "Can the North Atlantic Oscillation Speak? Climate and Empire in North Africa." Paper presented at the French Colonial History Society annual meeting, 2017.

———. "Historical (f)Actors: Environments and Histories in Modern North Africa." *History Compass* 16:e12509 (2018): 1–10. https://doi.org/10.1111/hic3.12509.

———. "'Water Mania!': Drought and the Rhetoric of Rule in Nineteenth-Century Algeria." *Journal of North African Studies* 19 (2014): 317–37.

Dalrymple, Theodore. "The Architect as Totalitarian: Le Corbusier's Baleful Influence." *City-Journal* (Autumn 2009). https://www.city-journal.org/html/architect-totalitarian-13246.html.

Dartois, Marie-France. *Agadir et le Sud marocain*. Paris: Éditions de Courcelles, 2008.

Davis, Diana. "Restoring Human Nature: French Identity and North African Environmental History." In *Environmental Imaginaries of the Middle East and North Africa*, edited by Diana Davis and Edmund Burke III, 60–86. Athens: Ohio University Press, 2011.

———. *Resurrecting the Granary of Rome: Environmental History and French Colonial Expansion in North Africa*. Athens: Ohio University Press, 2007.

Davis, Mike. *Late Victorian Holocausts: El Niño Famines and the Making of the Third World*. London: Verso, 2001.

Davis, Muriam Haleh. "Restaging Mise en Valeur: 'Postwar Imperialism' and the Plan de Constantine." *Review of Middle East Studies* 44 (2010): 176–86.

Debia, René Yves. *Orléansville: Naissance et destruction d'une ville: Sa résurrection*. Algiers: Éditions Baconnier, 1955.

Despeyroux, J. "The Agadir Earthquake of February 29th, 1960: Behavior of Modern Buildings during the Earthquake." *Proceedings of the World Conference on Earthquake Engineering* (1960): 521–41.

Dethier, Jean. "60 ans d'urbanisme au Maroc." *Bulletin économique et social du Maroc* 32, nos. 118–19 (1973): 1–56.

Dias, Jill R. "Famine and Disease in the History of Angola c. 1830–1930." *Journal of African History* 22 (1981): 349–78.

Djemai, Abdelkader. *Saison des pierres*. Algiers: Entreprise Nationale du Livre, 1986.

Donat, Olivier. *La tragédie Malpasset*. Mont-de-Marsan: Imprimerie Lacoste, 1990.

Drury, A. Cooper, Richard Stuart Olson, and Douglas A. Van Belle. "The Politics of Humanitarian Aid: U.S. Foreign Disaster Assistance, 1964–1995." *Journal of Politics* 67 (2005): 454–73.

Duffaut, Pierre. "The Traps Behind the Failure of Malpasset Arch Dam, France, in 1959." *Journal of Rock Mechanics and Geotechnical Engineering* 5 (2013): 335–41.

Dynes, Russell. "The Dialogue between Rousseau and Voltaire on the Lisbon Earthquake: The Emergence of a Social Science View." *University of Delaware Disaster Research Center Preliminary Paper* no. 293 (1999). http://udspace.udel.edu/handle /19716/435.

Edwards, M. Kathryn. *Contesting Indochina: French Remembrance between Decolonization and Cold War*. Berkeley: University of California Press, 2016.

Effros, Bonnie. *Incidental Archaeologists: French Officers and the Rediscovery of Roman North Africa*. Ithaca NY: Cornell University Press, 2018.

Eickelman, Dale F. "Is there an Islamic City? The Making of a Quarter in a Moroccan Town." *International Journal of Middle East Studies* 5, no. 3 (1974): 274–94.

Eldridge, Claire. *From Empire to Exile: History and Memory within the Pied-Noir and Harki Communities, 1962–2012*. Manchester: Manchester University Press, 2016.

Elyazghi, Mohamed. "Dialogues sur la ville: La genèse." In *Urbanités en recomposition: dialogues sur la ville, textes et références; Commémorations du discours royal adressé aux architectes à Marrakech*, 12–22. Rabat: Ministère de l'Aménagement du Territoire, de l'Urbanisme, de l'Habitat et de l'Environnement, 2013.

El Younssi, Anouar. "*Souffles-Anfas* and the Moroccan Avant-Garde Post-Independence." *Journal of North African Studies* 23 (2017): 34–52.

Entelis, John. *Algeria: The Revolution Institutionalized*. Boulder: Westview, 1986.

Erikson, Kai. *A New Species of Trouble: The Human Experience of Modern Disasters*. New York: Norton, 1995.

Evans, Martin. *Algeria: France's Undeclared War*. New York: Oxford University Press, 2012.

Evison, F. F. "Lessons from Agadir." *New Zealand Engineering* 18, no. 10 (1963): 369–71.

Fanon, Franzt. *A Dying Colonialism*. Translated by Haakon Chevalier. New York: Grove, 1967.

Faraj, Abdelmalek. "Historical Background to the Mass Poisoning in Morocco 1959." In *Triaryl-Phosphate Poisoning in Morocco 1959: Experiences and Findings*, edited by A. Albertini, D. Gross, and William M. Zinn, 5–15. New York: Intercontinental Medical, 1968.

———. "Social and Vocational Aspects." In *Triaryl-Phosphate Poisoning in Morocco 1959: Experiences and Findings*, edited by A. Albertini, D. Gross, and William M. Zinn, 156–59. New York: Intercontinental Medical, 1968.

Favier, René, and Anne-Marie Granet-Abisset. "Society and Natural Risks in France, 1500–2000: Changing Historical Perspectives." In *Natural Disasters, Cultural Responses: Case Studies Toward a Global Environmental History*, edited by Christof Mauch and Christian Pfister, 103–36. Lanham MD: Lexington, 2009.

Fletcher, Yaël. "The Politics of Solidarity: Radical French and Algerian Journalists and the 1954 Orléansville Earthquake." In *Algeria and France, 1800–2000: Identity, Memory, Nostalgia*, edited by Patricia Lorcin, 84–98. Syracuse NY: Syracuse University Press, 2006.

Flood, Christopher, and Hugo Frey. "Questions of Decolonization and Post-Colonialism in the Ideology of the French Extreme Right." *Journal of European Studies* 28 (1988): 69–89.

Fontaine, Darcie. *Decolonizing Christianity: Religion and the End of Empire in France and Algeria*. New York: Cambridge University Press, 2016.

Fordham, J. H. "Agadir." *Postgraduate Medical Journal* 36, no. 421 (November 1960): 652–57.

Foucou, Marcel. *Malpasset: Une tragédie déjà entrée dans l'histoire; Naissance, vie, mort d'un barrage*. Fréjus: self-published, 1978.

Gaillard, Jean-Christophe, Ilan Kelman, and Ma Florina Orillos. "US-Philippines Military Relations after the Mt. Pinatubo Eruption in 1991: A Disaster Diplomacy Perspective." *European Journal of East Asian Studies* 8, no. 2 (2009): 301–30.

Gaudefroy-Demombynes, Roger. *L'oeuvre française en matière d'enseignement au Maroc*. Paris: Librairie Orientaliste Paul Geuthner, 1928.

Gellner, Ernest, and Charles Micaud, eds. *Arabs and Berbers: From Tribe to Nation in North Africa*. Lexington MA: D. C. Heath, 1972.

Gershovich, Moshe. "The Long Shadow of Lyautey: Long-Term Effects of French Colonialism on Contemporary Morocco." Panel organized at Middle East Studies Association annual meeting, 2012.

Godfrey, C. M. "An Epidemic of Triorthocresylphosphate Poisoning." *Canadian Medical Association Journal* 85 (September 16, 1961): 689–91.

Goebel, Michael. "'The Capital of the Men without a Country': Migrants and Anticolonialism in Interwar Paris." *American Historical Review* 121 (2016): 1444–67.

Gold, John R. "Creating the Charter of Athens: CIAM and the Functional City, 1933–43." *Town Planning Review* 69 (1998): 225–47.

Gross, D. "Diagnosis and Sympomatology." In *Triaryl-Phosphate Poisoning in Morocco 1959: Experiences and Findings*, edited by A. Albertini, D. Gross, and William M. Zinn. New York: Intercontinental Medical, 1968.

———. "Results of Treatment." In *Triaryl-Phosphate Poisoning in Morocco 1959: Experiences and Findings*, edited by A. Albertini, D. Gross, and William M. Zinn. New York: Intercontinental Medical, 1968.

Gross, D., S. Robertson, and William M. Zinn. "Organisation and Contributions." In *Triaryl-Phosphate Poisoning in Morocco 1959: Experiences and Findings*, edited by A. Albertini, D. Gross, and William M. Zinn, 160–73. New York: Intercontinental Medical, 1968.

Grzeda, Paulina. "Trauma and Testimony: Autobiographical Writing in Post-Apartheid South Africa." In *Postcolonial Traumas: Memory, Narrative, Resistance*, edited by Abigail Ward, 65–82. New York: Palgrave Macmillan, 2015.

Guerin, Adam. "'Not a Drop for the Settlers': Reimagining Popular Protest and Anti-Colonial Nationalism in the Moroccan Protectorate." *Journal of North African Studies* 20 (2015): 225–46.

Halstead, John P. *Rebirth of a Nation: The Origins and Rise of Moroccan Nationalism, 1912–1944*. Cambridge MA: Harvard University Press, 1969.

Hardy, Georges. *L'âme marocaine d'après la littérature française*. Éditions du bulletin de l'enseignement public du Maroc 73. Paris: Émile Larose, 1926.

———. *Une conquête morale: l'enseignement en A.O.F.* Paris: Librairie Armand Colin, 1917.

Hardy, Georges, and Louis Brunot. *L'enfant marocain: Essai d'ethnographie scolaire*. Éditions du Bulletin de l'enseignement public du Maroc 63. Paris: Émile Larose, 1925.

Hassan II, King. "Message de S.M. Hassan II." *A + U: Revue africaine d'architecture et d'urbanisme* 4 (1966): 2.

———. "Texte du discours royal pronouncé devant le corps des architectes le 14/1/1986." *Al omrane* 5 (1986): 3–15.

Haut-Commissariat au Plan. *Recensement général de la population et de l'habitat de 2004: Population légale du Maroc*. Rabat: Royaume du Maroc, 2004.

Heggoy, Alf Andrew. *Insurgency and Counterinsurgency in Algeria*. Bloomington: Indiana University Press, 1972.

Hicks, David T. "Rebuilt Agadir." *Architectural Review* 142 (1967): 292–300.

Hoisington, William. *The Casablanca Connection: French Colonial Policy, 1936–1943*. Chapel Hill: University of North Carolina Press, 1984.

———. "The Selling of Agadir: French Business Promotion in the 1930s." *International Journal of African Historical Studies* 18 (1985): 315–24.

Horne, Alistair. *A Savage War of Peace: Algeria 1954–1962.* New York: Viking, 1978.

Horowitz, Andy. "The Complete Story of the Galveston Horror: Trauma, History, and the Great Storm of 1900." In *Environmental Disaster in the Gulf South*, edited by Cindy Ermus, 62–79. Baton Rouge: Louisiana State University Press, 2018.

House, Jim, and Neil MacMaster. *Paris 1961: Algerians, State Terror, and Memory.* Oxford: Oxford University Press, 2006.

Hughes, Christian. "Souvenir d'un jour tragique." In *Barrage de Malpasset: De sa conception à sa rupture*, edited by Vito Valenti and Alfred Bertini, 109–10. Le Pradet: Éditions du Lau/ Societé d'histoire de Fréjus et de sa région, 2003.

Ifowodo, Ogaga. *History, Trauma, and Healing in Postcolonial Narratives: Reconstructing Identities.* New York: Palgrave Macmillan, 2013.

Ighil, Ibn. "The Tale of Agadir." Translated from Tashelhit. In "Every Man's Disaster, the Earthquake of Agadir: A Berber (Tashelhit) Poem." *Maghreb Review* 5, nos. 5–6 (1980): 128.

Jennings, Eric. "Conservative Confluences, 'Nativist' Synergy: Reinscribing Vichy's National Revolution in Indochina, 1940–1945." *French Historical Studies* 27, no. 3 (2004): 601–35.

Johnson, Jennifer. *The Battle for Algeria: Sovereignty, Health Care, Humanitarianism.* Philadelphia: University of Pennsylvania Press, 2016.

Jones, Randolph. "Otto Passman and Foreign Aid: The Early Years." *Louisiana History* 26, no. 1 (1985): 53–62.

Julien, Charles-André. *Le Maroc face aux impérialismes 1415–1956.* Paris: Éditions J. A., 1978.

Kateb, Kamel. "La gestion administrative de l'émigration algérienne vers les pays musulmans au lendemain de la conquête de l'Algérie (1830–1914)." *Population* 52, no. 2 (1997): 410–11.

Kelman, Ilan. *Disaster Diplomacy: How Disasters Affect Peace and Conflict.* Abingdon: Routledge, 2011.

Khaïr-Eddine, Mohammed. *Agadir.* Rabat: Tarik Éditions, 2010 [1967].

Khalfa, Boualem, Henri Alleg, and Abdelhamid Benzine. *La grande aventure d'Alger républicain.* Paris: Éditions Messidor, 1987.

Kidou, Brahim. "Les changements démographiques d'une nouvelle grande ville au sud du Maroc: Le Grand Agadir." September 28, 2009. http://www.abhatoo.net.ma/index .php/fre/content/download/11814/195728/file/KidouBrahim.pdfMarrakech.

Kréa, Henri. *Le séisme: Tragédie.* Paris: Pierre Jean Oswald, 1958.

Krieger, Nancy. "Epidemiology and the Web of Causation: Has Anyone seen the Spider?" *Social Science and Medicine* 39, no. 7 (1994): 887–903.

LaCapra, Dominick. *History and Memory after Auschwitz.* Ithaca NY: Cornell University Press, 1998.

———. "Trauma, Absence, Loss." *Critical Inquiry* 25 (1999): 696–727.

Lacheraf, Mostefa. Preface to *Pierres et lumières: Souvenirs et digressions d'un médecin algérien, fils d'instituteur 'd'origine indigene*, by Belgacem Aït Ouyahia, 6–9. Algiers: Casbah, 1999.

Lagumina, Salvatore. *The Great Earthquake: America Comes to Messina's Rescue*. Youngstown NY: Teneo, 2008.

Lahbabi, Abderrafih. "Changement social et aliénation en architecture au Maroc." *Lamalif* 77 (1976): 50–54.

Landauer. "Outline of Master Plan for the Reconstruction of Agadir." In *The Reconstruction of Agadir, Translation no. 3118(60)*, 1–6. Rabat: United States of America Operations Mission to Morocco, 1960.

Laroui, Abdallah. *L'histoire du Maghreb: Un essai de synthèse*. Casablanca: Centre Culturel Arabe, 1995.

Laskier, Michael. *North African Jewry in the Twentieth Century: The Jews of Morocco, Tunisia, and Algeria*. 2nd ed. New York: New York University Press, 1997.

Lasky, Ahmed. "La renaissance d'Agadir." *Europe france outremer* 428 (1965): 51–55.

Lee, Joomi. "Urban Politics of the Bouregreg Project: The Integration of Rabat-Salé and Morocco's Monarchial State." Paper presented at Middle East Studies Association annual meeting, 2011.

Lehman. *The Reconstruction of Agadir*. Translated by Language Service Section. Rabat: United States Operations Mission to Morocco, April 1960.

Le Toullec, Roger. *Agadir 1960: Mémoire d'un séisme*. Nantes: Éditions Marines, 2002.

Lewis, James. "The Algerian Earthquakes of May 2003: Some Precedents for Reconstruction." *Radix* (May 2003). www.radixonline.org/algeria2.htm.

Lorcin, Patricia. *Imperial Identities: Stereotyping, Prejudice, and Race in Colonial Algeria*. New York: I. B. Tauris, 1995.

——— . "Women, Gender and Nation in Colonial Novels of Interwar Algeria." *Historical Reflections/Réflexions Historiques* (2002): 163–84.

Luckhurst, Roger. *The Trauma Question*. London: Taylor and Francis, 2008.

Lyons, Amelia H. *The Civilizing Mission in the Metropole: Algerian Families and the French Welfare State during Decolonization*. Stanford CA: Stanford University Press, 2013.

——— . "The Civilizing Mission in the Metropole: Algerian Immigrants in France and the Politics of Adaptation during Decolonization." *Geschichte und Gesellschaft* 32 (2006): 489–516.

Makdisi, Ussama, and Paul Silverstein, eds. *Memory and Violence in the Middle East and North Africa*. Bloomington: University of Indiana Press, 2006.

Malverti, Xavier. "Méditerranée, soleil, et modernité." In *Architectures françaises d'outre-Mer*, edited by Maurice Culot and Jean-Marie Thivea, 29–64. Liège: Mardaga, 1992.

Mann, Gregory. "Locating Colonial Histories: Between France and West Africa." *American Historical Review* 110, no. 2 (2005): 409–34.

Marsh, Kate, and Nicola Frith, eds. *France's Lost Empires: Fragmentation, Nostalgia, and La Fracture Coloniale*. New York: Rowman & Littlefield, 2011.

Mas, Pierre. "Plan directeur et plans d'aménagement." *A + U: Revue africaine d'architecture et d'urbanisme* 4 (1966): 6.

McDougall, James. *History and the Culture of Nationalism in Algeria*. Cambridge: Cambridge University Press, 2006.

———. *A History of Algeria*. Cambridge: Cambridge University Press, 2017.

———. "The Impossible Republic: The Reconquest of Algeria and the Decolonization of France, 1945–1962." *Journal of Modern History* (2011): 772–811.

———. "Martyrdom and Destiny: The Inscription and Imagination of Algerian History." In *Memory and Violence in the Middle East and North Africa*, edited by Ussama Makdisi and Paul Silverstein, 50–72. Bloomington: Indiana University Press, 2006.

———. "Savage Wars: Codes of Violence in Algeria, 1830s–1990s." *Third World Quarterly* 26, (2005): 117–31.

McNeece, Lucy. "Le jour de la très grande violence: *Agadir* ou l'écriture séismique de Mohammed Khaïr-Eddine." In *Francophonie plurielle: Actes du congrès mondial du Conseil international d'études francophones tenu à Casablanca (Maroc) du 10 au 17 juillet 1993*, edited by Ginette Adamson and Jean-Marc Gouanvic, 147–58. Quebec: Hurtubise, 1995.

Menant, Georges. "La vielle femme et le barrage." In *À Fréjus ce soir là*, edited by Gaston Bonheur, 35–79. Paris: Julliard, 1960.

Millecam, Jean-Pierre. "Apocalypses." In *An Algerian Childhood: A Collection of Autobiographical Narratives (Une enfance algérienne)*. Translated by Marjolijn de Jager. Edited by Leïla Sebbar, 161–74. St. Paul: Ruminator, 2001.

Ministère de l'Agriculture. *Final Report of the Investigating Committee of the Malpasset Dam, Paris 1960 (Commission d'enquête du barrage de Malpasset, rapport définitif)*. Translated by D. Ben-Yakov. Jerusalem: Israel Program for Scientific Translations, n.d.

Ministère du Développement Durable. "Rupture d'un barrage: Le 2 décembre Malpasset [Var], France." *Analyse, recherche et information sur les accidents* 29490 (April 2009).

Minkin, Shana. *Imperial Bodies: Empire and Death in Alexandria, Egypt*. Stanford CA: Stanford University Press, 2019.

Mitchell, Timothy. *Rule of Experts: Egypt, Techno-Politics, Modernity*. Berkeley: University of California Press, 2002.

Mitchell, William A. "Reconstruction after Disaster: The Gediz Earthquake of 1970." *Geographical Review* 66, no. 3 (1976): 296–313.

Molotch, Harvey, and Marilyn Lester. "Accidental News: The Great Oil Spill as Local Occurrence and National Event." *American Journal of Sociology* 81 (1975): 235–60.

Morin, Emilie. "Unspeakable Tragedies: Censorship and the New Political Theatre of the Algerian War of Independence." In *Theatre and Human Rights After 1945: Things Unspeakable*, edited by Mary Luckhurst and Emilie Morin, 21–38. New York: Palgrave Macmillan, 2015.

Mouline, Saïd, ed. *Urbanités en recomposition, dialogues sur la ville: Textes de références, commémorations du discours royal adressée aux architectes à Marrakech*. Rabat:

Ministère de l'Aménagement du Territoire, de l'Urbanisme, de l'Habitat et de l'Environnement, 2000.

Mulcahy, Matthew. *Hurricanes and Society in the British Greater Carribean, 1624–1783*. Baltimore: Johns Hopkins University Press, 2006.

Nadau, Thierry. "La reconstruction d'Agadir." In *Architectures françaises d'outre-mer*, edited by Maurice Culot and Jean-Marie Thivea, 146–66. Liège: Mardaga, 1992.

National Research Council, Committee on Natural Disasters. *El-Asnam, Algeria Earthquake October 10, 1980: A Reconnaissance and Engineering Report*. Edited by Arline Leeds. Washington DC: National Technical Information Service, 1983.

Nunez, Juliette. "La gestion publique des espaces confessionnels des cimetières de la ville de Paris: L'exemple du culte musulman (1857–1957)." *Le mouvement social* 4 (2011): 13–32.

Osgood, Kenneth. *Total Cold War: Eisenhower's Secret Propaganda Battle at Home and Abroad*. Lawrence: University Press of Kansas, 2006.

Ouahmed, Karima Yahia. "De la double origine à l'être-deux dans l'écriture de Nina Bouraoui." *Synergies Algérie* 7 (2009): 221–29.

Ouardi, Brahim. "Écriture, théâtre et engagement dans le théâtre d'Henri Kréa et Noureddine Aba." Doctoral dissertation, Université de Oran, 2009.

———. "Mythe, théâtre et oralité dans *Le séisme* d'Henri Kréa." *Synergies Algèrie* 3 (2008). http://gerflint.fr/Base/Algerie3/ouardi.pdf.

Pauty, Edmond. "Rapport sur la défense des villes et la restauration des monuments historiques." *Hespéris: Archives berbères et bulletin de l'institut des hautes-études marocaines* 2, no. 4 (1922): 449–62.

Paye, Lucien. "Introduction et évolution de l'enseignement moderne au Maroc." Thesis, Université de Paris Sorbonne, 1957.

Pelling, Mark, and Kathleen Dill. "Disaster Politics: Tipping Points for Change in the Adaptation of Sociopolitical Regimes." *Progress in Human Geography* 34, no. 1 (2010): 21–37.

———. "'Natural' Disasters as Catalysts of Political Action." ISP/NSC Briefing Paper 06/01 (2006), 4–6.

Pelosse, Valentin. "Évolution socio-professionnelle d'une ville algérienne (Orléansville-Esnam 1948–1966)." Thesis, Universitéde Paris I Panthéon-Sorbonne Institut de Géographie, 1967.

Pennell, C. R. *Morocco since 1830: A History*. London: Hurst, 2000.

Péré, M. "Agadir, ville nouvelle." *Revue de géographie du Maroc* 12 (1967): 43–90.

Pernoud, Géorges. "Le Reyran se jette dans la Méditerranée." In *À Fréjus ce soir là*, edited by Gaston Bonheur, 168–225. Paris: Julliard, 1960.

Peyréga, Jacques. Preface to *Orléansville: Naissance et destruction d'une ville: Sa résurrection*, edited by René Yves Debia, 7–11. Algiers: Éditions Baconnier, 1955.

Pfister, Christian. "Learning from Nature-Induced Disasters: Theoretical Considerations and Case Studies from Western Europe." In *Natural Disasters, Cultural*

Responses: Case Studies Toward a Global Environmental History, edited by Christof Mauch and Christian Pfister, 17–40. Lanham MD: Lexington, 2009.

Picard, Aleth. "Orléansville: La reconstruction après 1954." In *Architectures françaises d'outre-mer*, edited by Maurice Culot and Jean-Marie Thivea, 65–75. Liège: Mardaga, 1992.

Prado, Max. *L'imprévisible nature: Tragique guet-apens de Malpasset*. Castelnau-le-Lez: Max Prado, 1998.

Primeau, Bertrand. "The Rehabilitation of 10,000 Victims of Paralysis in Morocco by the League of Red Cross Societies." *Canadian Medical Association Journal* 85 (December 2, 1961): 1249–52.

Quarantelli, E. L. "What Is a Disaster? The Need for Clarification in Definition and Conceptualization in Research." *University of Delaware Disaster Research Center Article no. 177* (1985). Accessed June 25, 2018. http://udspace.udel.edu/handle/19716/1119.

Rabinow, Paul. *French Modern: Norms and Forms of the Social Environment*. Chicago: University of Chicago Press, 1995.

Raqbi, Ahmed. "Agadir dans les écrits de Tahar Benjelloun et de Mohammed Khaïr-Eddine." In *Le Grand Agadir: Memoire et defis du futur*, edited by Mohamed Ben Attou and Hassan Benhalima, 127–31. Agadir: Université Ibn Zohr, 2004.

Ray, Gene. "Reading the Lisbon Earthquake: Adorno, Lyotard, and the Contemporary Sublime." *Yale Journal of Criticism* 17, no. 1 (2004): 1–18.

Ritzi, Matthias, and Erich Schmidt-Eenboom. *Im Schatten des Dritten Reiches: Der BND und sein Agent Richard Christmann*. Berlin: Ch. Links Verlag, 2011.

Rivet, Daniel. "Lyautey et l'institution eu protectorat français au Maroc 1912–1925." *Histoire et perspectives méditerranéennes*. Vol. 2. Paris: L'Harmattan, 1988.

Roberson, Jennifer. "The Changing Face of Morocco under King Hassan II." *Mediterranean Studies* 22 (2014): 57–83.

Roberts, Priscilla H., and James N. Tul. "Moroccan Sultan Sidi Muhammad Ibn Abdallah's Diplomatic Initiatives toward the United States, 1777–1786." *Proceedings of the American Philosophical Society* 143 (1999): 233–65.

Robion, Louis, and Marcel Foucou. *Fréjus Ve–XXe siècle: Déclins et renaissances*. Centre régional de documentation pédagogique de Nice, 1987.

Rothé, Jean-Pierre. "Les tremblements de terre d'Orléansville (Septembre–Octobre 1954)." *Revue pour l'étude des calamités* 14, no. 32 (1955): 77–82, as abstracted in *Geophysical Abstracts* 163 (1955): 240.

Roussafi, Lahsen, Yazza Jafri, and Abdallah Kikr. *Dhakirat Agadir fi al qarn al ashrin* (Memories of Agadir in the Twentieth Century). 3 vols. Agadir: Raïssa, 2010.

———. *Mémoires d'Agadir au XXe siècle*. Translated by Ali Ahlallay. Vol. 1: 1901–1945. Rabat: Imprimerie RabatNet, 2013.

Rowe, J. Stan. "What on Earth Is Environment?" *The Trumpeter* 6, no. 4 (1989): 123–26, rev. ed. Accessed August 10, 2017. http://www.ecospherics.net/pages/RoWhatEarth.html.

Rozario, Kevin. "What Comes Down Must Go Up." In *American Disasters*, edited by Steven Biel, 72–102. New York: New York University Press, 2001.

Rubin, Eli. "The Athens Charter." *Themenportal europäische Geschichte*, 2009. www .europa.clio-online.de/essay/id/artikel-3486.

Russo, Fabienne. "Gaby." In *Souvenirs intimes: Malpasset*, edited by Fabienne Russo and Michel Suzzarini. Brignoles: Association du Cinquantième de la Catastrophe de Malpasset/Éditions Vivre Tout Simplement, 2009.

Ryang, Sonia. "The Great Kanto Earthquake and the Massacre of Koreans in 1923: Notes on Japan's Modern National Sovereignty." *Anthropological Quarterly* 76, no. 4 (2003): 731–48.

Saada, Emmanuelle. "More than a Turn? The 'Colonial' in French Studies." *French Politics, Culture, and Society* 32 (2014): 34–39.

Saïd, Edward. *Orientalism*. New York: Vintage, 1979.

Saïgh-Bousta, Rachida. "Une vie, un rêve, un homme toujours errant . . ." *Le Maghreb littéraire* 1, no. 1 (1997): 17–33.

Saoudi, Samira. "Le passé, le présent, et peut-être le futur du lotissement dans la ville d'Agadir." Fin d'études diplôme thesis, École Nationale d'Architecture, Rabat, 1991.

Sari, Djilali. "Le démantèlement de la propriété foncière." *Revue Historique* 249 (1973): 47–76.

Sbenter, Abderrahman. "Eléments d'articulation urbaine: Ville d'Agadir." Fin d'études diplôme thesis, École Nationale d'Architecture, Rabat, 1990.

Schaefer, Wolf. "Global Civilization and Local Cultures: A Crude Look at the Whole." *International Sociology* 16 (2001): 301–19.

———. "Global History and the Present Time." In *Wiring Prometheus: Globalisation, History and Technology*, edited by Peter Lyth and Helmut Trischler, 103–25. Denmark: Aarhus, 2004. www.stonybrook.edu/globalhistory/PDF/GHAndThePresentTime .pdf.

———. "The New Global History: Toward a Narrative for Pangaea Two." *Erwägen, Wissen, Ethik* 14 (2003): 75–88. http://www.stonybrook.edu/globalhistory/PDF /Hauptartikel.pdf.

Schencking, Charles J. "The Great Kanto Earthquake and the Culture of Catastrophe and Reconstruction in 1920s Japan." *Journal of Japanese Studies* 34 (2008): 295–331.

Seaman, David. "'A Jumping Joyous Urban Jumble': Jane Jacobs's *Death and Life of Great American Cities* as a Phenomenology of Urban Place." *Journal of Space Syntax* 3 (2012): 139–49.

Sebbar, Leïla, ed. *An Algerian Childhood: A Collection of Autobiographical Narratives* (Une enfance algérienne). Translated by Marjolijn de Jager. St. Paul: Ruminator, 2001.

Segalla, Spencer. *The Moroccan Soul: French Education, Colonial Ethnography, and Muslim Resistance, 1912–1956*. Lincoln: University of Nebraska Press, 2009.

Shepard, Todd. *The Invention of Decolonization: The Algerian War and the Remaking of France*. Ithaca NY: Cornell University Press, 2006.

Shrady, Nicholas. *The Last Day: Wrath, Ruin, and Reason in the Great Lisbon Earthquake of 1755*. New York: Penguin, 2009.

Smith, Honor V., and J. M. K. Spalding. "Outbreak of Paralysis in Morocco due to Ortho-Cresyl Phosphate Poisoning." In *The Challenge of Epidemiology*, edited by

Carol Buck, Alvaro Llopis, Enrique Najera, and Milton Terris, 442–45. Washington DC: Pan American Health Organization, 1989.

———. "Outbreak of Paralysis in Morocco due to Ortho-Cresyl Phosphate Poisoning." *Lancet* 274, no. 7110 (December 5, 1959): 1019–21.

Smith, Richard. *Ahmad Al-Mansur: Islamic Visionary*. New York: Longman, 2006.

Smith, William J., and Leif Sjöberg. Preface to *Agadir*, by Artur Lundkvist, ix–xiii. Translated by Leif Sjöberg and William J. Smith. Athens: Ohio University Press, 1976.

Spivak, Gayatri Chakravorty. "Can the Subaltern Speak?" In *The Post-Colonial Studies Reader*, edited by Bill Ashcroft, Gareth Griffiths, and Helen Tiffin, 24–28. New York: Routledge, 1995.

Steinberg, Ted. *Acts of God: The Unnatural History of Natural Disaster in America*. New York: Oxford University Press, 2006.

———. "Down to Earth: Nature, Agency, and Power in History." *American Historical Review* 107 (2002): 798–820.

Stoler, Ann Laura, and Frederick Cooper. "Between Metropole and Colony: Rethinking a Research Agenda." In *Tensions of Empire*, edited by Frederick Cooper and Ann Laura Stoler, 1–56. Berkeley: University of California Press, 1997.

Stora, Benjamin. *Algeria: 1830–2000: A Short History*. Translated by Jane Marie Todd. Ithaca NY: Cornell University Press, 2001.

———. "La différenciation entre le FLN et le courant messaliste (été 1954–décembre 1955)." *Cahiers de la méditerranée* 26, no. 1 (1983): 15–82.

Swearingen, Will D. *Moroccan Mirages: Agrarian Dreams and Deceptions, 1912–1986*. Princeton NJ: Princeton University Press, 1987.

Taithe, Bertrand. "Humanitarianism and Colonialism: Religious Responses to the Algerian Drought and Famine of 1866–1870." In *Natural Disasters, Cultural Responses: Case Studies Toward a Global Environmental History*, edited by Christof Mauch and Christian Pfister, 137–63. Lanham MD: Lexington, 2009.

Talbott, John. *The War without a Name: France in Algeria 1954–1962*. New York: Knopf, 1980.

Tengour, Habib. "Childhood." In *An Algerian Childhood: A Collection of Autobiographical Narratives (Une enfance algérienne)*. Translated by Marjolijn de Jager. Edited by Leïla Sebbar, 199–211. St. Paul: Ruminator, 2001.

Thenault, Sylvie, and Raphaele Branche. "Le secret sur la torture pendant la guerre d'Algérie." *Matériaux pour l'histoire de notre temps*, no. 58: *Le secret en histoire* (2000): 57–63. https://doi.org/10.3406/mat.2000.404251 and https://www.persee.fr/doc/mat _0769-3206_2000_num_58_1_404251.

Torres, Jacques. *L'Orléansvillois: Un essai sur l'histoire du département du Chéliff*. Saint Sympohrien: Micro et Logo, 2008.

Tosi, Luigi, Carlo Righetti, Carlo Adami, and Giampietro Zanette. "October 1942: A Strange Epidemic Paralysis in Saval, Verona, Italy." *Journal of Neurology, Neurosurgery, and Psychiatry* 57 (1994): 810–13.

Touaf, Larbi. "The Legacy of Dissent: Mohamed Khaïr-Eddine and the Ongoing Cultural Diversity Debate in Morocco." *Journal of North African Studies* 21 (2016): 50–59.

Touati, Hajar. "Le devenir et la qualité de vie des victimes de l'intoxication aux huiles frelatées de 1959." Doctoral thesis in medicine, Université Sidi Mohamed ben Abdellah, 2017. http://scolarite.fmp-usmba.ac.ma/cdim/mediatheque/e_theses/280-17.pdf.

Travers, P. R. "The Results of Intoxication with Orthocresyl Phosphate Absorbed from Contaminated Cooking Oil, as Seen in 4,029 Patients in Morocco." *Proceedings of the Royal Society of Medicine* 55 (January 1962): 57–60.

Tricou, Luc. "La création d'Orléansville." *L'Algèrianiste* 107 (September 2004): 88–93.

Trumbull, George, IV. "Body of Work: Water and Reimagining the Sahara in the Era of Decolonization." In *Environmental Imaginaries of the Middle East and North Africa*, edited by Diana Davis and Edmund Burke III, 87–112. Athens: Ohio University Press, 2011.

———. "The Environmental Turn in Middle East History." *International Journal of Middle East Studies* 49 (2017): 173–80.

Tuyns, Albert. "Conversation with Albert Tuyns." *Addiction* 91, no. 1 (January 1996): 17–23.

Valenti, Vito, and Alfred Bertini. *Barrage de Malpasset: De sa conception à sa rupture*. Le Pradet: Éditions du Lau/Societé d'histoire de Fréjus et de sa région, 2003.

Valeri, Odette. "La Vague." In *À Fréjus ce soir là*, edited by Gaston Bonheur, 85–130. Paris: Julliard, 1960.

Valette, Jacques. "Le maquis Kobus, une manipulation ratée durant la guerre d'Algérie (1957–1958)." *Guerres mondiales et conflits contemporains* (1998): 69–88.

Vickroy, Laurie. *Reading Trauma Narratives: The Contemporary Novel and the Psychology of Oppression*. Charlottesville: University of Virginia Press, 2015.

Ville de Fréjus. *Malpasset un an après*. Draguigan (Var): Imprimerie Riccobono, December 1960.

Wainscott, Ann. "Opposition Failure or Regime Success? Education, the Decline of the Left and the Rise of Islamism in Post-Independence Morocco." Unpublished paper, Florida Maghreb Workshop, Tampa 2013.

Walker, Charles F. *Shaky Colonialism: The 1746 Earthquake-Tsunami in Lima, Peru, and its Long Aftermath*. Durham NC: Duke University Press, 2008.

Wallet, Regina. *La nuit de Fréjus*. Draguignan: Éditions de la Tour, 1968.

Ward, Abigail, ed. *Postcolonial Traumas: Memory, Narrative, Resistance*. New York: Palgrave Macmillan, 2015.

Wiebelhaus-Brahm, Eric. "Globalization, Modernity, and their Discontents." *SSRN*, August 27, 2002. http://dx.doi.org/10.2139/ssrn.1666871.

Williford, Daniel. "Seismic Politics: Risk and Reconstruction after the 1960 Earthquake in Agadir, Morocco." *Technology and Culture* 58 (2017): 982–1016.

Wright, Gwendolyn. "Tradition in the Service of Modernity: Architecture and Urbanism in French Colonial Policy, 1900–1930." *Journal of Modern History* 59 (1987): 291–316.

Zartman, I. William. "The Moroccan-American Base Negotiations." *Middle East Journal* 18, no. 1 (1964): 27–40.

———. *Morocco: Problems of New Power.* New York: Prentice-Hall, 1964.

Zingg, Paul. "The Cold War in North Africa: American Foreign Policy and Postwar Muslim Nationalism, 1945–1962." *Historian* 39, no. 1 (1976): 40–61.

Zinn, William M. "Survey of Earlier Triaryl-Phosphate Intoxications." In *Triaryl-Phosphate Poisoning in Morocco 1959: Experiences and Findings*, edited by A. Albertini, D. Gross, and William M. Zinn, 1–4. New York: Intercontinental Medical, 1968.

———. "Treatment and Rehabilitation." In *Triaryl-Phosphate Poisoning in Morocco 1959: Experiences and Findings*, edited by A. Albertini, D. Gross, and William M. Zinn. New York: Intercontinental Medical, 1968.

Page numbers in italics indicate maps.

To order or obtain more information on these or other
University of Nebraska Press titles, visit nebraskapress.unl.edu

Lightning Source UK Ltd.
Milton Keynes UK
UKHW010151270521
384458UK00001B/37